COMPARATIVE
HOMELAND SECURITY

WILEY SERIES IN HOMELAND AND DEFENSE SECURITY

Series Editor

TED LEWIS   Professor, Naval Postgraduate School

Foundations of Homeland Security: Law and Policy / Marin J. Alperen, Esq.
Comparative Homeland Security: Global Lessons / Nadav Morag

COMPARATIVE HOMELAND SECURITY

Global Lessons

NADAV MORAG

A JOHN WILEY & SONS, INC., PUBLICATION

Published by John Wiley & Sons, Inc., Hoboken, New Jersey
Published simultaneously in Canada

For general information on our other products and services or for technical support, please contact our Customer Care Department within the United States at 877-762-2974, outside the United States at 317-572-3993 or fax 317- 572-4002.

Wiley also publishes its books in a variety of electronic formats. Some content that appears in print may not be available in electronic formats. For more information about Wiley products, visit our web site at www.wiley.com.

Library of Congress Cataloging-in-Publication Data:

Morag, Nadav, 1965-
    Comparative homeland security : global lessons / Nadav Morag.
        p. cm.
    Includes index.
    ISBN 978-0-470-49714-2 (cloth)
    1. National security. 2. Terrorism–Prevention. 3.   Law enforcement. 4.
Emergency management. 5.   Public safety. I. Title.
    UA10.5.M67 2011
    363.34–dc22

                                                                    2010050398

Printed in the United States of America

ePDF ISBN: 978-1-118-04868-9
oBook ISBN: 978-1-118-04825-2
ePub ISBN: 978-1-118-04827-6

10 9 8 7 6 5 4 3

For Galia, Adi, and Edan

CONTENTS

PREFACE

This volume is designed primarily as a textbook for students of the emerging academic and practitioner discipline of homeland security. Although no universally accepted definition of homeland security exists at present (or is ever likely to exist), in the introductory chapter I posit a working definition around which the other chapters are organized. The book is not designed to be an introductory text for students of homeland security, as there are a number of these. Instead, it is designed to serve as a text and resource for a subfield within the discipline of homeland security, that of *comparative homeland security*. This subfield is concerned primarily with analyzing and understanding the homeland security policies followed outside the United States (homeland security is a quintessentially American concept, as explained in the Introduction). Comparative homeland security accordingly mirrors the various subject areas within the broader field of homeland security, and hence I touch on most (or all, depending on one's definition of homeland security) of these issue areas.

Regrettably, a field as sweeping in scope as comparative homeland security cannot be comprehensively addressed in a single volume. As noted in the Introduction, the field of homeland security is extremely broad and covers issues as diverse as counterterrorism, law enforcement, emergency management and response, public health, strategic communications, and a host of other public policy issues. Adequate treatment of this topic solely within the American context would require, at a minimum, a shelf load of books, and doing so in the context of the handful of foreign countries addressed in this book would require several shelf loads of books. Moreover, aside from space issues, comprehensiveness is not really possible at this stage in the development of this subfield because only a small percentage of the information needed to address this issue in a truly thorough manner is publicly accessible.

Researchers who focus on homeland security strategies and policies in the domestic American context often do not have access to the materials that they need because these are either classified or otherwise held close and not made publicly available or, in some cases, are unwritten and can only be accessed through identifying the appropriate persons and obtaining their acquiescence to be interviewed. Nevertheless, a surprising amount of material is available in the public sphere, as many organizations and agencies produce reports, analyses, strategy papers, and other types of documentation, and there is also a growing body of academic studies in the field. Consequently, although researchers of domestic homeland security policy will sometimes still come up empty when looking for documentation on which to base their research, they also enjoy an extensive and expanding pool of materials with which to work.

The researcher interested in exploring the homeland security policies of other nations is, however, in a position of comparative disadvantage. This is because not all countries of interest tend to follow the American approach of, by and large, making strategy and policy publicly available as a way of ensuring governmental accountability to the public. Granted that materials produced by governmental agencies for public consumption are sanitized and always designed to portray the agency in question in as favorable a light as possible (many embellish their documents with photos of smiling agency personnel and members of the public). Nevertheless, much can still be learned from them if one reads between the lines and triangulates this information with data from other sources. In addition, academic studies and documents produced by various assessment entities (public and private) often provide a more critical view of policies. The culture of public accountability is quite strong in the United States and is shared by some of the countries of interest in this book, including the United Kingdom, Canada, Australia, and, in some cases, Germany, and these countries offer greater access to information about policy. In other cases, such as Israel, France, and Italy, there is little of the culture of public accountability, so far fewer materials are available to the public because there is less of a sense that the public has the "need to know."

In addition to the absence of materials with respect to many countries, there is also a linguistic barrier to the comprehensive study of homeland security policies overseas. An all-inclusive study of the publicly available materials in the handful of countries dealt with in this book would require the researcher to be fluent not only in English, but also in Hebrew, French, German, Italian, Dutch, and Japanese. Perhaps a few lucky (and brilliant!) people with this linguistic repertoire can be found somewhere, but I am definitely not among them (having true command of only two of these), and thus some documentation could not be analyzed. Consequently, this is somewhat similar to the story of the man who is found searching attentively for his car keys underneath a streetlight in the middle of the night. When asked where he dropped the keys, he points to his car, shrouded in darkness, down the street. When asked why he is searching for his car keys near the streetlight when he dropped them near his car, he replies: "because this is where the light is."

As the subfield of comparative homeland security is quite new, it also, at present, lacks the kind of comprehensive methodological base that other, more mature, areas of inquiry enjoy. Most works that deal with comparative analysis in fields and subfields such as comparative politics, comparative public health, and comparative policing do not integrate the data and analyze them but, rather, lay out different policies (followed by different nations or jurisdictions within a nation) side by side, although this methodology does help increase understanding of how and why things are done in different contexts (such as countries) through comparing and contrasting. As this is an introductory text designed to introduce the reader to the subfield of comparative homeland security, the goal here is not to produce a theoretical tome that will solve the problem of the absence of a good comparativist methodology that truly integrates data and analysis. Nevertheless, by breaking the book down by issue areas within homeland security and then looking at the approaches of different countries in those contexts, I have attempted to at least take one step in the direction of some sort of integrative approach to comparing across countries.

Finally, comparative homeland security is a very dynamic field, with homeland security laws, policies, and strategies overseas constantly evolving, and while I have attempted to provide as much up-to-date information as possible, changes are occurring constantly and no book in this area can be 100 percent current. In addition, the reader should also bear in mind that policy and strategy as expressed in documents and briefings is not necessarily what really happens. To understand what does really happen, researchers have to have worked in the various areas within homeland security in a senior capacity (in order to have a good overall view of policy and strategy) in all of the countries touched on in this book, and they need simultaneously to continue working for all of these agencies in all of these countries to make sure that the knowledge they have is indeed still relevant but clearly no such person exists.

Despite the above challenges—most of which, in one form or another, are common to all works in homeland security or other areas of public policy— this book will provide you with a strong grounding and basic understanding of the emerging subdiscipline of comparative homeland security. Since comprehensiveness is not an option, the focus here is on providing vignettes of information that are interesting and useful, so each chapter touches on a different mix of countries and different sets of issues. Hopefully, the book will stimulate your interest in this field and encourage you to look outside your national borders (although written primarily for an American audience, the book will be of use to others as well) for answers to homeland security problems. The more that policymakers and practitioners in different countries can learn from each others' strategies and approaches, the greater will be the shared pool of knowledge, and this knowledge will ultimately make people safer. That is reason enough to study comparative homeland security.

Acknowledgments

I would like to thank Kristin Darken and Melissa Lieurance for their invaluable assistance in creating the maps and graphics for this book.

NADAV MORAG

INTRODUCTION
Studying International
Homeland Security Policies

WHAT IS HOMELAND SECURITY?

Homeland security is a uniquely American concept. Although a number of other countries around the world have employed the term since its entrance into common usage in the wake of the monstrous terrorist attacks against the World Trade Center and Pentagon on September 11, 2001, (9/11), they have done so essentially because they were following America's lead. Despite the fact that many countries have partially adopted the term, they have yet to really internalize the emerging discipline of homeland security in the way that it is being developed in the United States. Of course, disciplines, both in terms of their practitioner and academic components, take at least several decades to become fully developed and accepted, so it is no surprise that homeland security is still evolving and that there is a wide range of definitions for this discipline. It is not our purpose in this volume to provide a definitive definition of homeland security but, rather, to focus on the approaches and policies followed by a select group of countries *within the realm* of homeland security. However, to do this we must begin with some sort of baseline working definition in order to determine which types of overseas policies should be surveyed and which should not.

As homeland security is an American concept, there is some logic in turning to the premier homeland security strategy document, the *National Strategy for Homeland Security*, a revised version of which was issued by then president George W. Bush and his Homeland Security Council in October 2007, to shed some light on the concept. According to the *National Strategy, homeland security* is defined as "... a concerted national effort to prevent terrorist attacks within the United States, reduce America's vulnerability to terrorism, and

Comparative Homeland Security: Global Lessons, First Edition. Nadav Morag.
© 2011 John Wiley & Sons, Inc. Published 2011 by John Wiley & Sons, Inc.

FIGURE I.1 Old Glory at Ground Zero. U.S. Navy photo by Journalist 1st Class Preston Keres. September 15, 2001. Image released by the U.S. Navy with the ID 010915-N-3995K-024 @ Wikimedia Commons.

minimize the damage and recover from the attacks that do occur" (Bush, 2007, p. 3). Based on this definition, homeland security would appear essentially to be focused on counterterrorism and thus recognizable overseas as a "national strategy for counterterrorism." Indeed, the British government issued just such a strategy in 2006, known as *Countering International Terrorism: The United Kingdom's Strategy*, which is discussed in Chapter 2.

All the countries surveyed in this book have either written strategies or unwritten approaches to dealing with terrorism and hence, in this context, the United States would appear to be just another country with just another counterterrorism strategy. Indeed, in the wake of 9/11, *homeland security* may have been viewed by many as an alternative term for *counterterrorism*. Nevertheless, on page 6 of the *National Strategy* it is noted that preparedness in a homeland security context also requires coping with "...future catastrophes—natural and man-made..." (Bush, 2007, p. 6), and page 10 of the document notes that catastrophic natural disasters and public health emergencies are part of the homeland security threat menu. The *National Strategy* goes on to refer to a broad range of other policy issues, including transportation security, policing, border security, critical infrastructure protection, countering radicalization, and cyber security, to mention a few. Looking at this document as a central reference point and viewing it holistically thus suggests that homeland security, as interpreted by the leadership of the executive branch of government in the United States, is an extremely broad field. It seems to involve most threats to the stability and normal operation of government and society at the local, state, and/or federal levels of government—perhaps barring strictly economic threats such as the collapse of the stock market or consumer spending, the breakdown of credit markets, and other such issues that are not brought about directly by disasters, health emergencies, or terrorism. This does not, of course, imply that the federal government has a monopoly on knowledge and understanding and is thus able to define homeland security in an unambiguous and correct manner. However, given the absence of a universally agreed-upon definition, using the broad definition developed by the federal government represents a reasonable topological compromise in view of the central role played by the federal government in defining homeland security policy.

In terms of actual policies and institutions, one of the most important outcomes of the *National Strategy* and a slew of other federal strategy documents that have been produced and updated over the years since 2001 is the creation of homeland security agencies (or homeland security functions within existing agencies) at all levels of American governance. The most significant of these institutional changes was, naturally, the creation of the Department of Homeland Security (DHS) in November 2002—although a much smaller Office of Homeland Security and a Homeland Security Council (modeled on the National Security Council) had been in existence previously and were established by presidential executive order in October 2001. Thus, to help define homeland security, and in addition to looking at the *National Strategy*, one can also look at the policy areas for which DHS is responsible, as DHS is the principal federal agency with homeland security duties. On June 6, 2002, then president George W. Bush, in an address to the nation, outlined the four essential missions of the newly proposed DHS: (1) border and transportation security, (2) emergency preparedness and response, (3) coping with the threat of weapons of mass destruction, and (4) intelligence gathering and analysis designed to create an integrated intelligence picture (DHS, 2008a, p. 5). Without getting into a survey of the convoluted process of organization and reorganization in DHS and the evolution of DHS missions and areas of responsibility since the creation of this mammoth department (except for the Department of Defense and the Department of Veterans Affairs, the largest, by employee numbers, in the federal government) suffice it to say that the current strategic goals of DHS, based on the department's strategic plan published in 2008, include (1) counterterrorism (including border security, enforcement of immigration laws, and procedures); (2) protection from weapons of mass destruction; (3) protection of critical infrastructures and key resources (including partnerships with private-sector providers, ensuring government continuity of operations during emergencies, cyber security, and protection of air, maritime, and surface transportation sectors); and

(4) ensuring emergency preparedness through strengthening response and recovery (DHS, 2008b, pp. 6–20).

In view of the above, and without attempting to produce a perfect and definitive definition of homeland security, a *functional* categorization of the policy areas that fall within the sphere of homeland security may be summarized as follows:

- Policies directed at mitigating the threat of terrorism and large-scale criminality (of the type that threatens social and economic stability), including:
  - o Counterterrorism strategy
  - o Intelligence sharing and coordination
  - o Policing strategies
  - o Countering homegrown radicalization
- Policies directed at enhancing security measures, including:
  - o Border security and immigration enforcement
  - o Transportation security (air, maritime, and surface)
  - o Critical infrastructure protection
- Policies directed at management of the immediate and long-term effects of acts of terrorism, natural disasters, and/or public health emergencies, including:
  - o Emergency and disaster preparedness and response
  - o Public health
  - o Development of political, social, and economic resiliency

Needless to say, these areas include a great deal of overlap, and all of them require the sharing of information, inter- and intraagency cooperation, and interfacing with the public. Although the list of policy areas above is no doubt imperfect, it does have the advantage of, more or less, covering those areas viewed in various governmental strategy papers and academic studies as part of homeland security, and these areas are, accordingly, addressed in this book in differing degrees of detail (based on the availability of information and space for analysis) with respect to approaches taken to them outside the United States.

Although we may view the policy areas above as constituting the field of homeland security in the United States, one would be hard-pressed to find a similar amalgamation of what would appear to be disparate policy domains in the rest of the world. Counterterrorism, security, and crisis management, broadly speaking, are not seen as part of the same discipline overseas, and there certainly is nothing equivalent to DHS in trying to bring these policy areas together under a single institutional framework (the Australians toyed briefly with the idea of creating their own version of DHS and then decided against it). Of course, there are long-standing cooperative relationships between intelligence agencies and police or between police and fire and emergency medical services since many challenges require a multidisciplinary response, but no one overseas has attempted to put so many different functions and activities within the same rubric and argue that they are all part of the same type of public policy issue.

HOMELAND SECURITY VERSUS NATIONAL SECURITY

The concept of homeland security is uniquely American largely because most other democratic countries do not distinguish as clearly between what in the United States was referred to by some as the "home game" versus the "away game." Historically, the United States benefited from its geographic isolation far from the wars and political machinations of the leading powers on the European and Asian continents. Consequently, there gradually developed a view that there was a distinct separation between domestic and international challenges and that policies (and their attendant institutions) employed overseas were largely irrelevant domestically, and vice versa. As the United States came to play a very large role on the world stage in the wake of World War II, the concept of national security, and its attendant institutions, including the Department of Defense, the Central Intelligence Agency, the National Security Council, and the National Security Agency, developed apace. The goal of national security policy was to protect and enhance the various elements of national power (military, diplomatic, economic, etc.) and to safeguard national interests overseas.

Since much of the development of the concept of national security occurred in the context of the Cold War, it is not surprising that the discipline of national security was focused on the Soviet threat and ensuring that the United States was able to contain and deter Soviet ambitions and actions worldwide. Many of the tools employed in safeguarding national security; such as fighting wars, espionage, paying off allies and potential allies, disinformation campaigns, the occasional assassination, and other measures, were seen as necessary and legitimate policies for use overseas but certainly fundamentally contrary to American laws and values in a domestic context. Consequently, apart from counterespionage operations and other limited spillovers into the domestic arena, national security was essentially an overseas endeavor.

With the effective end of the Cold War in 1989 and the collapse of the Soviet Union in December 1991, threats to national security were transformed and diminished. Although countries such as North Korea, Iraq under

FIGURE I.2 Soviet SS-X-15 mobile intercontinental ballistic missile. May 23, 1984, U.S. Department of Defense @ Wikimedia Commons.

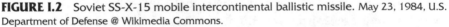

Saddam Hussein, and Iran (the states constituting George W. Bush's "Axis of Evil") were still problems (and, of course, the United States would go to war against Iraq in 2003) and the rise of China was certainly to pose challenges, the pervasive sense of fear of Soviet encroachment (and Soviet nuclear missiles) that propelled national security to the top of the governmental agenda came to an end with the faltering and then collapse of the Soviet empire. At the same time, new threats were developing, of which the primary one was to become that of international terrorism.

Terrorism, however, is a very different type of threat, in terms of both scope and modus operandi. Terrorists cannot, of course, command the elements of national power and thus do not constitute anything remotely similar to an advanced, capable, and belligerent nation-state. Nevertheless, with the advancement of military technology and the global media, terrorist groups were increasingly in a position to effectively attack vulnerable targets and populations and to create the impression that they were almost as dangerous and threatening as the Soviet Union once was. Since, in public affairs, impressions are more important than reality as people act based on their impressions (whether or not those impressions conform strictly to reality), terrorists were surprisingly well-placed to transform themselves into major threats to national security even though, in terms of the power that they were able to wield, they

FIGURE I.3 Poster of Osama Bin Laden. Al Qaeda propaganda poster found by U.S. Special Operations Forces in Afghanistan, n.d., public domain under terms of Title 17, Section 104 of the U.S. Code @ Wikimedia Commons.

really did not deserve to be viewed in such a way. Impressions may motivate people's actions, but they cannot magically transform terrorists from relatively weak actors into global superpowers. Consequently, terrorists must generally infiltrate their target societies in one manner or another and then strike at relatively undefended and vulnerable targets (usually those that involve public access of some sort). In democratic countries at least, this means that the terrorist threat cannot be addressed using exactly the same policies and institutions that were employed against national security threats outside the borders. Arguably, nowhere was this distinction starker than in the United States, with its long traditions of keeping the military and intelligence agencies largely out of domestic affairs. Consequently, in dealing with terrorist threats within American territory, a new concept and approach seemed to be needed.

FIGURE I.4 Aerial view of floodwaters in downtown New Orleans after Hurricane Katrina. September 1, 2005, U.S. Navy photo @ Wikimedia Commons.

After the shock of 9/11 gradually began to abate and in the absence of any additional successful terrorist attacks on U.S. soil coupled with a new kind of shock, the massive natural disaster of Hurricane Katrina and what was perceived as the impotence of governmental entities at the local, state and federal levels in dealing with its aftermath, the concept of homeland security evolved and began focusing on natural disasters as well as those of the human-made variety. There is, of course, some logic to viewing the threat from terrorism and that of natural disasters or pandemics as part of the same discipline. After all, many of the measures instituted to prepare for and recover from the aftermath of emergencies can be applied to terrorist threats, natural disasters, and public health threats. Also, many of the preventive techniques are also common to coping with all three categories of threats (e.g., intelligence gathering, analysis, and dissemination are critical elements of prevention regardless of whether the information pertains to the activities of a terrorist cell or the spread of a pandemic). Moreover, if one is to consider terrorism to be a major threat to the lives and livelihoods of Americans, natural disasters and pandemics certainly qualify as threats that are equally serious, if not more so (particularly in the case of a pandemic outbreak that may kill millions and bring about the collapse of the health system). Perhaps other countries did not evolve toward developing a concept of homeland security because they did not see national security as focused almost exclusively overseas and ending at the national borders and because, for many other democracies (with the exception of Japan), large-scale natural disasters are simply not as common or as threatening as they are for the United States.

The comments above, at least to some degree, are philosophical musings regarding the concepts of national security and homeland security (although the reader should bear in mind that philosophical musings are usually the first step and basis for strategic policymaking). The upshot is that homeland security is a concept and discipline that is of American origin and still largely alien to the rest of the world (even if they may use the term to describe counterterrorism policy). The reader may thus inquire as to the utility of a book that focuses on international homeland security policies if homeland security does not technically exist internationally, at least not as a discipline. The answer to this is that although homeland security as a discipline is alien to other countries, there is a wealth of experience and tested policies and approaches employed overseas in the various areas that constitute homeland security, and it would not make sense for Americans to remain blissfully unaware of them. Reinventing the wheel and repeating the mistakes of others are activities that are both wasteful and counterproductive.

INTERNATIONAL COMPARATIVE HOMELAND SECURITY

Learning from other countries' experiences and approaches is important not only because it makes sense for American decision makers to learn from the experiences of foreign governments but also because in many cases the threats are transnational, and consequently, although homeland security may appear to be a fundamentally domestic concept, safeguarding it requires cooperation with other countries. Whether the threat emanates from radicalized Europeans accessing the United States under the visa waiver program in order to execute terrorist attacks or aircraft passengers flying into the United States from an Asian city carrying the latest viral mutation with them, many homeland security threats emanate from abroad. Examples of such threats abound. In the terrorism sphere, in addition to the 9/11 attackers, Ahmed Resam (the "millennium bomber"), arrested in 1999, used Canada as a staging area for his plot to bomb the Los Angeles International Airport, and Richard Reid (the "shoe bomber") boarded a Miami-bound flight in Paris in December 2001. In addition, the 2006 transatlantic liquid explosives plot (the "Overt plot") was hatched and prepared in the United Kingdom (UK), and Umar Farouk Abdulmutallab (the "underwear bomber" or "Christmas bomber") boarded his Detroit-bound flight in Amsterdam in December 2009. The spillover of Mexican criminal violence into the United States has also been an issue of concern for some time. In the pandemic sphere, the SARS outbreak in China led to some outbreaks in the United States, with the public health system being put on alert in December 2003 and the outbreaks of avian influenza and swine flu in Southeast Asia and Mexico, respectively, led to pandemic concerns in the United States. In short, there is no lack of examples of homeland security threats emanating from overseas.

It therefore follows that addressing them will not only require international cooperation but also an understanding of how other countries, particularly allied democratic nations, address these issues within their own borders. To be able to do this, one must have some baseline knowledge of the governmental and institutional framework and legal basis under which these countries operate. An additional advantage to conducting comparisons is that they help identify options that may otherwise be overlooked as well as the manner in which various policies that have not, thus far, been adopted in the United States might play out here (Watts, R. L., 1999, p. 2).

The focus of this book is on a handful of democratic countries, for three primary reasons: (1) as noted in the Preface, time and data limitations do not allow for an across-the-board survey of policies followed by countries worldwide; (2) there is little point in looking at nondemocratic countries since part of the goal of this analysis is to provide information and ideas that might be used by students of homeland security to improve policies in the United States, and nondemocratic countries are simply less relevant because their policies and practices are usually considerably less applicable to democratic states; and (3) there is little point in looking at significantly dysfunctional countries, countries without significant homeland security–related policies in place or those with such policies that clearly do not work. In view of the above, in this book we focus primarily on Israel, the UK, France, Germany, Canada, and Australia, with a more cursory discussion of additional countries when their policies are of particular interest, including Japan, the Netherlands, and Italy. Also as noted in the Preface, it was not possible to obtain sufficient data on the policies of each of these countries with respect to each policy area of homeland security surveyed. Because of the time constraints and, in many cases, the sensitivity of the data, it is simply not feasible to create a neat matrix in which each area of homeland security can be laid out and complete data on each country can be filled in. Accordingly, the focus will be on the more significant strategic policies (with some detailed examples) of the various countries where information is available. This means that not every country is covered in every chapter, and thus each chapter involves a different mix of countries with differing levels of emphasis on each. The choice of country to be addressed per homeland security issue area will depend not only on the availability of information but also on the degree to which a particular country has a particularly interesting or useful set of policies in a given issue area. It will be left to future researchers in this field to write definitive accounts of the policies of each of the countries touched upon here in each issue area of homeland security.

This volume fits into the general literature dealing with comparative government (although, of course, the focus is on homeland security–related issues). Consequently, a few words with respect to the comparative method are in order.

THE COMPARATIVE METHOD AND COMPARATIVE HOMELAND SECURITY

As with any other social science research methodology, the comparative method has its advantages and its disadvantages. If we confine ourselves to focusing on policy-oriented research and analysis, we find that the comparative method does not provide us with a means of measuring the degree of efficacy of policies followed by different entities in different contexts (in our case, countries) and designed to achieve goals that may be slightly different from one another. This is because the comparative method is not designed to be able to provide meaningful measurements of differing policies operating in differing contexts. To use a fairly simplistic example, the adversarial legal system, the system generally in place in common law countries such as the United States, UK, Canada, and Australia puts a judge in the role of impartial arbiter, with the court's primary role being to ensure that the proceedings adhere to the law and provide due process. On the other hand, an inquisitorial legal system, the system in place in many civil law countries, such as France, Italy, and Spain, gives the court the role of determining the facts of the case and assigning guilt. Common law systems are focused on judges (and juries) and allow considerable scope to ad hoc decisions by courts with respect to specific legal cases, whereas civil law systems tend to leave little room for judicial discretion and to focus more on a codified body of generalized principles (Slapper and Kelly, 2009, pp. 1869–1876). In reality, there is a considerable degree of crossover between these approaches, and adversarial systems can act inquisitorially in some ways, and vice versa. Nevertheless, these two systems are different in their fundamental principles as well as in the way in which they operate.

Although it may be possible to measure the overall efficacy of the judicial system in France and then compare that to the efficacy of the system in the United States (by looking in both cases at variables such as conviction rates, the number of unresolved cases in the system in a given year, the length of legal proceedings, and the cost of legal proceedings), it is not possible to measure the effectiveness of the judicial system in the United States by looking at how things are done in France and then use French measures of success, based as they are on the French system, to determine U.S. success. This is equivalent to the proverbial comparison of apples to oranges. Rather than measuring things, the comparative method is designed to discover "empirical relationships among variables" (Lijphart, 1971, p. 683), and this means that the comparative method allows us to understand how processes work and thus increases our understanding of policy issues and our range of conceivable policy options. Looking at the approaches, policies, and experiences of other countries with respect to homeland security policy (whether or not they view it as that) makes it possible to gain a greater understanding of the options available to U.S. policymakers, a sense of how policies should be selected and evaluated, and an understanding of the options available to overseas partners as well as how they operate and cope with their own threats—many of which are, as noted earlier, "transferrable" to the United States. In other words, the comparative methodology gives us the framework in which to study different policies and policy contexts but not really the ability to translate and apply one country's policies in another.

To understand homeland security–related approaches, strategies and policies followed in the primary countries surveyed in this book: Israel, the UK, France, Germany, Canada, and Australia (and to a lesser degree, Japan, the Netherlands, and Italy), it will first be important to understand the historical, political, and institutional contexts in which these countries operate (more on this later).

STRUCTURE OF THE BOOK

In this book we address each area of homeland security based on the categories comprising the field of homeland security but with a focus on the overarching approaches, legal bases, institutions, and some of the specific policies followed by the countries noted above as well as, in certain contexts, the supranational European Union. Each chapter, however, focuses on a different mix of countries and issues within the general topic of each chapter because the goal of the book is to provide some interesting perspectives of policies and approaches followed overseas rather than an exhaustive catalog of countries and their respective laws, institutions, and policies. In addition, sidebars that provide snapshots of particular practices or issues pertaining to topics being addressed within each chapter are interspersed within the text to provide the reader with some examples and a more concrete sense of how the issues being discussed are addressed by one or another of the countries in this survey.

Chapter 1 consists of a brief overview of the political institutions and judicial systems of Israel, the UK, Canada, Australia, France, Germany, Italy, the Netherlands, and Japan. The objective of this chapter is to provide the reader with the general political and institutional context in which these countries operate. Chapter 2 is a survey of counterterrorism laws, strategies, institutions, and examples of specific policies followed by a number of countries. Chapter 3 focuses on policing and law enforcement institutions and strategies, which are an important facet of homeland security given that traditional policing plays an integral part in counterterrorism efforts. This is the case both because in all the countries surveyed as well as in the United States, the first line of defense and most ubiquitous counterterrorism actor is the local law enforcement official and because a significant component of terrorist activity involves a criminal nexus of one sort or another. Chapter 4 focuses on the status of Muslims in Europe and the counter-radicalization strategies followed by some European countries given that homegrown radicalization is considered a growing problem not only in the United States, but especially in Europe. Particularly as international travel (especially to areas of known Jihadi activity such as Pakistan, Yemen, or Somalia) has come under greater scrutiny by the authorities, domestic radicalization has increasingly afforded global Jihadi groups an alternative to traveling to their target countries. Moreover, radicalized individuals from Europe and other areas that enjoy visa-free travel to the United States can pose a significant threat to homeland security.

Chapter 5 focuses on the role played by military forces in a number of countries in the provision of domestic security and support for civilian authorities. The issue of the military's role in domestic security is often quite controversial in the United States, but as shown in this chapter, many countries have few qualms about employing their respective military establishments for domestic security missions, particularly when conditions are not normal. Chapter 6 focuses on border security, immigration policy, and survey border management approaches followed with respect to the supranational European Union (a model that could conceivably be applicable at some future date to the North American Free Trade Area or any other North American combined border security regime). Chapter 7 focuses on security strategies with respect to protecting potential terrorist targets. The chapter begins with a brief discussion of critical infrastructure protection and governmental partnerships with the private sector—which in all the countries surveyed currently make up the bulk of critical infrastructure operators. Chapter 7 then moves on to focus on transportation security and look at the air, maritime, and surface transportation sectors. The ready access and mass use of transportation networks (and their criticality for economic and social interaction and activity) have made them prime targets the world over for terrorist attacks, as attested to by the large number of attacks against buses in Jerusalem, Tel Aviv, and other Israeli cities between 2000 and 2004; the

Madrid rail network in 2004; the 7/7 and 20/7 attacks against public transportation in London in 2005; and of course, the Al Qaeda attack against four U.S. airliners on 9/11.

In Chapter 8 we look at strategies, institutions, and policies followed by the countries surveyed that are designed to respond to emergencies (whether terrorism-, natural disaster-, or public health–related) and to manage the intermediate- and long-term impact of such emergencies, including emergency preparedness and emergency response and management. The focus is also on the approaches taken by several countries toward crisis communications, with an eye to fostering public resiliency. Finally, in Chapter 9 we look at a number of public health strategies, institutions, laws, and policies followed by a variety of countries surveyed.

Although none of the countries and policies to be surveyed represent perfect policy approaches and solutions to all homeland security problems, many of them have proven useful and comparatively successful in achieving specific policy objectives and thus should be of interest to others in analyzing and improving upon homeland security laws, strategies, and policies in the United States and other countries as well as building a foundational knowledge base for students of homeland security.

ISSUES TO CONSIDER

- Why is homeland security a uniquely American concept?
- How can homeland security be defined?
- How has the concept of homeland security evolved?
- How does homeland security differ from national security?

CHAPTER 1

COUNTRY OVERVIEW

Each of the countries surveyed in this book should first be understood in the context of their governance systems. This means looking at the constitutional underpinnings, the relationship between the executive, legislative, and judicial branches and the nature of territorial governance (federalism, centralism, and other models). There are a number of excellent texts focusing on particular countries' governance systems that can provide the reader with a comprehensive understanding of these countries. The goal here is not to repeat those efforts but, rather, to focus on homeland security–related issues. At the same time, to provide the context mentioned above it is important to establish some very basic knowledge of the governance systems of the countries to be surveyed. The following is therefore a highly abridged overview of the countries focused on in the survey. These are, as noted earlier, Israel, the UK, Canada, Australia, Germany, France, the Netherlands, Italy, and Japan.

THE STATE OF ISRAEL (MEDINAT YISRAEL)

Israel is a small country with a total area of 22,072 square kilometers (approximately the size of the American states of New Jersey or Massachusetts). It has a population of 7 million inhabitants (80 percent Jews and 20 percent Arabs). It also controls a large section of the West Bank and has annexed the northern, southern, and eastern sections of the city of Jerusalem and the Golan Heights (all of these territories were conquered during the Six-Day War of 1967). Most of the population lives in the temperate northern two-thirds of the country (which enjoys a Mediterranean climate), with most of the desert regions of the south being sparsely populated. The topography varies from rocky and partially wooded hills in the north and east to sandy coastal

Comparative Homeland Security: Global Lessons, First Edition. Nadav Morag.
© 2011 John Wiley & Sons, Inc. Published 2011 by John Wiley & Sons, Inc.

FIGURE 1.1 Israel.

plains in the west, to rugged desert hills in the south. Israel is a highly urbanized country, with 92 percent of the population living in towns or cities and 82 percent of the workforce employed in service industries, 16 percent in heavy industry, and only 2 percent employed in the agricultural sector. The leading sector in the economy is the high-tech sector; Israel is one of the world's leading producers of computer software, communications technology, avionics, and medical electronics.

The State of Israel was declared on May 14, 1948 upon expiration of the British Mandate for Palestine. The new state, however, did not appear in a vacuum and was established upon a foundation of three decades of nation building and institution building by a largely autonomous Jewish community (known in Hebrew as the Yishuv) operating under the administration of the British Mandate for Palestine. This incubatory period made it possible for the new state to come into existence with surprisingly robust and tested democratic institutions and traditions. In fact, it is quite remarkable that Israel was able to maintain an unbroken record of democratic rule throughout the years given the significant security challenges that it faced, including no less than seven full-scale wars as well as several additional

FIGURE 1.2   Israeli parliament (Knesset) building. This Wikipedia and Wikimedia Commons image is from the user Joshua Paquin and is freely available at http://commons.wikimedia.org/wiki/ File:Knesset_building.jpg under the Creative Commons Attribution 2.0 license.

significant military operations and long periods of dealing with intensive terrorist campaigns.

Israel is a parliamentary democracy and thus follows the principle of *responsible government* (in that the executive branch, known as the "government," is responsible to parliament and can be replaced by it). This means that the government must enjoy the support of the majority of the parliament (or, at the very least, avoid being voted out by a majority of the parliament), and the parliament has the power to unseat the prime minister and the rest of the cabinet if they lose majority support in the parliament (usually via the parliamentary procedure known as a *vote of no confidence*). The upshot is that in such systems, the parliament is not only responsible for passing legislation, but is also responsible for creating governments (cabinets). All of the countries surveyed in this book are parliamentary democracies of one sort or another, the only exception being France, which has a hybrid, or semipresidential, system. Indeed, although it may seem strange to American readers, the presidential system employed by the United States (in which the executive branch is independent of the legislative branch) is rare among democracies and is largely confined to the Western Hemisphere. In a parliamentary system such as Israel's, the government—that is, the ministerial level of the executive branch (the cabinet)—is created from the legislature (the parliament), so that the prime minister and the other cabinet ministers are also members of parliament [in some systems all cabinet members must be MPs (members of parliament) in others only some are MPs, and in yet others cabinet ministers cannot be MPs]. In the Israeli case, at a minimum, the prime minister and half of the cabinet must be MPs, but in practice the vast majority of (and often, all) government ministers are also MPs. In a parliamentary system, the prime minister is not elected directly but is elected to parliament (either by representing a voting district or, as in the Israeli case, by running at the head of a party list of candidates), and those cabinet ministers who are also MPs are similarly elected to parliament (with the non-MP ministers appointed by the prime minister). Consequently, there is no constitutional separation between the executive and legislative branches in a parliamentary democracy. Most parliamentary systems have a bicameral parliament (two legislative houses), but Israel has a unicameral parliament—the Knesset. The Knesset consists of one house with 120 MPs (known as MKs, members of Knesset), and the prime minister and the vast majority of his/her cabinet are among those 120 members (each having one vote).

As noted above, all Israeli MKs are voted in by party list, as there are no voting districts in Israel (or rather, the country is one voting district). This system of election, known as *proportional representation*, is quite rare among parliamentary systems, most of which employ some version of the U.S. *winner-take-all system*, in which the candidate with the most votes (although not necessarily a majority of votes) in any given voting district is elected to represent that district (the British, using a horse-racing metaphor, refer to this as *first-past-the-post*). In many ways the proportional representation system is very democratic, in that it essentially means that the leaders of smaller parties that represent only a fraction of the voters are able to achieve parliamentary office and thus, at least theoretically, represent the views and preferences of those voters. Thus, entire swaths of minority opinion can enjoy representation, whereas in a winner-take-all voting system such as that of the United States, voters who supported candidates and parties that garner only a fraction of the votes are essentially ignored. This is one of the reasons that in such systems fewer candidates from non-mainstream parties tend to achieve a place in the legislature. If the United States were, hypothetically, to institute a proportional representation voting system, one can be certain that Congress would include a wide variety of parties and that the effective two-party monopoly of power that exists today would be challenged and would probably break down over time. One of the downsides of this voting system, however, is that it often affords to small parties and their leaders, who represent a political minority of one kind or another, the power to impose themselves on the majority (something that is not terribly democratic).

As a result of the proportional representation voting system, and in view of the deep divisions in the Israeli body politic, elections for the Knesset produce a very large number of parties. At present, the Knesset members (the eighteenth Knesset, voted in on February 24, 2009) belong to no fewer than 12 separate political parties, with the largest party, Kadima, with 28 seats in the opposition, and the second-largest party, the Likud, with 27 seats, forming the core of the current Israeli government. Since a government (i.e., the prime minister and the other members of the cabinet, who are collectively tasked with running the executive branch) can only be voted in with a majority in the Knesset, this means that the Likud was 34 seats shy of enjoying a slight majority in the Knesset (61 seats, of course, being needed for a tiny majority).

This current distribution of seats in the Israeli parliament is not unique. No Israeli political party has ever come close to enjoying a majority in the parliament, and consequently, all Israeli governments are formed through an alliance (or *coalition*) of parties elected to the Knesset. The current government (Israel's thirty-second), is made up of six parties, the largest and central one being the party of Prime Minister Benyamin Netanyahu, the Likud. Since the Likud is far from enjoying a majority in parliament, however, Netanyahu must ensure the integrity of his coalition, and this means that he, or any Israeli prime minister for that matter, must compromise and share power in a manner that would be quite foreign to a U.S. president. In fact, in parliamentary systems, the cabinet as a whole makes policy and the prime minister is not the chief executive and commander-in-chief, as is the U.S. president, but rather *primus inter pares* (first among equals) in the collective decision-making of the cabinet. In parliamentary systems in which one party enjoys a majority in the parliament, the prime minister (who is head of his/her party) is in a much more powerful position than in countries such as Israel in which rule is by coalitions of parties. Nevertheless, even in systems in which one party enjoys a clear majority in parliament, the prime minister does not enjoy a separate status (as does the president of the United States) since prime ministers are not voted in directly, and their status is dependent on the maintenance of the domination of their party over the parliament. Moreover, prime ministers must act in the context of the cabinet, with a majority vote in the cabinet for all important policy issues.

Parliamentary democracies also maintain a separation between the functions of "head of state" and "head of government" (whereas in the United States, these functions are amalgamated in the person of the president of the United States). As Israel is a republic, the head of state is the president, whose role is almost entirely ceremonial, with substantive powers being confined largely to the right to commute the sentences of convicted criminals or pardon them (and this only at the recommendation of the Ministry of Justice). The president is supposed to be "above" politics and act as a unifying figure— although this never really happens, as most Israelis do not put much stock in the Israeli presidency and usually ignore it.

The realities of coalition politics sometimes makes Israeli cabinets chaotic, and Israeli prime ministers often have to act more as consensus builders than

FIGURE 1.3 Israel Supreme Court building. This Wikipedia and Wikimedia Commons image is from the user Adiel lo and is freely available at http://commons.wikimedia.org/wiki/File:Israeli_ supreme_court_building_nightshot.JPG under the Creative Commons Attribution-Share Alike 3.0 Unported license.

as leaders, in order to keep together coalitions of parties with different agendas and ideologies. One of the repercussions of this need to maintain coalitions is that long-range planning is highly difficult, as Israeli cabinets do not always last for their entire four-year term (when they do not it is usually because coalitions disintegrate, and this leads to a loss of support in the Knesset, which usually results in the calling of early elections), and the prime minister must be careful not to be seen as supporting positions that might irrevocably alienate his/her coalition partners in the cabinet, causing them to leave the government and vote against it in the Knesset. This also means that the prime minister cannot use the cabinet as a true decision-making and deliberation body because the cabinet is stacked with his/her political rivals, both in the prime minister's own party and among the prime minister's coalition allies (Freilich, 2006, pp. 639–640, 645–646).

Unlike the linkage between the executive and legislative branches that exists in Israel and other parliamentary democracies, the court system in Israel is independent of these other institutions. Although Israel has a number of specialty courts that deal with things such as municipal issues, labor disputes, juvenile criminality, personal law matters that fall under the purview of religious courts, and a military justice system (more on this in a subsequent chapter), the primary court system has three tiers and is responsible for dealing with both criminal and civil cases. The lowest level of courts in this system are magistrate courts (Betei Mishpat Ha'shalom), of which there are currently 26, which generally deal with criminal offenses punishable by incarceration of up to seven years and a range of civil issues. These courts are overseen by a single judge, and there are no juries in this or any other court in Israel. The next level is the district courts (Betei Mishpat Mehozi'im), of which there are five. These deal with more serious criminal cases and more monetarily significant civil cases and also act as an appellate court for cases previously tried in magistrate courts. Many of the cases heard in these courts are presided over by a single judge, but appeals and very serious cases are handled by a panel of three judges. The highest legal body in Israel is the Supreme Court (Beit Mishpat Ha'elyon), which usually consists of 12 to 14 justices (the number is set by the Knesset). The Supreme Court acts as the supreme appellate court (cases are usually heard by a panel of three justices, although the president of the Supreme Court can create a larger odd-numbered panel for specific cases). In addition, the Israeli Supreme Court acts as a high court of justice (known in this context by the acronym Bagatz—Beit Mishpat Gavoha Le'tzedek) in exercising judicial review of government policies and the actions of official bodies—and it has even been argued that it has the right to review legislation passed by the Knesset, although that has yet to happen and will probably cause a constitutional crisis if and when it does. Unlike the U.S. Supreme Court, the Israeli Supreme Court receives petitions from citizens and noncitizens requesting rulings on matters related to public policy independent of specific judicial cases and frequently intervenes and issues rulings forcing the government to modify or abandon certain policies. For example, in the counterterrorism and security context, the court ruled on two separate occasions (in 2004 and 2005) that the government must change the route of the fence and wall security barrier that Israel built in the West Bank to lessen the negative impact on Palestinian civilians living near specific sections of the fence—despite arguments made by government attorneys with respect to the importance, from a security perspective, of maintaining the existing routes of the fence.

In this context, the Israeli Supreme Court may be thought of as one of the most powerful courts in the world and one of the primary guarantors of civil liberties in Israel. This is particularly so given the fact that like the UK, Israel, lacks a formal constitution against which legislation or the policies of government can be compared. While incorporating elements of other legal traditions, Israel's court system is still based fundamentally on common law, and consequently, precedents established in higher courts are binding on lower courts (a principle known as *stare decisis*).

Israel's small geographic size also causes it to be unique among the countries surveyed in terms of the manner in which governance occurs across the national territory. Among the countries surveyed, Israel, France, Italy, the Netherlands, and Japan all have a primarily centralized form of government, but the other countries, which are significantly larger and more populous than Israel, divide their respective territories into administrative regions of various kinds. Israel has no such administrative divisions, as the national territory is only the size of New Jersey or Massachusetts, so appointing governors, prefects, county supervisors, and the like would make little sense. Accordingly, Israel has a central government (based largely in the capital, Jerusalem, although the Ministry of Defense is based in Tel Aviv and all other cabinet ministries have branch offices in Tel Aviv) which holds considerable power and comparatively weak local governments (in the form of munici-palities for cities, local councils for towns, and regional councils for rural areas). The vast majority of policing functions, for example, are centralized, with one national police agency under the direct control of the central government having law enforcement authority throughout the country. In this sense one could argue that Israel is an example of the most centralized country in our survey (with Germany arguably being the least centralized, although Canada and Australia have highly federalized systems as well). While the role of local government has grown in Israel over recent years [in matters of policing, to use the previous example, municipal inspectors have been given some limited police powers (see Chapter 3)], the lion's share of policy issues are still handled at the central government level. The small geographic size and small population (7 million persons) of the State of Israel means that only national-level agencies have the budget, personnel, and clout to design and implement most homeland security policies.

THE UNITED KINGDOM OF GREAT BRITAIN AND NORTHERN IRELAND

The United Kingdom (UK) has a territorial area of 243,610 square kilometers (roughly the size of the state of Oregon) and a population of just over 61 million persons. Approximately 90 percent of the population lives in cities or towns. Close to 74 percent of the UK population is English, with close to 9 percent Scottish, 5 percent Welsh, and 3 percent Northern Irish. In addition, approximately 8 percent of the population originally hails from areas outside the British Isles. The climate is generally wet and overcast, and the topogra-phy varies from rugged mountains and hills in the north and west to rolling plains in the south and east. Industrial activity takes up some 24 percent of the economy, but the bulk (75 percent) of economic activity is in the service sector, the UK being one of the world's leading financial centers and enjoying one of the four largest economies in Europe.

Of the countries surveyed in this book, the UK has the longest tradition of parliamentary rule. It is not, however, a republic (as are Israel, France, Italy,

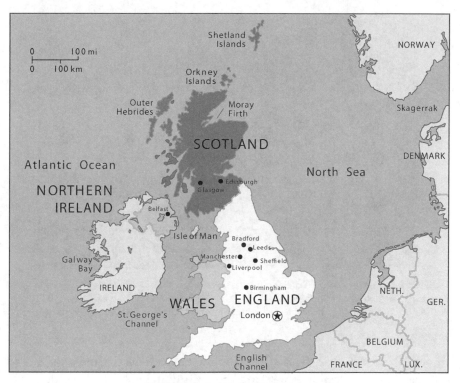

FIGURE 1.4 United Kingdom.

and Germany) but rather, as its name suggests, a constitutional monarchy in which the head of state is a hereditary monarch (Queen Elizabeth II, who also reigns over Canada and Australia). Much of English history involved a tug-of-war between the Crown, desiring to maintain royal prerogatives, and Parliament, desiring to increase its share of power. Ultimately, Parliament was largely victorious in this contest, but the Crown was able to retain some significant residual powers (known as the *royal prerogative*). Those powers include the power to enter into international treaties, the power to declare war and peace, the power to summon and to dissolve Parliament, the appointment of a government (a cabinet), and the power to commute sentences or grant pardons (Barnett, 2002, pp. 8–9). Perhaps even more significantly, the Crown must assent to any bill passed by Parliament before it can become law. Although these powers appear very impressive on paper and, indeed, appear to be at odds with many of the principles of democracy, there are strong conventions in place that regulate these powers. These conventions, although they do not enjoy the status of legal requirements, are extremely binding nonetheless. For example, while the Queen could theoretically reject a bill passed by Parliament, this would in fact be the first time

FIGURE 1.5 British Houses of Parliament. This Wikipedia and Wikimedia Commons image is from the user Maurice and is freely available at http://commons.wikimedia.org/wiki/File:British_Houses_of_Parliament.jpg under the Creative Commons Attribution 2.0 Generic license.

since 1707 that a British monarch would have done so, and this would unquestionably precipitate a serious constitutional crisis that would probably result in the significant curtailment of royal powers. Consequently, the monarchy cannot really exercise many of the significant powers that it theoretically enjoys. Moreover, most of these prerogative powers are no longer exercised by the Crown but rather by the government in the name of the Crown (e.g., powers of war and peace, the signing of international treaties, decisions on dissolving Parliament in order to call new elections), and it is generally understood that the Crown will assent to whatever the government requests of it. Beyond this, it is generally accepted that if Parliament passes a law regarding a particular matter, that issue will then be dealt with according to that act of Parliament rather than by government determining how the issue will be handled based on the powers it enjoys under the royal prerogative (Barnett, 2002, p. 10).

Unlike Israel, the UK has a bicameral parliament with two houses: a lower house (the House of Commons) and an upper house (the House of Lords). This bifurcation of Parliament (not unlike the rationale behind the creation of the more exclusive upper house of the U.S. Congress) was originally designed, at least in part, to allow the nobility, operating through the House of Lords (whose membership was once largely hereditary but is now largely appointed), to maintain their historic prerogatives and to act as a limitation on the "excitability of the masses" as reflected through the

House of Commons. The poet Samuel Taylor Coleridge reflected this view-point when he noted:

> You see how this House of Commons has begun to verify all the ill prophecies that were made of it—low, vulgar, meddling with everything, assuming universal competency, and flattering every base passion—and sneering at everything noble refined and truly national. The direct tyranny will come on by and by, after it shall have gratified the multitude with the spoil and ruin of the old institutions of the land.
>
> —Coleridge, 1833

As of 2005, there were 736 peers in the House of Lords (membership fluctuates). Fewer than 100 of those with voting rights are still hereditary peers, and the institution still contains 26 senior clergy of the Church of England as well as a large number of "life peers," appointed for life by the prime minister. Given that the House of Lords also plays an important judicial role, it includes in its membership up to 28 senior judges (including the Lord Chancellor, who heads the judicial branch, is also a member of the cabinet, and acts as Speaker of the House of Lords). The House of Lords is thus a nonrepresentative parliamentary body. Interestingly, members do not receive a salary for serving on this body, and consequently, it is not a professional body, with attendance ultimately being at the discretion of the individual. In the legislative process, the Lords have the role of scrutinizing legislation passed by the Commons and often improve upon legislation that is sometimes passed hurriedly by the Commons (Watts, D., 2006, p. 70). They can also delay the passage of a bill from the Commons (although only temporarily) and can also generate bills for the consideration of the Commons (approximately one-fourth of the bills passed by Parliament were drafted initially in the Lords). Consequently, while the basis for membership of the House of Lords is undemocratic in the sense that the Lords are neither elected nor directly accountable to the voters, the fairly limited powers of this institution in the legislative process ensure that most of the power and authority lie with the elected members of Parliament (MPs) in the Commons.

As with other parliamentary democracies, the leadership of the executive branch (the cabinet) is formed through the creation of an elected majority in Parliament (in this case, the House of Commons). Members of the House of Commons are elected to represent 659 constituencies in the UK and its overseas territories. Unlike the Israeli case, virtually all governments in the UK have been formed from one party, which has been able to gain a majority in the House of Commons in the wake of a national election (something that is much more possible with a voting system based on candidates running in voting districts rather than party lists elected via proportional representation). While coalition governments are not unknown, they are quite rare—although the present British government is, in fact, a coalition government, but it is the first one in several decades. Consequently, a British prime minister generally has the luxury of not having to deal with fractious coalition partners—although he or she may be the recipient of considerable grief from party backbenchers—not unlike the position that U.S. Presidents find themselves in from time to time with respect to members of Congress from their own party. As in other parliamentary systems, the prime minister is a member of the House of Commons and the other ministers are also MPs (four from the Lords and the rest from the Commons) and the cabinet makes national policy decisions as a collective body. One other point that is interesting about the role of Parliament in the British system is that Parliament is sovereign, meaning that no court or other entity has the authority to overturn an act of Parliament, and only Parliament can overturn its legislation. Since the UK, like Israel, does not have a written constitution, there is no document to which laws must conform. Courts have the authority to rule on the manner in which the government implements legislation, and thus the principle of judicial review exists and is acted upon in the UK, but they do not have the authority to review legislation passed by Parliament (although they can review bylaws passed by local authorities).

Unlike Israel, there is no clear separation in the UK between the executive and legislative branches on the one hand and the judicial branch on the other. As noted above, the person who effectively heads the judicial branch, the Lord Chancellor, is both a member of the cabinet and a peer in the House of Lords (although there are restrictions, by convention, on the Lord Chancellor's powers when fulfilling one of these roles with respect to the other functions of the office). Moreover, the 26 judges who are peers in the House of Lords (known as the Law Lords) act as the country's highest court of appeals, causing further intertwining of the relationship between the legislature and the courts.

The United Kingdom of Great Britain and Northern Ireland consists of four "countries" (in addition to overseas dependencies): England, Wales, Scotland, and Northern Ireland, and these form three jurisdictions with their own court systems: (1) England and Wales, (2) Scotland, and (3) Northern Ireland. England and Wales were formally united in 1536 (although English law had been applied to Wales since 1284, two years after the country was conquered by England). England and Wales were united with Scotland (creating the Kingdom of Great Britain) in 1707 and with Ireland in 1800 (creating the United Kingdom of Great Britain and Ireland), but at present only six northern counties still remain part of the UK and comprise the Province of Northern Ireland. Although there are differences in the terminology and function of various courts in the three legal jurisdictions within the UK, it is still possible to summarize the system in general terms. At the lowest level of the court system are magistrate courts, each of which is presided over by a volunteer, nonlegal professional known as a justice of the peace (of which there are approximately 30,000). In addition, there are 140 district judges and 170 deputy district judges: experienced lawyers who sit in magistrate courts as salaried justices. Magistrate courts deal with minor offenses (known as *summary offenses*), such as assault, vandalism, family disputes, youth issues, and public drunkenness. The maximum penalty that can be handed down by a magistrate court is a level 5 fine (currently a maximum of £5000 and/or a 12-month prison sentence). Serious cases (known as *indictable offenses*) are heard in crown courts, and crown courts also hear appeals from magistrate courts. Crown court trials on serious offenses involve jury trials, whereas most magistrate trials do not involve juries. Minor civil cases (e.g., small claims) are dealt with initially in county courts, and more serious ones are heard by the high court (which also hears appeals from the county courts). The high court is divided into divisions dealing with various civil issues. Civil and criminal matters may be appealed from the high court or the crown courts, respectively, to the court of appeal (which contains both a civil and a criminal division). The UK Supreme Court acts as the highest court of appeal. As with Israel, the principle of *stare decisis* applies to

FIGURE 1.6 Scottish parliament building. This Wikipedia and Wikimedia Commons image is from the user Ron Almog and is freely available at http://commons.wikimedia.org/wiki/File:The_Scottish_Parliament.jpg under the Creative Commons Attribution 2.0 Generic license.

English adjudication, and thus decisions by higher courts are binding on lower courts, with rulings by the Supreme Court binding on all courts in the legal system (Slapper and Kelly, 2009, pp. 4555–4570).

While the UK is an amalgamation of England, Wales, Scotland, and Northern Ireland, it does not have a federal system of government—although it is also not a centralized state in the manner of Israel or France. In fact, the UK incorporates both very significant elements of local autonomy, separate jurisdictions, separate laws, and separate institutions while maintaining a strong central government. In this respect, and in comparison to the other countries surveyed, it is somewhat of an anomaly. In addition, the relative influence of each of these countries differs, with England, which contains some 87 percent of the British population, being much larger, more populous, and wealthier than the other UK countries. In terms of ultimate power, Parliament is sovereign and its ability to legislate for the entire country is not in question (and in this sense the UK is a centralized state), but there has been a significant divestment of central government powers over the years. Moreover, MPs representing constituencies in Scotland, Wales, and Northern Ireland are able to influence national policy from the center. At present, Northern Ireland, Scotland, and Wales all have their own devolved legislatures with varying degrees of local power (England does not have such an assembly, and at present there seems to be little popular desire for such a body). Wales has the lowest degree of local autonomy, with the Welsh Assembly having primarily administrative and executive responsibilities, and it can only legislate with respect to the manner of implementation of legislation passed at Westminster (the district of London that contains the Houses of Parliament). Scotland and Northern Ireland enjoy far greater autonomy; their own devolved legislatures have the power to tax and to pass legislation with respect to certain matters. Finally, greater London has, since 2000, had its own mayor and regional assembly with some degree of autonomy. In addition, there are other forms of regional governance. As with more centralized countries, local authorities enjoy the power to tax, pass bylaws, and otherwise enjoy some limited autonomy, and policing is regionalized in the form of 52 police district–based policing organizations (more on this later). The UK thus has a system of regional governance that is neither completely centralized nor truly federal.

THE DOMINION OF CANADA

Canada is the second largest country in the world (after Russia), with a territorial scope of almost 10 million square kilometers. Its population is, however, rather small, being just under 34 million people (just over half the

FIGURE 1.7 Canada.

size of the UK population). Given the harsh and intemperate climate of most of the country, 90 percent of Canada's population is clustered in the south of the country (within 160 kilometers of the U.S. border). Most of the center of the country is wooded wilderness, while most of the northern third is desolate tundra with subarctic and arctic climates. Most of the population is of European origin (approximately 66 percent), an additional 26 percent of the population is of mixed ethnic background, 2 percent is indigenous (known in Canada as "First Nations"), and the remainder are from Asia, Africa, and the Arab world. In terms of its labor force, 71 percent are employed in the service industries and 26 percent in heavy industries. Canada has an affluent economy and lifestyle, with vast reserves of natural resources. Its primary trading partner is the United States, to which it sends 80 percent of its exports.

Canada shares much of the basis of its system of government with the UK. This is not surprising given that Canada's process of detachment from the British Empire occurred very slowly and in a piecemeal fashion. What had been six separate British colonies united and became a dominion (a self-governing member of the British Empire) in 1867 with the promulgation by

London of a Canadian constitution known as the Constitution Act. The separate dominion of Newfoundland subsequently joined the Canadian federation in 1949. However, Canada only became legislatively independent of the mother country with the passage of the Statute of Westminster in 1931, and the British Parliament maintained the exclusive right to amend the Canadian constitution until the passage of the Constitution Act in 1982, at which point Canada is considered to have achieved complete independence from Britain in the full legal sense. Symbolically, however, the Canadians have not cut their ties to the UK completely and remain a constitutional monarchy, with Queen Elizabeth II also serving as Canada's head of state.

Canada's constitution is thus based on both the 1867 and 1982 acts. The 1982 act also included a bill of rights known as the Canadian Charter of Rights and Freedoms. As Canada is a federal state, each of Canada's 10 provinces has veto power over amendments to the constitution. Canada also has three territories, which derive their powers from the federal government—unlike provinces, which derive their legal authority from the Constitution Act of 1867 and thus are not legally beholden to the federal government in terms of their respective spheres of authority. The most contentious constitutional issue in Canada is that of the status of the province of Quebec and whether or not it has the legal right to secede from the Canadian federation. In 1995, a referendum held in Quebec nearly gave a victory to separatist political forces, and the Canadian Supreme Court subsequently ruled that while it was not legal for any province or territory to secede from Canada if a "clear majority" of Quebeckers voted in favor of secession, the federal government would be obligated to enter into negations on Quebec's secession. The court also ruled that it was up to the federal government to determine what a clear majority was, and this led Parliament in 2000 to promulgate the Clarity Act, which gave Parliament the authority to determine what constitutes a clear majority (Malcolmson and Myers, 2005, p. 44). Any future attempt at secession by Quebec is thus likely to be highly complicated, but that prospect seems to be receding, as support for independence among Quebeckers has dropped sharply since the mid-1990s.

Canada has a parliamentary regime with a bicameral parliament modeled on the British Parliament. It consists of an elected lower house (the House of Commons) and an appointed upper house (the Senate). As in the UK, the Canadian House of Commons is equated with the term *parliament* because it holds the virtually exclusive power to legislate. The Commons consists of 308 MPs, each representing a territorially based constituency (known as a *riding*), meaning that the lion's share of parliamentarians come from the most populous provinces: Ontario and Quebec. Like the British House of Lords, the Senate was an elite "club" designed to act as a break on the democratic

FIGURE 1.8   Canadian parliament. This Wikipedia and Wikimedia Commons image is from the user Matthew Samuel Spurrell and is freely available at http://commons.wikimedia.org/wiki/File: Canada_Parliament2.jpg under the GNU Free Documentation License, Version 1.2.

power represented by the Commons. Unlike the House of Lords, the Senate was not made up of hereditary peers, church officials, and the like but, rather, senators representing provinces. At present there are 105 senators, with the seats distributed by the population size of each province (Ontario and Quebec each have 24 senators, the western provinces each have six, etc.). The senators are nominally appointed by the governor-general (the Queen's representative), but, in practice, are chosen by the prime minister. The Senate has legal powers similar to those of the Commons, but because it is an unelected body, it almost never makes use of its full powers, and thus its primary role is to review bills and make suggestions for changes. Since the middle of the twentieth century, it has been the convention that the Senate will not oppose bills that enjoy the support of the Commons (Malcolmson and Myers, 2005, pp. 132–133). As in other parliamentary regimes, the government (cabinet) comes to power through obtaining the support of a majority of members of the House of Commons in the wake of parliamentary elections, and as in the UK, coalition governments are rare and usually formed during times of national crisis. As with other parliamentary systems, the government determines policy as a collective body, and the prime minister and other government

FIGURE 1.9 Supreme Court of Canada. This Wikipedia and Wikimedia Commons image is from the user Peregrine981 and is freely available at http://commons.wikimedia.org/wiki/File: Supreme_Court_of_Canada.jpg and has been released into the public domain.

ministers are members of the parliament. Since the British monarch is also Canada's monarch, the Canadian Crown exercises similar powers (again, primarily symbolic and ceremonial) in the Canadian context. However, since the Queen resides in the UK, she is represented on an ongoing basis by a governor-general, who exercises the royal prerogatives in the name of the Queen and is appointed by her (although on the recommendation of Canada's prime minister).

Since Canada has a written constitution, the judiciary, as in the United States, has the power to review legislation to determine its constitutionality, and consequently, unlike the UK, Canada's parliament does not have unchallenged sovereignty and the Canadian judicial branch is independent of the other branches of government. As in the United States, the court system in Canada is divided between federal courts and, in the Canadian case, provincial ones, with the Canadian constitution (specifically the Constitution Act of 1867) empowering the provinces to establish courts. Canada's provinces thus have courts that hear both civil and criminal cases and are divided into inferior courts (such as traffic, family, and small claims courts) and superior courts, which function either as trial courts for serious offenses and significant civil litigation or as appellate courts that receive appeals from the inferior courts (also known as provincial courts) (Malcolmson and Myers, 2005, p. 150). Since a good deal of law (including much criminal law) is based on federal legislation, the provincial court system can issue rulings based on both provincial and federal law. Given the unique nature of Quebec, it should not be surprising that while all the courts in Canada's other provinces operate under common law principles, Quebec's courts adjudicate based on the French tradition of civil law. While provincial courts are independent of federal courts, the federal government is given the authority to appoint all superior court judges, including those in provincial courts.

As in the United States, Canada also has a federal court to adjudicate certain matters of federal law (which range from antigang legislation to maritime law to intellectual property laws). It should be noted, however, that unlike the United States, the federal government has exclusive legislative power with respect to criminal law and procedure and, consequently, provincial courts can adjudicate over criminal matters but provincial parliaments cannot pass laws with respect to such matters (UK Foreign Office, 2005, p. 6). The Supreme Court of Canada has the authority to hear appeals from the provincial superior courts and the federal court and, as noted above, possesses the power of judicial review of legislation.

Unlike the UK, Canada is a federal country created, as was the United States, through the amalgamation of individual colonies. Naturally, the circumstances surrounding the creation of the Canadian federation were very different from those relating to the creation of the American federation. In addition to having a very different attitude and relationship with Britain compared to their American cousins, Canadians (with British guidance) also had to create a system that would make it possible for English-Canadians and French-Canadians to cohabit. This necessitated a federal system in which power would be shared between the central government and the provinces (which jealously guarded their autonomy). Unlike the U.S. federation, where power is more evenly balanced between state and federal authorities, the Canadian federal government based in Ottawa is considered to be more powerful in relation to the provinces.

The Canadian constitution gives the federal parliament exclusive control over trade and commerce and criminal law and, unlike the United States, where residual powers (those not specifically granted to the federal government) are in the hands of states, in Canada the opposite is true. Whereas in the United States federal laws are considered to supersede state law when there is a contradiction between the two, in Canada the federal government has the power of *disallowance*, whereby it can annul legislation passed by the provinces (Malcolmson and Myers, 2005, pp. 71–72). Despite the constitutional dominance of the federal government, in practice governance in Canada is based largely on consultation and buy-in and, more often than not, involves cooperative ventures between the federal government and the provinces rather than a federal *diktat*, and in this sense, Canada has become more decentralized over time. With respect to Canada's three territories (the Yukon, Northwest Territories, and Nunavut), however, their powers emanate from the federal government rather than the constitution, and consequently, they enjoy less autonomy than do the provinces. Thus, while the federal government does not generally interfere in areas that are seen as being clearly within the competence of provincial governments (such as education), it does play this role with respect to territories. Provinces are headed by a premier and possess a unicameral legislature. Local government is run via municipalities (which can be based on cities, counties, or districts), and local governments are created by the provinces and can be dissolved by them. Thus, unlike the United States, local government does not exist as an independent power base but rather is beholden to the provincial government.

THE COMMONWEALTH OF AUSTRALIA

Australia has a total area of over 7.7 million square kilometers, making it the sixth largest country in the world and slightly smaller than the continental United States. Australia is primarily a country of low desert plateaus with huge internal expanses (known as the "Outback") consisting of parched desert, with small temperate zones in the south and east and a tropical climate in the far north of the country. Not surprisingly, the country does not support a large population; with some 21.5 million inhabitants, Australia has

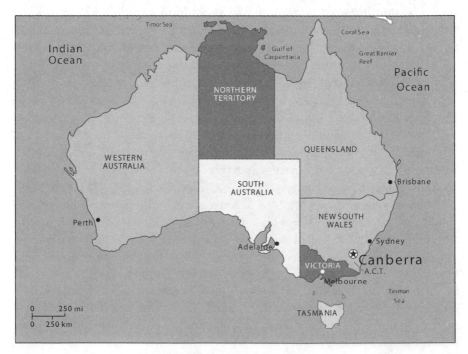

FIGURE 1.10 Australia.

a population significantly smaller than Canada's. Australia is highly urban-ized, with 89 percent of the population living in towns or cities. In terms of ethnicity, 92 percent of the population is of European origin, 7 percent is Asian (including Middle Eastern), and the aboriginal population consists of less than 1 percent of the overall population. In terms of the workforce, some 70 percent are employed in the service industries and 26 percent in heavy industries. Like Canada, Australia has abundant natural resources (although, unlike Canada, it has severe water problems), and it enjoys a high standard of living.

There are many similarities between Canada and Australia but some important differences as well. As in the Canadian case, Australia's gover-nance and legal systems are also based on the British model and, as with Canada, Australia did not suddenly break with the mother country and go its own way. Like Canada, Australia was also formed through the creation of a federation between separate British colonies (in this case, six), although the Australians did not need to reconcile pronounced differences in language and culture between colonies as there was and is no Australian equivalent to Quebec. The Commonwealth of Australia was formed on January 1, 1901 in the wake of the passage in the British Parliament of the Commonwealth of Australia Act in 1900. Prior to this, there had been six self-governing colonies

FIGURE 1.11   Australian parliament. This Wikipedia and Wikimedia Commons image is from the user Andrea Schaffer and is freely available at http://commons.wikimedia.org/wiki/File: Parliament_House_Canberra_(281004929).jpg under the Creative Commons Attribution 2.0 Generic license.

on the Australian continent, and each colony's legislature had to vote in favor of joining the commonwealth and to hold popular referenda on the issue.

Like the Canadians, the Australians have a written constitution (promulgated in July 1900, which can only be revised by popular referendum), which again as with Canada, establishes Australia as a constitutional monarchy with a parliamentary form of government and a federalist system of governance. The Australians share the same monarch with the UK, and the Queen is represented in Australia, as in Canada, by a governor-general appointed on the recommendation of the prime minister. Finally, as with the Canadians, Australia's parliament gained full law-making authority in 1931 under the Statute of Westminster. The Australian parliament is also a bicameral legislature with a lower house (the House of Representatives) and an upper house (the Senate). Members of parliament in the House of Representatives are elected to represent 150 constituencies that are geographically disbursed based on population. In a departure from the British system, however, the Senate is also an elected body representing the six Australian states equally, with 12 senators being elected for each state, two from the Northern Territory and two from the Australian Capital Territory (Canberra and its environs). Voting for the Senate is on the basis of proportional representation, as in Israel (although in Australia, this voting occurs within each state and territory and Australians can vote either for party lists or for specific candidates). This means that small parties have a greater chance for representation and that it is very rare for one party to enjoy a majority in the Senate. Legislation must be passed by both houses before it can become law, but most bills originate in the House of Representatives, and the Senate does not have the power to block appropriations bills. The Australian legislative system thus represents a significant variance from the British Westminster model and, in fact, contains clear elements of the American model. Nevertheless, as in other parliamentary systems, the government (cabinet) is created from the lower house and must maintain its majority in that body in order to govern. Moreover, unlike any of the other countries surveyed here, voting for federal and state/territorial legislatures is compulsory in Australia and a small fine is levied against those who fail to vote and do not have a justifiable excuse.

Like its parent British judicial system, the Australian judicial system is based on common law. However, the federal nature of the state and the fact that there are three categories of law—state, concurrent, and federal—has necessitated the creation of two different court systems: state and commonwealth (federal). In the state court system, the lowest courts, the magistrate courts (also known as local courts), handle infractions and small-scale civil issues, and magistrates are not legal professionals. Serious criminal and civil cases are handled at the state level by district courts (usually with jury trials), and each state has a supreme court that acts as the highest appellate court.

FIGURE 1.12   High Court of Australia. This Wikipedia and Wikimedia Commons image is from the user Nicholas Brown and is freely available at http://commons.wikimedia.org/wiki/File:High_Court_of_Australia_from_lake_cropped_.jpg under the Creative Commons Attribution 3.0 Unported license.

At the federal (commonwealth) level, the High Court of Australia oversees the federal court system and acts as the highest appellate court, in addition to fulfilling the function of judicial review with respect to the constitutionality of legislation passed by Parliament. The High Court also rules on appeals from state supreme courts and can overturn decisions made by state courts. Below the High Court within the commonwealth court system are the Federal Court of Australia, the Federal Magistrates Court of Australia, and the Family Court of Australia (these courts deal primarily with matters of bankruptcy, discrimination, trade practices, privacy rights, industrial law in the case of the first two, and family-related matters with respect to the latter). State courts deal with matters arising under both state and federal law.

In terms of federalism, the state governments are granted law-making rights by the constitution, but these can be superseded by federal legislation. Moreover, the federal government has exclusive powers to make laws in matters of trade and commerce, taxation, immigration and citizenship, and so on. There is also a third category of powers, concurrent powers, in which both levels of government are empowered to enact legislation (examples include laws relating to insurance, banking, and regulation of businesses). Any powers not granted specifically to the states or deemed concurrent powers are reserved for the federal government. States differ from territories in terms of their degree of legislative and administrative autonomy, with territories more dependent on federal legislation and governance. The Northern Territory has full state powers, however, and has simply retained the term "territory" in its name. Each state has its own institutions that mirror the federal ones, with a governor (who acts as the local governor-general representing the Queen), an upper and lower house of Parliament (except for Queensland, which has a unicameral parliament), and a premier and government.

THE FEDERAL REPUBLIC OF GERMANY (BUNDESREPUBLIK DEUTSCHLAND)

Germany is just over 357,000 square kilometers in size, making it one of the largest countries in Europe and slightly smaller than the state of Montana. Its topography consists primarily of fertile plains and woodlands in the center and north and the Alps in the south, and its climate is temperate and marine. As with the UK, this hospitable climate has resulted in a large population in relation to its size, and Germany today has just over 82 million inhabitants (thus making it the most populous country in Europe). The population, of which 74 percent live in cities or towns, consists primarily of persons of German ethnic origin (approximately 92 percent), with the remainder of

FIGURE 1.13 Germany.

southern or eastern European origin and just over 2 percent of Turkish origin. Germany has a significant industrial sector (almost 30 percent of the workforce is employed in heavy industries) with close to 68 percent in the services sector. Germany is the economic powerhouse of Europe and possesses the world's fifth largest economy, a highly skilled workforce, and a sophisticated, export-based industrial infrastructure.

Although every country's political and legal systems are a product of its past, that of Germany is particularly so given the profoundly negative effect that the totalitarian Nazi regime had on Germany (not to mention the rest of Europe). Moreover, the Allied occupation of Germany at the end of World War II resulted in the imposition of a democratic regime in western Germany with the creation of the Federal Republic of Germany in 1949, as opposed to the organic development of law and government that occurred in most of the other countries in our survey (with the additional exceptions of Italy and Japan). Germany, of course, had its own democratic traditions, dating back to the period of the ill-fated Weimar Republic (1918–1933), but in terms of lessons learned, many of these were relevant for the failure of democracy rather than being a strong basis for the formation of a stable and long-lived democracy. Germany had a longer history with respect to the antecedents for federalism and, in particular, local independence, as the German state had only come into existence in 1871 and was preceded by a wide range of independent German states of various kinds that had existed since the onset of the medieval period.

The current German political system and constitutional order (which was extended to the formerly Marxist East Germany with the unification of the two German nations in October 1990) is based on the German constitution, the Grundgesetz (Basic Law). The Basic Law enshrines the principles of democracy, federalism, and constitutional rule in which all governing bodies are subject to judicial review and control. The creation of a judiciary with the power to oversee actions by the executive and legislative branches and that of a strong federalist system were seen as important balancers that would prevent the aggrandizement of power at the executive level in a way that had brought about the collapse of the Weimar Republic and the rise of the Third Reich. One feature of the Basic Law is the "eternal character" of its primary principles, which are basic democratic rights, federalism, and the welfare state. This means that amendments to the Basic Law, or even the writing of a new constitution at some future date, cannot abrogate these basic principles (*Facts About Germany*, 2009). The Basic Law also requires that political parties that run for public office accept democratic values. This means that the federal government can petition the Federal Constitutional Court for a ban disallowing any party to run for public office on the grounds of that party adhering to antidemocratic principles.

FIGURE 1.14 Bundestag building (Reichstag). This Wikipedia and Wikimedia Commons image is from the user Matthew Field and is freely available at http://commons.wikimedia.org/wiki/File:Berlin_reichstag_west_panorama.jpg under the GNU Free Documentation License, Version 1.2.

As is the case with the other countries surveyed above, Germany has a parliamentary form of government (although it is a republic, like Israel, rather than a constitutional monarchy). Also, similar to Israel, German governments are formed through coalitions of parties rather than by one party with a majority in parliament and there has only been one case in the last 56 years in which one party was able to obtain a majority in parliament. Germany, however, has a bicameral parliament with a lower house (the Bundestag) and an upper house (the Bundesrat), with the upper house, as in Australia, reflecting the federal nature of the country. As with other bicameral parliamentary regimes, the government (cabinet), including the prime minister (known in Germany as the chancellor) are members of the lower house and must enjoy a majority in the lower house in order to govern (although bringing down the government is more difficult in Germany because the opposition must first muster a majority for an alternative government before they can pass a vote of no confidence to bring down the existing government—this is called a *constructive vote of no confidence*). Unlike other parliamentary systems surveyed here, however, the chancellor is elected by majority vote in the Bundestag rather than serving only as head of the party that forms the coalition. This is because the Basic Law invests the chancellor with executive authority, and thus his/her authority is derived from the constitution and not only by virtue of enjoying majority support in the parliament. This also means that the chancellor can determine the size of his/her cabinet and appoints the cabinet ministers without needing to receive the approval of the Bundestag. The reason that governments in Germany are almost always the result of coalition building is that half the seats in the 598-seat Bundestag are allocated on the basis of proportional representation, with voters selecting party lists (as is done in Israel with respect to the entire parliament), and this, of course, means that smaller parties have an easier time gaining representation at the expense of potential majorities for the larger parties. The other half of the Bundestag membership is elected on the basis of winner-take-all candidate-based elections in 299 territorially based constituencies.

The upper house of the German parliament, the Bundesrat, consists of 69 members who are appointed by state (Länder—plural, or Land—singular) governments and represent state interests at the federal level. Each of Germany's 16 states has a minimum of three Bundesrat delegates, and the other seats are allocated on the basis of state populations. Approximately 50 percent of the bills passed by the lower house, the Bundestag, must also be passed by the Bundesrat in order to become law (these are bills in which the states enjoy concurrent powers with the federal government or bills that must be enforced or administered by the states). Also, the federal government, which produces the lion's share of bill initiatives, must first consult with the Bundesrat before it submits the bill to the Bundestag for consideration. This system ensures that critical legislation will have the buy-in of the majority of state governments because the upper house is, in effect, a state entity operating within the federal government (much more so than the U.S. Senate because the Bundesrat members are appointees of state governments, whereas U.S. senators are not). This system also allows the state governments the ability to veto legislative initiatives at the federal level that they believe are inimical to their interests. Germany, as a republic, also has a federal president (Bundespräsident) as head of state, whose powers, as in Israel, are largely ceremonial.

Unlike the other countries discussed above, the German legal system is based not on the common law tradition but on the civil law system (which does not rely on prior rulings but, rather, on a broad system of legal codes). The highest court in Germany is the Federal Constitutional Court (the Bundesverfassungsgericht or BVerfG), which enjoys the right of judicial review and also, as with the Israeli Supreme Court, receives petitions on which it can hold hearings and make rulings (but it does not act as an appellate court). Half of its judges are elected by the Bundestag and the other half by the Bundesrat. Germany also has specialized courts that deal with administrative issues, labor disputes, intellectual property disputes, and so on. The primary court system is organized into four levels. At the lowest level are local courts (Amtsgericht), which deal with minor criminal offenses punishable by two years' incarceration or less and with small-scale civil suits and are overseen by a professional judge in minor cases and a judge assisted by lay judges in criminal cases involving penalties of one year or more in prison. The second level consists of regional courts (Landesgericht) which have sections that hear major civil cases and others that hear major criminal

Bundesarchiv, B 145 Bild-F083310-0001
Foto: Schaack, Lothar | 19. Dezember 1989

FIGURE 1.15 German federal constitutional court. This Wikipedia and Wikimedia Commons image is from the German Federal Archive, Photo B 145 Bild-F080597-004 and is freely available at http://en.wikipedia.org/wiki/File:Bundesarchiv_B_145_Bild-F083310-0001,_Karlsruhe,_Bundes-verfassungsgericht.jpg.

cases with trials presided over by a panel of judges. Regional courts also act as courts of appeal for local courts and have basic jurisdiction over most criminal and civil matters. Particularly serious cases (such as murder) are heard by regional courts with a panel of three professional judges and two lay judges. The third level consists of state appellate courts (Oberlandesgericht) that deal primarily with appeals from lower courts and also preside over trials dealing with acts of treason or antidemocratic activity. The top tier court is the Federal Court of Justice (Bundesgerichtshof), which is a final court of appeal for cases from regional and appellate courts (Federal Research Division of the Library of Congress, 1995).

Federalism in Germany is very deeply ingrained, and state power and autonomy are seen, as with other federal states, as a guarantee against the concentration of power in a central government. But this need to balance the power of the central government is seen as particularly critical given Germany's history of despotic, unrestrained centralized control. Moreover, the Basic Law allows for three categories of legislation, as with Australia: federal law, concurrent law, and state law. The states also enjoy residual powers, as in the United States, in that any powers not explicitly granted to the federal government become state powers by default. However, as in the United States, when there is a conflict between federal and state law, federal law takes precedence.

Ultimately, the German solution to the need to balance power, as noted above, is not only to provide for state autonomy under a federal system but also to give the state governments a significant role, via the Bundesrat, in the federal government. German states have the power of taxation and are obligated to implement federal policies and enforce federal laws when implementation of those policies and enforcement of those laws are not within the purview of federal agencies, and the states usually have a wide degree of autonomy in choosing the manner of implementation of federal policies or enforcement of federal laws. In fact, one of the things that makes the German system unique is that implementation of the vast majority of federal policies and the enforcement of most federal laws are in the hands of the states (Roberts, 2000, pp. 101–102). State governments are responsible for education, internal security (including policing), the four-tier court system described above, the provision of social services, most transport systems, and the organization of municipal and other local governments. At the same time, Articles 72 and 74 of the German constitution state that when common legislation is necessary in all states, the federal parliament has the power to pass such legislation—and these articles have been used by the federal authorities over the years to expand their legal and institutional powers at the expense of the states (Konze, 2009, p. 102).

State governments have their own constitutions and are organized on the basis of a parliamentary system with a cabinet led by a prime minister (Ministerpräsident) and a unicameral legislature (state parliament—Landtag) from which the prime minister is elected by majority vote. The terminology for the minister-president and state parliament differs in some of the states, but this system is essentially the same across the various state governments in Germany. The states themselves are subdivided into districts, boroughs, municipalities, and other divisions (each state has a somewhat different system), but the state governments maintain a monopoly of control over the local level of government.

THE FRENCH REPUBLIC (RÉPUBLIQUE FRANÇAISE)

France has a territorial area of just over 643,000 square kilometers, making it slightly smaller than Texas and larger than Germany. In fact, France is the largest country in Europe except for European Russia. Its topography and climate are highly diverse: from plains and rolling hills and a marine climate

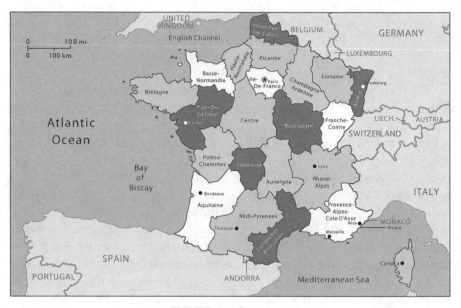

FIGURE 1.16 France.

in the north and west to a central plateau to the Alps in the southeast and a hilly southern region with a Mediterranean climate. The population is just slightly larger than that of the UK, with just over 64 million inhabitants. The majority of the population is of French ethnicity (although some are from families originally from Italy, Spain, or Portugal) and between 5 and 10 percent of the population is of North African and Middle Eastern ethnic origins. France, too, is largely urbanized (although slightly less so than some of the other countries in the survey), with 77 percent of the population living in urban areas. Approximately 79 percent of the workforce is employed in the service industries and 19 percent in heavy industry (although the agricultural sector enjoys considerable political patronage). France enjoys a highly developed economy that is one of the four largest in Europe.

France's political and administrative system was heavily influenced by the rule of Napoleon Bonaparte (crowned emperor in 1804), particularly with respect to the centralized nature of the French state. Whereas Canada and the UK chose, in differing degrees, to reflect the ethnic, social, and linguistic aspects of their societies through various measures of decentralization, France ultimately chose to approach the challenge of social differences through strong principles of centralization and a strong national identity based on the principles of the Republic. France traditionally was a very diverse country with different cultures and languages (French only became the most widely spoken language in France in the wake of the public education efforts begun in the 1880s that worked to effectively eradicate other dialects). France has experimented with a number of different constitutional orders since the revolution that began in 1789. The current constitutional order (based on the constitution promulgated in 1958), known as the Fifth Republic, represents a sort of middle ground between the American presidential system and the traditional parliamentary system adhered to by the other countries in this survey. The system maintains the distinction between head of state and head of government, with a president as head of state and a prime minister as head of government. However, unlike parliamentary republics such as Israel and Germany in which the president has a primarily symbolic role (and is elected by the legislature), France's president is elected directly by the voters and has wide powers. As this matter is not spelled out clearly in the constitution, the distribution of powers between the president and the prime minister is complex and dependent, at least to a point, on the degree of ambition and governing styles of the people filling these positions—and also on whether or not they are members of the same political party (which is usually the case, but not always). The president is considered to be the most powerful executive branch figure, and it is up to the president to appoint the prime minister and the other cabinet ministers. The prime minister and the rest of the cabinet ministers must, however, enjoy a majority in the parliament in order to govern.

When the president, prime minister, and a majority of members of Parliament are all from the same party (and assuming the president has a strong hold on his party), the French president's powers are quite substantial, and some refer to such situations as an *elected monarchy*. On the other hand, when the prime minister and his parliamentary majority are from a different party (a power-sharing arrangement known in French as *cohabitation*), the president's power, at least over domestic affairs, can be significantly limited. In very simplistic terms, then, the president is seen as the preeminent actor with respect to foreign affairs and defense, whereas the prime minister is primarily responsible for domestic policy, but reality can be more complex, particularly when the president is very ambitious and active (as is the current president, Nicholas Sarkozy). The French president also enjoys de facto law-making powers in the form of presidential decrees (décrets), which are similar to presidential executive orders in the United States. The president's power is balanced, however, by the constitutional requirement that the prime minister countersign most types of presidential decrees and other orders. This requirement for countersignature ensures that the president is accountable to the government (cabinet), which is, in turn, accountable to the legislative branch. Countersignature by the prime minister acting for the government as a whole also ensures that the government will take responsibility for implementing policy (Stevens, 2003, p. 68). The president also has the authority to delay legislation by

FIGURE 1.17 French National Assembly. This Wikipedia and Wikimedia Commons image is from the user David Monniaux and is freely available at http://commons.wikimedia.org/wiki/File:Paris_Assemblee_Nationale_DSC00074.jpg under the GNU Free Documentation License, Version 1.2.

asking the parliament to review it and is designated the commander of the armed forces.

The French parliament is bicameral and comprises a lower house, the National Assembly, and an upper house, the Senate. The National Assembly (Assemblée Nationale) consists of 577 delegates (députées) and the Senate (Sénat) consists of 343 members. Elections to the National Assembly are based on voting districts, with run-offs if no single candidate was able to acquire 50 percent of the vote. Elections to the Senate are based on an electoral constituency within each *department* (the administrative districts in which France is divided—the term is roughly equivalent to "county") or overseas territory, and some senators are elected to represent French citizens abroad. Senators are chosen by an electoral college comprised mainly of elected municipal councils and thus, to some degree like the German Bundesrat, they represent more localized governmental interests. Quite a few senators are also mayors or members of municipal or regional councils. The Senate is, however, clearly the weaker body in comparison with the National Assembly. Not only does the National Assembly (in traditional parliamentary fashion) form the government (cabinet) and gives it the authority to govern, by virtue of its majority in the National Assembly, but it is also the primary legislative body since the Senate's powers, in most categories of legislation, are limited to reviewing bills and providing input to the National Assembly. The National Assembly must make an effort to ensure that bills are acceptable to the Senate, including, when necessary, undertaking mediation between the two houses. Only in the fairly rare event that the Senate rejects a bill supported by the National Assembly will the National Assembly override the Senate and pass the legislation nonetheless (France, Senate, 2009).

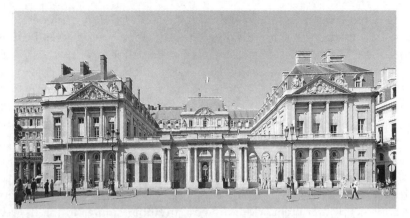

FIGURE 1.18   French Council of State. This Wikipedia and Wikimedia Commons image is from the user Jastrow and is freely available at http://commons.wikimedia.org/wiki/File:Conseil_d%27Etat_Paris_WA.jpg and has been released into the public domain.

As with the German legal system, French law is based on civil law rather than common law, and thus rulings are based on statutes and legislation as opposed to case law. The French judicial system is inquisitorial, meaning that the role of the judge is to actively discover the truth rather than to preside over a courtroom and ensure that a fair trial ensues. The role of the defense counsel (avocat) is primarily to ensure that correct procedures have been carried out by judges and is thus more of an auxiliary to the judge than a major actor in the legal process as in the adversarial system (Hodgson, 2006, p. 32). There are two separate French court systems: administrative courts and judicial courts. Administrative courts handle disputes between citizens and state entities. The supreme administrative court is the Council of State (Conseil d'Etat), below it are a number of specialized administrative courts and administrative appeals courts. The judicial courts deal with disputes between people and offenses against people, property, or society. The supreme court in this system is the Court of Cassation (Cour de Cassation), which acts as the highest court of appeal in criminal and civil cases. Below it are appellate courts, and below the appellate courts are a range of specialized civil courts, each dealing with a different area of civil litigation: between disputants, on commercial matters, with regard to employment disputes, and so on. There are also a range of specialized criminal courts, including police courts (infractions and misdemeanors), regional criminal courts, juvenile courts, and assize courts, the latter which involve jury trials and deal with serious offenses. In keeping with the inquisitorial nature of their judicial system, the French also employ investigative magistrates (juges d'instruction) to conduct investigations prior to cases being heard in court (see Chapter 2).

As noted earlier, like Israel, France is a centralized, not a federal state. Unlike Israel, however, France is divided into regions (regions) and counties (départments) in addition to possessing municipalities (communes). France has 26 regions, each administered by a regional council and a chairperson. Most of the regional powers relate to budgetary issues and economic development. The 100 departments serve as the most important administrative units. The "county seat" in each department is known as the *prefecture* and is generally the city or town closest to the geographic center of the county. The leadership of these counties consists of an elected General Council (Conseil Général), with the president of the council serving as the chief executive (prior to 1871, departments were headed by prefects appointed by the national government). Departments do not enjoy anything similar to the degree of autonomy enjoyed by states in federal systems but play the role of territorial units for the organization of central government activities. Departments deal with social welfare issues, public health, maintenance of roads and public buildings, support for education, and similar activities (Stevens, 2003, pp. 145–147). As most local issues are handled by departments, the authority of municipalities is quite limited in contrast to some other countries and deals primarily with communal order, oversight of public services, and some other matters. Ultimately, then, while France does have locally elected leaderships at the regional, county, and municipal level and while the French Senate does represent local interests, the bulk of public policy matters are developed and administered from Paris.

THE NETHERLANDS (KONINKRIJK DER NEDERLANDEN)

The Netherlands is a small country of just under 42,000 square kilometers, about twice the size of the state of New Jersey (and hence only twice the size of Israel). Its topography consists of fertile lowland plains (some of them reclaimed from the North Sea), and most of the country lies at or under sea level (the highest point in the country is only 322 meters high). The climate is primarily marine and thus mostly wet and overcast. The population of the country, at almost 17 million persons, is also small, although not for the country's physical size. In terms of ethnicity, almost 81 percent of the population is of Dutch ethnicity and over 6 percent of Indonesian, Middle Eastern, or North African ethnic origins. The Netherlands is a highly urbanized country, with 82 percent of the population living in cities or towns. Nearly 75 percent of the workforce is employed in service industries and almost 24 percent in heavy industries. Overall, the Netherlands is a prosperous country with an important role as a transportation hub for Europe (Rotterdam, the country's main port, is the largest port in Europe).

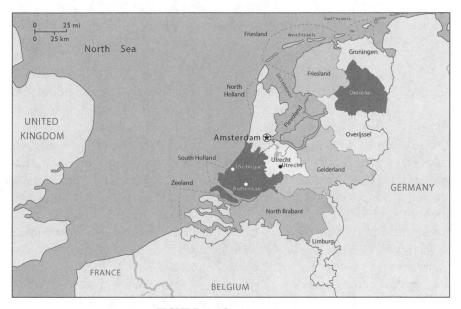

FIGURE 1.19 Netherlands.

FIGURE 1.20 States-General. This Wikipedia and Wikimedia Commons image is from the user Patrick Rasenberg and is freely available at http://en.wikipedia.org/wiki/File:The_hague_hofvijver.jpg and has been released into the public domain.

The population of the Netherlands, like the landscape bisected by rivers, dikes, and estuaries, is divided and consists of four principal groups: Catholics, Protestants (following a number of denominations), Socialists, and Liberals (chiefly the middle classes). To allow for the inclusion and representation of each group (none of which ever consisted of a population large enough to claim a majority), it was necessary to develop, as in Israel, inclusive institutions. Accordingly, again as is the case in Israel, the Netherlands has a proportional representation voting system in which citizens vote for political parties for parliament. As with other proportional representation systems, the Dutch parliament is made up of a number of parties, none of which have a majority (in fact, it is rare for any of the parties to possess even one-third of the seats in a given parliament) and, consequently, governments are formed through the creation of coalitions. The Dutch parliament, known as the States-General (Staten-Generaal), consists of two houses. The lower chamber, the House of Representatives, is elected directly through national proportional representation elections for parties, whereas the upper house, the Senate, is elected by the provincial legislatures. The House of Representatives consists of 150 seats and the Senate consists of 75 seats. Similar to upper houses in countries such as the UK, Canada, and Australia, the Dutch Senate has very limited powers. Legislation is initiated and passed in the House of Representatives. The Senate has no powers to amend legislation, and its role is primarily to review legislation for errors and to either pass or veto the legislation. The House of Representatives, in typical parliamentary fashion, initiates, amends, and passes legislation and forms the government—which can be voted out of power if the government loses majority support in the lower house. Like the UK, Canada, and Australia, the Netherlands is a constitutional monarchy. Although the cabinet is not appointed by the Queen and does not serve at her pleasure, the Queen is called upon to play the role of arbiter since the election outcome never produces a clear winner with a majority of seats in the House of Representatives (Andeweg and Irwin, 2009, p. 126). The Queen fulfills this role by appointing a politician to oversee the coalition negotiations and the subsequent formation of a new government. Dutch prime ministers tend to be weak compared to their counterparts in other parliamentary democracies and have very few formal powers (they lack, for example, the power to dismiss members of the cabinet), and virtually all executive decisions are taken by the cabinet as a whole, with each minister having significant power to run his/her ministry autonomously.

The Dutch legal system follows the French inquisitorial judicial model. There are no jury trials in the Netherlands; cases are heard by a single judge in subdistrict courts (which deal with minor cases) or a panel of three to five judges in district courts (Rechtbank). There are 19 district courts (each with

FIGURE 1.21 Netherlands Supreme Court. This Wikipedia and Wikimedia Commons image is from the user M. Minderhoud and is freely available at http://commons.wikimedia.org/wiki/File:Hoge_raad_gebouw_lange_voorhout.jpg under the GNU Free Documentation License, Version 1.2.

seven subdistrict courts), five appellate courts (Gerechtshof), and a Supreme Court (Hoge Raad, which acts as an appellate court and focuses on supervising the administration of justice). The Supreme Court is not allowed to rule on the constitutionality of legislation and must restrict its activities to judicial appeals and oversight.

The Netherlands essentially follows the model of a centralized state with some local autonomy (closer to the French model of territorial governance than to federal systems). The country is divided into 12 provinces, each with its own directly elected legislature and government. However, with the exception of a few areas such as infrastructure and environmental protection, provincial control over policy is limited (Andeweg and Irwin, 2009, p. 192). Municipal and other local governments are more powerful than provincial governments and have, among other duties, responsibility for education, urban planning, and the provision of social services.

THE REPUBLIC OF ITALY (REPUBBLICA ITALIANA)

Italy's territorial area is just over 301,000 square kilometers, making it slightly larger than the state of Arizona. Its territory comprises a long peninsula, two large islands, and several smaller islands, and most of the land area is rugged and mountainous, with some plains and coastal lowlands. The climate is chiefly Mediterranean, although the far north is alpine and the farther south one goes, the hotter and dryer it becomes. Most of the population is of Italian ethnic origin (with less than 5 percent of African, North African, and Middle Eastern origins), but there are significant cultural, linguistic, and economic differences between the northern and southern halves of the country. Italy has a comparatively larger agricultural sector that employs just over 4 percent of the workforce, with traditional industries employing 25 percent of workers and the service sector employing 65 percent of the workforce. Italy's north is highly developed and generally on a par with northern European living standards, but its south is considerably poorer and less developed.

Like contemporary Germany, contemporary Italy was profoundly affected by dictatorial rule and the desire to prevent a return to fascism and the attendant centralized concentration of power. Like Germany, Italy is a relatively new country and was not unified until the mid-nineteenth century (with the completion in 1870 of an 11-year unification process). Like the other countries surveyed (with the exception of Israel and the UK), Italy has a

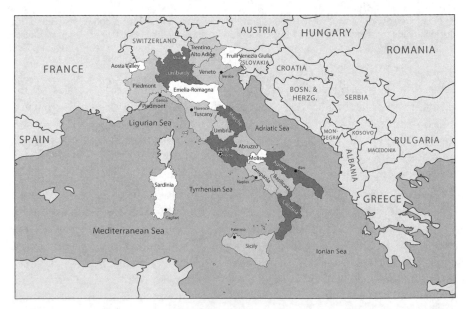

FIGURE 1.22   Italy.

FIGURE 1.23   Italian Chamber of Deputies. This Wikipedia and Wikimedia Commons image is from the user Hadi and is freely available at http://commons.wikimedia.org/wiki/File: Parlament_italien.JPG and has been released into the public domain.

constitution that sets out the role of the branches of government, and like the other countries surveyed, has a parliamentary system of government. The government (cabinet), known as the Council of Ministers (Consiglio dei Ministri), is headed by the prime minister. As in the case of the other countries surveyed here, the Italian government can remain in office only if it enjoys the confidence of the parliament (i.e., if it is not voted out of power). In a departure from the parliamentary norm, however, the Italian government must ensure the support of both houses of its bicameral parliament (the Chamber of Deputies and the Senate) in order to govern.

Unlike the other countries surveyed here, Italy has a truly bicameral parliament, in that both the lower house, the Chamber of Deputies (with 630 members), and the upper house, the Senate (with 315 members), have the same authority and role (Bull and Newell, 2005, p. 116). This not only means that the government must ensure the support of both houses but also that bills in Italy must be approved by both houses in order to become law. Accordingly, unlike the far less representative (but also less powerful) upper houses in many of the other countries surveyed here, Italian senators (senatori) are elected in a very similar manner as Italian deputies (deputati). As in Israel, elections for parliament are carried out on the basis of voting for party lists. But unlike Israel and the Netherlands, but similar to the other countries in the survey, elections are regionally based. There are 26 voting districts for the

Chamber of Deputies and 20 for the Senate (the latter corresponding to Italy's 20 administrative districts).

Between 1991 and 1993, Italian electoral law was reformed in an attempt to achieve political stability. For decades Italy was infamous for having one of the most unstable (if not *the* most unstable) political systems among democracies, due to the fact that its proportional representation voting system made it very easy for small parties to enter parliament, thus creating governments dependent on broad and unstable coalitions. The instability of these coalitions resulted in governments being brought down on a regular basis, so that, up to the time of these political reforms, postwar Italian governments lasted an average of nine months. Under the present system, elections are still held under a proportional representation system with voters voting for party lists, but the largest coalition of parties in the Chamber of Deputies is automatically granted enough seats (if it has not obtained these in the actual elections) to gain a 55 percent majority in the lower house. Similarly, elections for the Senate allow for what is termed a "majority prize," in that the largest coalition of parties per region is granted a 55 percent majority of the seats representing that particular region. Without getting into the intricacies of the Italian electoral system, the bottom line is that the new system has reduced the power and number of smaller parties, giving governments a greater chance of serving out their terms.

As a parliamentary republic, Italy also has a president, who, as in Israel, is elected by parliament, but the Italian president plays a more substantive constitutional role. The Italian president enjoys the power to dissolve the parliament and can decide to refrain from dissolving a given parliament in the wake of the collapse of a government, to encourage the existing parliament to produce a new coalition government and thus to attempt to ensure political stability and continuity (Bull and Newell, 2005, p. 129). The president appoints the prime minister and government from the dominant coalition of parties, but the Italian president has more discretion regarding the choice from among competing groups of parties in parliament. The Italian president also enjoys the power to veto legislation, albeit only temporarily. Finally, in Italy, uniquely among the republics surveyed here, outgoing presidents are automatically entitled to a senatorial seat for life and have the right to nominate five lifetime senators from among Italians who have distinguished themselves in social, scientific, or artistic fields (Bull and Newell, 2005, p. 117).

The judicial branch in Italy is independent of the other two branches of government and, to assure this independence, the judicial branch has nearly complete control over the appointment and dismissal of judges, this being carried out by the High Council of the Judiciary (Consiglio Superiore della Magistratura). Italian law is based on Napoleonic law and the civil law

FIGURE 1.24 Italian Court of Cassation. This Wikipedia and Wikimedia Commons image is from the user Zavijavah and is freely available at http://commons.wikimedia.org/wiki/File: Palazzo_di_Giustizia_a_Roma_102_2049.jpg under the Creative Commons Attribution-Share Alike 3.0 Unported license.

system. However, unlike France, a 1990 reform of the legal system resulted in the phasing out of inquisitorial judicial practice with an adversarial system similar to that of the common law countries.

With respect to the organization of the judicial system, minor legal matters are handled by magistrates (conciliatori), who are lay judges. At the next level, the courts of the first instance for more serious cases (civil and criminal) are tribunals (tribunale) and three other specialized courts. Tribunals usually consist of one judge, but with important cases heard by three judges (there being159 tribunals, each with jurisdiction over its own district). Tribunals are default courts for general civil and criminal disputes, but other more specialized courts operate at this level of the judicial system, including 90 criminal courts (Corte d'Assise) that deal with felonies as well as courts that deal with disputes concerning minors and courts that deal with traffic matters, minor civil claims, and so on. At the next level, there are courts of the second instance that hear appeals from these lower courts: tribunal appellate courts (also known as tribunale), appeals courts (Corte di Apello), and criminal appeals courts (Corte d'Assise d'Apello). At the top tier of the Italian judicial system lies the Court of Cassation (Corte di Cassazione), which acts as the highest court of appeal. Constitutional matters, however, are ruled upon by a separate 15-member Constitutional Court (Corte Constituzionale).

Like France and Israel, Italy is a centralized state. However, the nature of the country's history and geography (including the fact, as noted earlier, that Italy was not even a unified country for most of its medieval and a significant portion of its modern history) has meant that there are stark localized differences in dialect, mentality, and culture. This is pronounced enough that regional political parties, such as the Northern League (which has alternated between calls for a federalist system and demands for complete separatism), have been able to make political inroads. Italy has 20 regions (regioni), each of which enjoys a significant level of autonomy. Each region is governed by an elected regional council (Consiglio Regionale), a government (Giunta Regionale), and a president. The constitution guarantees the regions' financial autonomy although their power of taxation is limited and can be overridden under certain conditions by the central government. Each region is broken down into provinces (province), and these, in turn, are divided into communes (comuni). Provinces and communes (municipalities of various sizes) also have legislatures, executives, and chief executives, and consequently their structure largely mirrors that of regional governing institutions. The regions are responsible for local policing, environmental protection, public services, and economic development. Provinces in general have few powers, as commune powers have increased over time. Aside from some planning issues and a few other matters, provincial power is quite limited. Communes focus largely on public works, transportation, commerce, utilities, hygiene, and cultural events.

JAPAN (NIPPON-KOKU)

Like the UK, Japan is an island nation comprising four main islands and several smaller ones, with a total area of almost 378,000 square kilometers which makes it slightly smaller than the state of California. Its climate is primarily marine and temperate, with a subtropical climate in the far south and a cold climate in the far north. Most of the terrain is rugged and mountainous, necessitating the judicious use of land. Japan has a population of almost 127 million persons, making it the tenth most populous country in the world (while being the sixty-first largest). Japan is the most ethnically homogeneous country in this survey and one of the most ethnically homogeneous countries in the world. Fully 98.5 percent of the population is of Japanese ethnicity, and the largest minority group, Koreans, comprises only half of 1 percent of the population. Surprisingly for such a crowded country, the rate of urbanization is comparatively low, with 66 percent of the population inhabiting cities or towns and 4 percent of the workforce involved in agricultural pursuits such as rice farming. A further 28 percent are

FIGURE 1.25 Japan.

employed in Japan's vaunted industries and 68 percent in the service sector. Japan is an economic superpower, and despite some two decades of economic stagnation, Japan is still the world's third largest economy.

As with Germany and Italy, Japan's governmental structure represents a reaction to the dictatorship and aggression of the prewar and wartime regime, overthrown by the U.S. occupation of the country in 1945. The American authorities, in part to build legitimacy for the new Japanese institutions and to maintain some degree of continuity with the past, allowed the Emperor to continue to reign, thus instituting a constitutional monarchy that remains in force to this day. The Emperor was forced to renounce his "divine" status in 1945 but was allowed to continue to function as head of state. As with many other constitutional monarchs, the Emperor's role is to represent the country, to act as its symbolic embodiment, and to fulfill various formal ceremonial functions of state, such as appointing the prime minister (although the actual determination as to who will be prime minister depends on majority support in the parliament). Like most other democracies, Japan also possesses a formal constitution (this document, like so many other aspects of postwar Japanese governance) was a product of the U.S. occupation. It includes a section (Article 9) that renounces war as the sovereign right of the Japanese nation in perpetuity, although this has been interpreted over the years as renouncing Japan's right to wage aggressive war but not its right to self-defense.

The Japanese parliament (the Diet) consists of two houses, the lower house (House of Representatives—Shugi-in) with 480 members and the

FIGURE 1.26 Japanese Diet. This Wikipedia and Wikimedia Commons image is from the user Chris 73 and is freely available at http://commons.wikimedia.org/wiki/File:Japanese_diet_outside.jpg under the creative commons cc-by-sa 2.5 license.

upper house (House of Councillors—Sangi-in) with 242 members. Unlike some of the other bicameral parliaments discussed above, the upper house is elected directly, although its constituencies mirror Japanese prefectures and thus are conceived of as representative of local interests. For elections to the lower house, the country is divided into 300 single-seat constituencies and 180 seats are allocated on the basis of proportional representation. Each voter votes once for a candidate to represent his/her local constituency and once for a party list. Representatives are voted in for four-year terms, although this can be shortened if early elections are called. Elections to the upper house are also divided between constituencies (although these are often multiseat constituencies) and proportional representation (with 146 councillors voted in by constituency and 96 by proportional representation via party lists). Elections to the House of Councillors are for six-year terms. With respect to matters of power and authority, the House of Representatives is considered more powerful than the House of Councillors. The prime minister and government are chosen by the Diet as a whole and must be members of the Diet, but if both houses disagree as to the appropriate choice, the lower house has ultimate authority to choose. In terms of legislation, a bill passed by the lower house but rejected by the upper house can be made into law if the House of Representatives passes it again with a two-thirds majority.

FIGURE 1.27 Japanese Supreme Court. This Wikipedia and Wikimedia Commons image is from the user Wiiii and is freely available at http://commons.wikimedia.org/wiki/File:Supreme_Court_of_Japan_2010.jpg under the Creative Commons Attribution-Share Alike 3.0 Unported, 2.5 Generic, and 1.0 Generic license.

Like that of France and the Netherlands, the Japanese judicial system is inquisitorial and is based on five types of courts: the Supreme Court, high courts, district courts, family courts, and summary courts. The Supreme Court (which consists of a grand bench with 15 justices and three petty benches made up of five justices each) is the country's highest appellate court. The court also has the authority to adjudicate legislation. The high courts handle appeals from the other courts, with appeals usually heard by a panel of three professional judges. Most criminal and civil cases are first adjudicated in district courts (the exceptions being family matters and juvenile delinquency cases, which are handled by family courts, and minor civil claims, which are handled by summary courts). Most district court cases are handled by a single professional judge. Investigations in Japan are handled by the public prosecutor's office with different levels of prosecutors dealing with the different levels of courts (Secretariat of the Judicial Reform Council, 1999). Strictly speaking, Japan does not have a jury system for trials, but it recently enacted a law that added "lay judges" to the existing panels of professional judges, with lay judges being chosen randomly from among the members of the public. In district courts, cases are now heard by a panel consisting of three professional judges and six lay judges and, in keeping with the principles of the inquisitorial legal approach, all the judges having authority to initiate and oversee investigations.

As with the Western Allies in Germany at the end of World War II, U.S. occupation forces in Japan viewed decentralization of the Japanese state as an important guarantor of democracy that would prevent the concentration of power and the return of authoritarian rule. Article 92 of the Japanese constitution affirms the decentralization of power and the safeguarding of local autonomy. Japan is divided into 47 prefectures (Todofuken) with different degrees of autonomy and size (including three municipal prefectures: Tokyo, Osaka, and Kyoto). In addition, there are close to 200 local governments with their own degrees of autonomy. Prefectures (headed by governors) and local governments (headed by mayors) enjoy authority over administration, budgets, planning, and bylaws, and they fulfill functions that in other countries would be within the purview of central government (Stevens, 2008). For example, provision of welfare and other social insurance as well as health care is a shared responsibility between municipal and prefectural governments (with municipalities providing basic health care and prefectures running hospitals) on the one hand and central government on the other. Nevertheless, unlike Germany, the prefectures are not autonomous states, and Japan can still be considered to be a centralized system with regional administration (much like France).

CONCLUSION

As this chapter has shown, the countries addressed in this book have a wide variety of political and legal systems and institutions. They also differ dramatically in terms of the types of homeland security challenges they face. In the terrorism realm, all the countries in our study have experienced terrorism of one kind or another over the years, but the scope and intensity of the terrorist threat has been very different. Clearly, Israel and the UK have been the countries that have had to cope with the greatest amount of terrorism in the most intense form (Israel, of course, more so than the UK), but at one point or another in their fairly recent past, Canada, Japan, Germany, France, Italy, and the Netherlands have had to cope with terrorism problems, and Australians have been the subject of terrorist attacks overseas. Moreover, all of these countries view terrorism as a significant threat that must be guarded against. In the areas of natural disasters, the Canadians and Japanese have extensive experience with these, and the British, Germans, and others have had to deal with wide-scale flooding and other problems. All of the countries surveyed have had to cope with immigration and integration issues and with public health problems (with Canada having had to deal with the SARS outbreak). Additionally, all of the countries surveyed have had to beef up airport and seaport security and develop policies to strengthen critical infrastructures. In short, although there are significant differences in institutions and political and legal systems (and we have not even broached the issue of cultural differences) across these countries, many of the threats that they must prepare for are similar and are also common to those facing homeland security policy-makers in the United States.

ISSUES TO CONSIDER

- How is a country's historical experience reflected in its constitution and/or institutions?
- What are the primary differences between common law/adversarial legal systems on the one hand and civil law/inquisitorial systems on the other?
- How do the versions of federalism differ among the federal systems surveyed in this chapter?
- How do the versions of centralism differ across the centralized systems surveyed in this chapter?

CHAPTER 2

COUNTERTERRORISM STRATEGIES, LAWS, AND INSTITUTIONS

> Mrs. Thatcher will now have to realize that Britain cannot occupy our country and torture our prisoners and shoot our people in their own streets and get away with it. Today we were unlucky, but remember that we only have to get lucky once. You will have to be lucky always....
>
> —IRA communiqué in the wake of the bombing of the Grand Hotel in Brighton in October 1984 during the Conservative Party Conference. The attack was an assassination attempt against the then British prime minister.

In this chapter we focus on a survey of counterterrorism strategies, laws and institutions of relevance to many of the countries in the survey. As we show below, there is a considerable degree of variability in counterterrorism approaches taken by different countries.

Counterterrorism strategy may be thought of as a continuum. On one end, the policy is based on a law enforcement approach, whereas on the other, it is based on a war-fighting approach. Democratic countries facing serious terrorist threats must decide—based on political traditions, institutional makeup, or other factors—which approach to emphasize. All democracies will tend to array themselves somewhere in the center of this continuum, although some may be closer to one or another of the ends. Countries that favor the war-fighting approach view the conflict with terrorists essentially as they would a war with a hostile country and, indeed, at times the terrorists

Comparative Homeland Security: Global Lessons, First Edition. Nadav Morag.
© 2011 John Wiley & Sons, Inc. Published 2011 by John Wiley & Sons, Inc.

FIGURE 2.1   Grand Hotel in Brighton, England, following the IRA bombing of October 12, 1984. This Wikipedia and Wikimedia Commons image is from the user D4444n and is freely available at http://commons.wikimedia.org/wiki/File:Grand-Hotel-Following-Bomb-Attack-19Grand-Hotel-Following-Bomb-Attack-1984-10-12.jpg under the GNU Free Documentation license.

operate from one or more countries whose rulers collude actively with the terrorists—as in the case of Al Qaeda and the Taliban in Afghanistan prior to late 2001. As with more conventional forms of warfare, fighting terrorism, according to this school of thought, requires the use of military or quasimilitary tools, the goal being the maximum devastation of enemy personnel and infrastructure in order to destroy the enemy's capacity to wage war (or, in this case, terrorism). Moreover, this strategy allows for the use of any and all means of intelligence gathering without the need to safeguard rights to privacy or other civil liberties, as it usually targets noncitizens in the territory of foreign countries. The use of violence is also frequently intensive and tempered only by the principle of proportionality (i.e., only the amount of force needed to accomplish a mission should be applied, and that use of force should be comparable to the nature of the threat). In other words, force should be used in a measured and relevant fashion and never gratuitously. In the context of this approach, victory against terrorism means, as with conventional warfare, the complete cessation of all hostile activity through the destruction of the enemy's capacity to engage in hostile acts.

The law enforcement approach, at the other end of the continuum, is based on an entirely different philosophy and modus operandi. The goal here is not to kill the enemy to deny them the ability to conduct hostile acts. In fact, the law enforcement approach, unlike the war-fighting approach, does not, at least for purposes of prosecution, generally view the "enemy" collectively, but rather as individuals. These individuals may or may not be organized into terrorist groups, but in any case they are not treated as a collective threatening the state, but rather as distinct individuals carrying out specific criminal acts for which each has to be prosecuted independently. Those taking the law enforcement approach to combating terrorism generally spend most of their time operating within the borders of the democratic state and thus are subject to legal restrictions designed to safeguard the basic rights of the population in those countries. In the law enforcement context, victory against terrorism means, at a minimum, the successful maintenance of law and social order in the face of the terrorist threat and, at a maximum, the complete cessation of terrorist activities and the successful prosecution and incarceration of all terrorists.

In some ways, the law enforcement approach contains a more flexible definition of success than the war-fighting approach. Since the proponents of the law enforcement approach view terrorism as a form of crime, and given the fact that crime is never something that can be truly eradicated, the traditional focus on the maintenance of law and order and reduction of crime to "acceptable levels" means that success in counterterrorism does not necessarily imply total eradication of the phenomenon. The war-fighting approach, on the other hand, emphasizes the total destruction of the enemy's capacity to engage in hostile acts, and accordingly, this results in the expectation that all hostile activity can and will be ended once the adversary understands that it no longer possesses the capacity to wage war. Clearly, no democratic state facing a terrorist threat will employ only one approach to the complete exclusion of the other. However, the mix of measures differs very markedly among democracies. Countries such as Germany, France, and to a lesser extent Britain (at least with respect to its past actions in Northern Ireland) tend to employ more of a law enforcement approach to countering terrorism, whereas Israel provides the opposite example, being a country that leans more heavily toward the war-fighting end of the continuum. Particularly in the wake of 9/11, the United States appears to have placed itself close to the center of this continuum, heavily employing both war-fighting and law enforcement strategies to combat terrorism.

For a liberal democracy, the counterterrorism strategy chosen must, regardless of its placement along the war-fighting/law enforcement continuum, not lose sight of what is being protected: namely, the very existence of the liberal-democratic state. According to several seventeenth- and eighteenth-century political theorists, the state was necessary to prevent men from succumbing to the passions of others, to defend the society against external enemies, and to regulate commerce and trade. The state thus represents a limitation on freedom, and freedom, if unchecked, as Thomas Hobbes suggests, will only lead to social chaos as well as institution of the "law of the jungle." According to John Locke, the *liberal-democratic state* thus represents a kind of "social contract" between the population and the state whereby the state will provide law, order, and security in exchange for some of its inhabitants' basic freedoms (Benoit-Smullyan, 1938, p. 105). Naturally, this "arrangement" between people and the state requires, according to John Stuart Mill, that "everyone who receives the protection of society owes a return for the benefit...in not injuring the interests of one another... which...ought to be considered as rights; and secondly in each person's bearing his share...of the labors and sacrifices incurred for defending society or its members from injury and molestation. These conditions society is justified in enforcing at all costs to those who endeavor to withhold fulfillment" (Mill, 1998). In other words, the liberal-democratic state is justified in enforcing its laws and requiring that its citizens fulfill their obligations to it regardless of whether particular citizens desire to respect those laws and/or fulfill those obligations. Consequently, according to Max Weber, the state must enjoy a "monopoly on the legitimate use of force" within its territory.

The liberal-democratic state is also founded on the principle of the rule of law. Locke believed that the legislative assembly should act as the lynchpin of the rule of law, and Alexander Hamilton argued that an independent court system is also an essential component of the rule of law. The liberal-democratic state must also endeavor to provide for the maximum number and type of rights and freedoms for its inhabitants, such as freedom of speech, assembly, and religion; the right to due process; and the right to equality before the law. Moreover, to buttress the smooth functioning of society within the framework of these rights, the liberal-democratic state should make all efforts to safeguard and promote societal tolerance and pluralism (Wilkinson, 1979, pp. 16–17).

Given all of the above, the central dilemma faced by liberal democracies in attempting to combat terrorism effectively has to do with the fact that combating a serious terrorist threat effectively requires measures that strengthen the power of government over the individual and that, in one way or another, reduce the freedoms and protections that individuals enjoy. The trade-off is one which, at best, represents an exchange of physical security (lives and property) at the expense of a greater degree of fundamental individual rights. The willingness of citizens of democracies to accept this trade-off will be a function of the level of threat that they perceive themselves to face from terrorism and the degree to which they feel that their fundamental democratic rights are in danger.

When liberal democracies face concerted and ongoing terrorist threats, they must endeavor to resist calls for "throwing out the baby with the bathwater." In other words, in cracking down on terrorists and otherwise heightening antiterrorism efforts, the state must not, in the process, lose its basic identity as a liberal democracy. A victory against terrorism that results in the liberal democracy transforming itself into an authoritarian state is arguably not a victory but a defeat. In fighting terrorism, the leadership of the liberal-democratic state must decide precisely what it is that it is fighting for: the preservation of the life of every last citizen or the preservation of the way of life. As disturbing and potentially damaging this may be at the individual level, the foremost goal of the liberal-democratic state is to preserve its system of government and the way of life of its citizenry. Even in liberal democracies that do not face a substantial terrorist threat, the preservation of freedoms and rights invariably results in higher levels of crime (as unsavory individuals take advantage of personal and collective freedoms for their own warped aims) than is the case with stable authoritarian or totalitarian regimes. Nevertheless, these higher levels of crime are considered to be an acceptable price to pay for the innumerable personal and societal benefits afforded by the liberal-democratic way of life.

One may view terrorism in a similar manner, although the terrorist threat, almost by definition, creates a greater sense of fear and vulnerability than the classical criminal threat. Accordingly, in fighting terrorism, liberal-democratic governments must show that their measures are designed solely to prevent terrorism and protect society and should not form a pretext for a general decrease in citizens' rights or other form of undermining of the democratic order. This is important not only because maintenance of the democratic order is a central goal, but also because repression and over-reactive policies often play into terrorists' hands, particularly with respect to "homegrown" terrorists, by strengthening their claim that democracy is a sham and that they are the true defenders of the people. Hence, counterterrorism policy must operate within the law even if this requires the use of more flexible legislation and intelligence; law enforcement and military agencies must remain accountable to the elected civilian leadership. As Paul Wilkinson wrote in his seminal work, *Terrorism and the Liberal State*: "To believe that it is worth snuffing out all individual rights and sacrificing liberal values for the sake of 'order' is to fall into the error of the terrorists themselves, the folly of believing that the end justifies the means" (Wilkinson, 1979, p. 122).

As noted above, since liberal democracies operate under the rule of law and must endeavor to act, at all times, within the legal framework, the question of defining terrorism becomes critical because laws operate on the basis of language and distinguish between different types of activities, legal and illegal, on the basis of terminology.

DEFINING AND CATEGORIZING TERRORISM AND THE EVOLUTION OF COUNTERTERRORISM LEGISLATION

The issue of defining terrorism is critical not only because democratic countries must work within the law and the law relies on specific terms but also because international cooperation between democratic countries in the counterterrorism realm is dependent on similar ways of defining terrorism. Otherwise, for example, terrorists being pursued by one country may be able to escape arrest in another if they, the organization with which they are affiliated, or the activities in which they are engaged, are not defined by the host country as terrorist. The reader is likely to be familiar with the well-worn academic debate over the definition of terrorism and platitudes such as "one man's terrorist is another man's freedom fighter." The reality is that this argument is of true importance only in the context of the manner in which it influences legislators to draft and ratify laws using one or another set of criteria. In any case, it is less important to determine *what* terrorism is than to determine *who* the terrorists are because, clearly, legal sanctions cannot be applied to the phenomenon of terrorism but can be applied to terrorist organizations and individual terrorists. It should, however, be noted that there is by no means a total consensus as to whether laws defining terrorism and prosecuting terrorist acts need to be promulgated in the first place.

In a report presented to the British Parliament in March 2007, the independent reviewer of terrorism legislation, Lord Carlile, notes that there are four different options or models for handling terrorism offenses. The first option is to abstain from providing a specific definition of terrorism or instituting specific procedures to handle such offenses—the argument being that terrorist acts can be prosecuted using existing laws and that the creation of a definition of terrorism and special procedures for handling terrorist offenses is more likely to lead to a reduction in personal freedoms and arbitrary actions on the part of the state that may impinge on democratic principles. The second option is to provide a definition of terrorism but to refrain from instituting special procedures to handle terrorism crimes. The claim with respect to the second model is that while acts of terrorism are addressed by existing laws, the nature of terrorism is such that it should be defined and subsequently recognized as a qualitatively different type of crime, and therefore the sentencing judge should be able to increase the sentence accordingly (similar to the manner in which hate crimes are often dealt with). The third approach holds that a narrow and very specific definition of terrorism is required and that special laws need to be instituted to punish terrorism crimes. According to this option, terrorism should indeed be recognized as a unique type of criminal activity requiring special legislation but that, to safeguard civil liberties, only the core activities of terrorists should be criminalized. The fourth and final option is to define terrorism in a broader-based way and to promulgate specific procedures and punishments for terrorism offenses—the argument being that the authorities need to be proactive rather than just reactive and consequently need to be able to employ a more generalized definition of terrorism and enjoy broader preemptive powers in dealing with terrorism plots and pre-attack activities (Lord Carlile, 2007, p. 19). Ultimately, all of the countries surveyed here employ a mix of broad definitions of terrorism, specific counterterrorism legislation (or provisions for aggravated sentencing of terrorist acts), and traditional criminal statutes in order to cope with the terrorist threat. In other words, not surprisingly, governments have shown a preference for providing themselves with a maximum of powers and flexibility in addressing terrorism.

All the countries surveyed in this chapter define terrorist acts, individuals, and organizations in terms of the nature of the actions and their objectives rather than in terms of the ideology or overarching aims of the organization behind the activities. It is, of course, legal in democracies to argue publicly and to organize politically to achieve goals that terrorist organizations desire to achieve—with the caveat that these rights do not extend to propagandizing or fund-raising for terrorist organizations: actions that are, in fact, illegal. It is not illegal in the UK to lobby for withdrawal from Northern Ireland, in Israel to push for the dissolution of Israel as it is presently constituted (indeed, several members of the Israeli parliament support this goal), or in Canada to support the implementation of Shari'a law and the creation of Islamic rule. Illegality comes into play when those individuals and organizations that advocate these solutions take steps, beyond their exercise of free speech and right to initiate change through the political system, to undermine the existing constitutional system, threaten or cause the loss of life, or attempt to bring about significant disruption of the economy or society. Next we provide a brief overview of definitional approaches among the countries surveyed.

Israel

Israel has chosen to forgo an elaborate and standardized definition of terrorism, opting instead for a minimalist and more flexible definition. This approach is, in fact, quite characteristic of Israel, in that flexibility and ad hoc solutions are generally preferred to more rigid and formalistic approaches. Under Israel's Prevention of Terrorism Ordinance, a terrorist organization is classified as having "committed acts of violence calculated to cause death or injury to a person or made threats of such acts of violence" (Israel, Prevention of Terrorism Ordinance, No. 33, Section 7, 1948). This definition also extends to organizations that take responsibility for acts of violence or whose members take responsibility on its behalf. Beyond this very basic definition, Section 8 of the Ordinance grants the government of Israel the power to classify groups (and therefore the members of these groups) as terrorist, indicating that "if the Government, by notice in the Official Gazette, declares that a particular body of persons is a terrorist organization, the notice shall serve, in any legal proceeding, as proof that that body of persons is a terrorist organization, unless the contrary is proved" (Israel, Prevention of Terrorism Ordinance, No. 33, Section 8, 1948). This nonspecific definition provides the government with significant latitude in being able to classify groups and individuals as terrorist and puts the burden of proof on those groups and individuals to prove in court that they are not, in fact, engaged in terrorism.

The government of Israel has used its powers to classify terrorist groups as such on an ongoing basis with respect not only to Palestinian groups but also to Israeli groups. For example, after the massacre by a right-wing extremist of 29 Palestinians in the West Bank city of Hebron in February 1994, the government declared the extreme right-wing Jewish groups Kach and Kahane Chai to be terrorist organizations and thus outlawed them. Finally, if in the course of a terrorism trial the court rules that a particular body of people constitutes a terrorist organization, this will be taken as prima facie evidence that the organization is a terrorist organization (Israel, Prevention of Terrorism Ordinance, No. 33, Section 11, 1948). Israel's Penal Law treats terrorist organizations as "unlawful associations" and expands upon the definitions of the Prevention of Terrorism Ordinance, focusing more on specific goals and methodologies. Under the Penal Law, unlawful associations are associations of persons (whether formally established or informal) who, among other things, attempt or encourage (1) the destruction of the legal and institutional system, (2) the violent overthrow of the government of Israel or any other government, and (3) destruction of the property of the Israeli government or any other government (Israel, Penal Law, Section 145, 1977). Moreover, many of the basic provisions of the Penal Law, such as causing the death of a person, assault, and so on, can also be used in prosecuting terrorist acts (State of Israel, 2001, p. 16).

The last major category of legislation that can be used in carrying out legal action against terrorist groups is an emergency law known as the Defense (Emergency) Regulations of 1945. This was a British Mandatory law that was effectively incorporated into Israeli law via the Government and Law Arrangements Ordinance (1948), although over the years it was amended a number of times. The Defense Regulations do not define terrorism but, rather, list a series of crimes involving the use of firearms and explosives, the carrying out of unauthorized military training, or the provision of protection or other support for individuals or groups that carry out such acts [Israel, Defense (Emergency) Regulations, Sections 58, 59, 62, 64, and 66, 1945]. The regulations also outlaw membership in groups that carry out such acts and engage in terrorist activity, with the minister of defense given authority to determine which groups are illegal [Israel, Defense (Emergency) Regulations, Section 84, 1945]. Overall, then, Israeli law is fairly flexible in its definition of terrorism and essentially allows the authorities to determine which behaviors constitute terrorism and which organizations are to be classified as terrorist and outlawed. Israel also employs a mix of traditional criminal legislation, specialized legislation dealing with terrorism, and emergency legislation to ban groups and prosecute their members. The definitions used are either dependent on the government or on a court determining that a particular organization is a terrorist group or on legislation that cites very broadly defined activities as terrorist activities.

United Kingdom

Given its long experience with terrorism, the UK, not surprisingly, has a range of laws to cope with the problem, including some laws that date back to the nineteenth century. The genesis of UK counterterrorism laws and therefore of definitions of terrorism lies in the conflict with Irish nationalism. In the nineteenth century, terrorism had yet to be viewed as a unique phenomenon designed to destabilize the state, economy, and society. Instead, terrorism was viewed primarily as a series of heinous murders rather than as a threat to the political and social order. Accordingly, UK counterterrorism legislation originally focused on defining and determining the penalties for violent acts as opposed to defining those acts as terrorism and treating them as a distinct category. Thus, the Explosive Substances Act outlaws the possession and/or use of explosives with the intent of causing loss of life or destruction of property (Explosive Substances Act, Sections 2 and 3, 1883) rather than referring specifically to the carrying out of terrorist acts. This piece of legislation, and others relating to sedition and murder along with a series of emergency laws, were used against Irish nationalist terrorism (and insurgency) well into the twentieth century.

During the early 1970s, however, in response to the dramatic increase in the terrorism in Northern Ireland and on the UK mainland and the apparent failure of British military operations in Northern Ireland (using emergency powers), Parliament passed a series of laws designed to deal with terrorism within the context of the civilian legal system (although conceived of as temporary emergency legislation). The Emergency Provisions Act of 1973 created a set of special police powers and a parallel court system for dealing with terrorism offences and was followed by the first formal counterterrorism law, the Prevention of Terrorism Act of 1974. These pieces of legislation were renewed a number of times due to sunset provisions, and while also designed (after 1974) to deal with a growing IRA terrorist campaign on the British mainland, these laws were geared primarily toward coping with terrorism in Northern Ireland. Gradually, however, laws were promulgated that expanded the purview of such legislation beyond Northern Ireland, as the Prevention of Terrorism Act (1984), which extended arrest and detention powers to international terrorism [Prevention of Terrorism Act, Section 12(3)(a), 1984]. By the late 1990s, however, progress on the Northern Ireland peace process, the newly developing problem of Islamic radicalization on the UK mainland, and the passage of the Human Rights Act in 1998 made it necessary to review and revise existing counterterrorism legislation (Walker, 2002, pp. 2–3).

The Terrorism Act of 2000 was designed to serve as permanent legislation that would further codify British counterterrorism laws and make it unnecessary periodically to renew what was ostensibly a piece of emergency legislation of a temporary nature, the Prevention of Terrorism Act. The Terrorism Act defines terrorism as a threat or act that involves violence (against persons or property) or the threat of violence, or threats to public health and safety or to the "electronic system" designed to intimidate the public or government, in order to advance a political or religious cause (Terrorism Act, Section 1, 2000). This law gives the Home Secretary authority to declare an organization to be a terrorist group and to outlaw (proscribe) it, thus (as with the Israeli approach) effectively giving the government the power to determine who the terrorists are. At the same time, there is a provision that allows organizations to appeal the Home Secretary's decision to a judicial body, especially created by the act, known as the Proscribed Organizations Appeals Committee (Terrorism Act, Section 5, 2000).

British law, similar to Israeli law, thus traditionally represented a mix of regular criminal legislation and ostensibly temporary emergency legislation, with very generalized and flexible definitions of terrorism that grant the executive the power to designate groups as terrorists. The British further created a specialized court to hear appeals against such designations, whereas in Israel there is no formalized system, but appeals against such designations can be lodged with the country's Supreme Court.

Canada

Canada has been faced with both domestic and international terrorism over the years. Between 1963 and 1970, a domestic terrorist group, the Front de Libération du Québec, carried out bombing attacks and kidnappings that culminated in the enactment of a state of emergency, the deployment of the military, and the suspension of civil liberties in the Province of Quebec in late 1970. In June 1985, a suitcase bomb planted by Sikh terrorists in Vancouver blew up in the cargo hold of an Air India flight over the Atlantic killing 329 persons. More recently, Canada was the staging area for "millennium bomber" Ahmed Ressam (who was caught trying to smuggle explosives into the United States in order to bomb the Los Angeles International Airport). In June 2006, 17 Canadian residents were arrested and charged with plotting to carry out bombings in southern Ontario, with the arrests representing one of the largest counterterrorist operations carried out in North America since 9/11 (Austen and Johnston, 2006).

Despite its experience with terrorism, Canada did not have any specific counterterrorism legislation prior to 9/11, relying instead on UN terrorism conventions that were integrated into the criminal code (St-Pierre, 2008). In December 2001, Canada enacted the Anti-Terrorism Act. The Anti-Terrorism Act amended a number of federal statues, including the Criminal Code and a number of other laws dealing with state secrets, the use of evidence in court, money laundering, and national defense (Council of Europe, 2008, p. 1). The Anti-Terrorism Act added a section in the Criminal Code that deals with terrorism and provides a list of offenses relating to issues such as hijacking, attacks on ports and airports, bombings, terrorism financing, the carrying out of hate crimes, and other areas of terrorist activity. The Anti-Terrorism Act also revised nine additional statutes, including the National Defense Act and the Canada Evidence Act. The Anti-Terrorism Act provided for stricter penalties for terrorism offences and new powers for preventive arrest, although these specific powers were revoked when they failed to be renewed in 2007 (the original legislation had included sunset provisions). The Criminal Code further outlaws participation in terrorist group activities, the facilitation of terrorist activity, acting to benefit a terrorist group, or harboring terrorists. Many of these activities are specifically defined as illegal whether or not terrorist activity is actually carried out and whether or not the person involved with the organization is aware of any specific attacks being planned by the organization, thus enabling the authorities to use the law in a preventive fashion in addition to punishment after the fact (Council of Europe, 2008). The Criminal Code defines terrorism as activity carried out for "political, religious or ideological purpose, objective or cause" designed to "intimidate[...] the public, or a segment of the public with regard to its security, including its economic security" and can involve the compelling of a person or a governmental entity to carry out or refrain from specific acts [Canada, Criminal Code, Part 11.1, Section B(i) a–e, 1985]. The Criminal Code further defines terrorism as acts carried out in order to cause death, endanger life, destroy property, and cause serious disruption of essential services [Canada, Criminal Code, Part 11.1, Section B(i) a–e,

1985]. As noted above, some clauses in the Anti-Terrorism Act expired in 2007, but the fundamental offenses and definition of terrorism have remained law. Unlike Israel and the UK, Canada thus employs a fundamentally criminal justice–based legal approach, eschewing the use of emergency legislation and also providing specific definitions of terrorism appended to the basic criminal legislation.

Australia

Of the countries surveyed in this section, only Australia has been relatively free of terrorism activities—although in 2004, Muslim extremists Jack Thomas and Faheem Lodhi were convicted of terrorism offenses. As in the case of many other countries, 9/11 served as a watershed in terms of counterterrorism laws and policies, and in the Australian case, the impact of the attacks in the United States was reinforced just over a year later by the Bali bombings. The primary vehicle for counterterrorism legislation is an amended federal Criminal Code. Between 1998 and 2008, the Criminal Code was amended 54 times, with many of those amendments relating to counterterrorism and some coming under a series of amendments passed in 2004 and 2005 and known as the Anti-Terrorism Act (parts 1, 2, and 3). The Criminal Code defines terrorism as "an act or threat of an act for a political, religious or ideological cause which kills or harms people or damages or interferes with property or systems for the purpose of intimidation of the government or the public" (Australia, Criminal Code, Section 100.1, 2008). The amendments of the Anti-Terrorism Act also outlaw organizations that advocate terrorism and update the definition of sedition to include support for terrorism (Australian Government, n.d.). At the same time, the Criminal Code makes it clear that "advocacy, protest, dissent or industrial action" are not terrorist acts as long as they are not designed to threaten public safety. The law does, however, provide allowance for preventive prosecution by outlawing preparations for committing a crime and, in some cases, shifts the burden of proof so that defendants must prove their innocence (Williams, 2006). Overall, then, the Australian approach is similar to the Canadian approach in that terrorism is essentially defined as a special subject of criminal activity and thus is dealt with as a criminal justice issue.

Germany

Germany has had to grapple with terrorist threats, both domestic and international, since the 1970s, including the Red Army Faction and Palestinian and Kurdish groups. However, the 9/11 attacks and the discovery that three of the hijackers had lived and plotted in Germany brought about significant changes in German law and policy. In the post-9/11 period, Germany has had to grapple with Al Qaeda–related threats, including the attempted bombing of two German passenger trains in July 2006 and the conspiracy to attack U.S. bases in Germany by a group of Muslim radicals who had 750 kilograms of hydrogen peroxide in their possession when arrested.

In November 2001, the German parliament passed a series of amendments to a range of laws designed to better address threats (known as Security Package I), and in December 2001 the legislature passed a series of amendments designed to enhance preventive efforts and increase the powers of the authorities in the area of security and immigration (known as Security Package II). German legislative initiatives and policy changes have, however, been tempered by two important considerations: first, a reluctance to increase state power substantially because the maintenance of diffuse state power and strong civil liberties are seen as critical safeguards in light of the country's Nazi past; and second, by a desire to base its laws and policies on the broader European consensus (also a reflection of attempts to avoid associations with past attempts by Germany to act unilaterally and dominate the European continent), something that is manifested in the use of European Union (EU) definition of terrorism (see Sidebar 2-1) and EU lists of terrorist organizations.

German law essentially eschews providing a specific definition of terrorism, yet provides aggravating penalties for terrorist acts. The German Criminal Code (Strafgesetzbuch) makes it illegal to form, participate in, or support a terrorist organization, even though it does not define terrorism as terrorism but, rather, focuses on traditional offences such as murder, causing physical or mental harm, and committing offenses against the environment (Germany, Criminal Code, Section 129a, 2008). Without defining terrorism, the Criminal Code does, however, outlaw the formation, recruitment for, or support of terrorist organizations. Moreover, judicial rulings have determined that a terrorist organization must consist of at least three persons, must exist over a period of time, and its members must subordinate themselves to the common objectives of the organization (Zoller, 2004, p. 477). Hence, German criminal legislation does not define terrorism and essentially employs a definition of a terrorist group that is so general as to be essentially meaningless, and thus nonconstraining—something that affords the government maximum flexibility in defining particular groups or activities as terrorist. Whereas many countries use a definition of terrorism that implies an intent to disrupt or create fear, German law is free of such distinctions, thus allowing the authorities to employ the broader investigative powers and treatment of prisoners permissible under criminal procedure and the various more recent amendments to criminal law to groups that engage in

Sidebar 2-1: EU Definition of Terrorism

In the wake of the attacks of September 11, 2001, the European Union attempted to create a comprehensive definition of terrorism for its use and the use of its member states, although they are not obligated to use the EU definition since EU authority does not yet extend to most security matters (with a few exceptions discussed in later chapters). Article 1 of the EU's *Council Framework Decision* of June 2002 defines terrorism and terrorist offenses as:

...offences...which, given their nature or context, may seriously damage a country or an international organization where committed with the aim of:

—seriously intimidating a population, or

—unduly compelling a Government or international organization to perform or abstain from performing any act, or

—seriously destabilizing or destroying the fundamental political, constitutional, economic or social structures of a country or an international organization,

shall be deemed to be terrorist offences:

a   attacks upon a person's life which may cause death;

b   attacks upon the physical integrity of a person;

c   kidnapping or hostage taking;

d   causing extensive destruction to a Government or public facility, a transport system, an infrastructure facility, including an information system, a fixed platform located on the continental shelf, a public place or private property likely to endanger human life or result in major economic loss;

e   seizure of aircraft, ships or other means of public or goods transport;

f   manufacture, possession, acquisition, transport, supply or use of weapons, explosives or of nuclear, biological or chemical weapons, as well as research into, and development of, biological and chemical weapons;

g   release of dangerous substances, or causing fires, floods or explosions the effect of which is to endanger human life;

h   interfering with or disrupting the supply of water, power or any other fundamental natural resource the effect of which is to endanger human life;

i   threatening to commit any of the acts listed in (a) to (h). (European Council Framework Decision 2002/475/JHA, 2002) criminal activities but do not necessarily have a terrorism nexus (Zoller, 2004, p. 478).

France

Like the UK, France was spurred into developing counterterrorism legislation in the wake of a campaign of terrorist attacks, but in the French case, this was as a result of the activities of Middle Eastern groups in the early and mid-1980s. The most important piece of legislation, the Act of 9 September 1986, as well as a series of other laws passed in the wake of the 9/11 terrorist attacks in the United States, have acted to reinforce existing criminal law. Like the Australians and Canadians, the French prefer the approach of defining terrorism and using existing law, particularly the Code of Criminal Procedure, with additional penalties for terrorism. Article 421-1 of the Code of Criminal Procedure defines as criminal offenses attacks upon persons, damage to property, computer offences, offenses using firearms, money laundering, endangerment of public health, and a handful of other activities (France, Code of Criminal Procedure, Articles 421-1 and 421-2, 1997). Under Article 421-3, terrorist acts have the legal status of aggravated criminal acts carried out by individuals or groups when the goal is disturbing the public order through intimidation and terror (France, Code of Criminal Procedure, Article 421-3, 1997). The determination of when an illegal act constitutes a terrorist act is left to the judicial and administrative authorities.

The French treat terrorism essentially as a criminal activity although, as noted, their laws make allowance for the provision of more severe punishments for terrorism, particularly in the wake of the 2006 promulgation of Law 2006-64, which enhanced sentences for terrorism, detention periods, and police powers. In 1986, in the wake of Middle Eastern and Armenian terrorist attacks during the earlier part of the decade, the French parliament passed a statute that made it illegal to conspire to commit a terrorist act. Most people detained in France for terrorist offenses today are initially arrested under this act (Dutheillet de Lamothe, 2006, pp. 5–6). Although the French do not have special counterterrorism legislation, they did (as we note later in the chapter) establish special procedural rules for terrorism investigations that provide the authorities with greater latitude than in the case of ordinary criminal investigations. The Code of Criminal Procedure defines terrorist acts as those "intentionally committed by an individual entity or by a collective entity in order to seriously disturb law and order by intimidation or terror" and lists a range of activities as terrorist acts, including attempted murder, kidnapping, hijacking of all means of transport, production or handling of weapons of mass destruction (WMDs) or illegal explosive substances, money laundering, and even willfully harming the environment (Tiefenbrun, 2003, p. 377).

France thus has the most expansive definition of terrorist acts of any of the countries surveyed—something that may perhaps be more of a reflection of France's civic law model with its codification of law than any unique thoroughness.

COUNTERTERRORISM LAWS

All of the countries discussed in this chapter employ a range of counterterrorism laws. These include both standard legislation applicable at all times and emergency legislation and/or emergency executive orders available to be used in times of acute national crisis (although in some cases it has been argued that these have been used in the absence of a condition of appropriate acuteness). The laws employed by the countries in the survey differ based on legal tradition, perceived counterterrorism needs, and the degree to which they incorporate emergency legislation or executive orders.

Israel

In light of the ongoing terrorism threat faced by the country, Israel has made extensive use of both standard and emergency legislation and emergency executive orders and employs what is, in effect, a parallel military tribunal system alongside the traditional civilian court system in order to cope with terrorism threats. Israel has, in fact, been in a legal state of emergency since its inception in May 1948. Since the establishment of the State of Israel, there have been three methods for using emergency legal measures: (1) through executive orders inherited from the British, (2) through emergency legislation promulgated by the Provisional State Council of Israel upon establishment of the state in 1948, and (3) via emergency legislation passed by the Knesset.

The state of emergency was first declared by the British authorities in Mandatory Palestine in 1937 as a result of the Arab uprising and was maintained during World War II and continued almost until the termination of the Mandate. The Defense Regulations (state of emergency), a 1945 British executive order, allows for the imposition of martial law in Israel (and, after 1967, was extended to the West Bank and Gaza Strip as well as other territories conquered by Israel). Under the Defense Regulations, military commanders have the power to demolish houses, impose curfews, arrest, search and detain without judicial authorization, censor the press, prohibit meetings, open and confiscate mail, and so on. In theory, the local military commander can use these powers, but in practice, no significant measures have been taken (especially not within Israel proper) without the authorization of elected political authorities (i.e., the cabinet or the minister of defense).

Upon the establishment of the State of Israel, the Provisional State Council (which was an appointed legislative body, in other words a provisional parliament) proclaimed a state of emergency and the country has been under this state of emergency (which is renewed annually) ever since. Section 9 of the Law and Administration Ordinance of 1948, promulgated by the Provisional State Council, authorized the government or a minister to promulgate emergency regulations (executive legislation) that can revoke, suspend, or alter the enactment of any legislation during a state of emergency. These orders are limited in time but can be renewed on an indefinite basis. One of the principal pieces of emergency legislation promulgated by the Provisional State Council was the Prevention of Terrorism Ordinance, which also dates from 1948 and allows for the prosecution of anyone suspected of membership in a terrorist organization, publishing propaganda for the organization, managing the organization, fund-raising, or transferring funds for the organization. The ordinance can be in effect only as long as Israel remains within a legal state of emergency and it empowers military tribunals (composed of three military officers, with the president of the court being an officer who is a member of the Israel bar association and including at least one other member with extensive legal knowledge who has been certified by the attorney general) to hear cases and make rulings [Israel, Prevention of Terrorism Ordinance, Section 12(a–c), 1948]. The ordinance also gives considerable legal power to the minister of defense, who, among other things, is granted authority by the cabinet over the military. After a judgment is handed down by a military court, the minister of defense must confirm the judgment and also enjoys the authority to reduce the sentence, quash the judgment, and either order a retrial by the same, or a different, military court, or acquit the defendant. In any case, the minister of defense must obtain a statement of opinion from an officer who serves as a president of a military court but who did not preside over this particular case [Israel, Prevention of Terrorism Ordinance, Section 15(a–b), 1948]. Even with the proviso of obtaining a legal opinion, the minister of defense still enjoys considerable powers and as long as emergency laws remain in effect, essentially has the authority to authorize the arrest and detention of virtually anyone under Israeli jurisdiction. In practice, however, this executive power is infrequently invoked within Israel itself and against Israeli citizens but is commonly used in the West Bank and the Gaza Strip (in the context of military operations in the latter since Israel ceased its occupation of the Strip in the summer of 2005).

In addition to the British-era executive orders and the emergency legislation from the provisional legislature, the Knesset has the legal power to transform emergency executive orders and emergency legislation into regular permanent legislation. The Knesset has transformed a number of such

FIGURE 2.2   Shata prison, northern Israel. This Wikipedia and Wikimedia Commons image is from the user Ori and is freely available at http://commons.wikimedia.org/wiki/File:Xatta137.jpg and has been released into the public domain.

orders and laws into parliamentary emergency legislation, and the process is ongoing, among other reasons because it is recognized that legal challenges and changing legal norms will not make it possible to maintain indefinitely emergency executive orders and emergency laws promulgated back in 1948 by the provisional legislature. The government is, however, wary of replacing old emergency orders and laws with Knesset-authored emergency legislation because this would bring the powers afforded by emergency orders and provisional laws under greater scrutiny. Moreover, new laws would have to be passed that allow powers similar to those under emergency regulations and laws, including regulations regarding travel abroad, regulations allowing the army to commandeer private property in crisis situations, border and maritime security laws, and even laws regulating patents and bakeries (Yoaz, 2005). Israeli society and the Israeli legal system have evolved considerably since 1948 and it is certainly conceivable that today's Knesset might balk at legislating such powers to the executive and it is also conceivable that the Supreme Court, on the strength of more recent legislation enshrining civil rights, may contest such legislation, thus leading to a constitutional crisis. All in all, then, it has been easier for the Israeli cabinet to try and maintain its emergency powers on the grounds of their utility for counterterrorism efforts than to try to end the somewhat anomalous situation of a mature and stable state continuing to operate under emergency executive orders.

Israel also sometimes employs standard law in its counterterrorism efforts and, specifically, the Penal Law (updated in 1996), which allows for prosecution of terrorism suspects under ordinary criminal procedures. Some auxiliary pieces of nonemergency criminal legislation, such as the Prohibition of Money Laundering Law (from 2000), the Combating Criminal Organizations Law (from 2003), or the 2001 amendment to the Extradition Law, may also be useful for the conduct of standard criminal prosecutions against terrorism suspects, or for facilitating extradition of suspected terrorists to other countries. The Penal Law, in addition to allowing prosecution for ordinary crimes, allows for prosecution of suspects trying to subvert the political order in Israel through attempts to overthrow the lawful government by force or employment of sabotage against state property. The Penal Law remains in force regardless of whether or not Israel is in a state of emergency.

Despite the problematic nature of this significant empowerment of the executive, the emergency executive orders and emergency legislation have nevertheless been upheld by the Israeli Supreme Court. In the words of former president (chief justice) of the Court Aharon Barak:

> There is no avoiding—in a democracy aspiring to freedom and security—a balance between freedom and dignity on the one hand, and security on the other. Human rights must not become a tool for denying security to the public and the State. A balance is required—a sensitive and difficult balance—between the freedom and dignity of the individual, and national security and public security.
>
> —*Anonymous v. Minister of Defense* 54(1), 1997

Hence, rather than challenge the existence of emergency executive orders and emergency legislation, the Supreme Court has thus chosen for itself the role of ensuring that the basic human rights of detainees are respected (Henry, 2009). The Supreme Court also serves as the court of first instance for detainees or others who wish to challenge the legality of their detention or the application of emergency orders and laws in a broader sense. The court also recognizes its authority to receive petitions from individuals detained outside Israel or those who carried out attacks outside Israel (Israel Supreme Court, 2005, p. 15). Consequently, and despite the far-reaching powers under emergency orders and legislation, Israel's highest court has chosen not to challenge them in a fundamental way because of their efficacy as a counterterrorism tool, yet it reserves for itself the right to alter their outcome on a case-by-case basis.

Nevertheless, pressure to normalize counterterrorism legislation has recently led the Israeli government to introduce a new bill that will replace the 1948 Prevention of Terrorism Ordinance, the Law Prohibiting the Funding of Terrorism, and parts of the Defense Regulations (State of Emergency). This bill, once passed into law, will serve not just as a compendium of previous legislation but will include clearer definitions of illegal activity pertaining to terrorism, including leading terrorist organizations, membership in terrorist organizations, support for terrorism, incitement to terrorism, failure to prevent a terrorist act, threatening to perpetrate a terrorist act, and providing training for terrorism. The bill also proposes to strengthen punishments for these acts, making criminal acts defined as terrorism punishable by periods of incarceration that are twice as long as those for similar ordinary criminal acts, up to a ceiling of 30 years in prison. Moreover, a person convicted of a terrorist crime will be eligible to have his/her sentence commuted only after 15 years incarceration, whereas those serving life sentences for non-terrorism-related crimes can have their sentence commuted after seven years (Izbenberg, 2010).

United Kingdom

Like Israel, the United Kingdom traditionally employed a mix of regular legislation, emergency legislation, and military-based emergency executive orders (some of the latter indeed formed the basis of Israel's emergency executive orders, as noted above). The conflict in Ireland represented the only real terrorist threat to the UK up to the last two decades of the twentieth century; consequently, British legislation and executive policy were focused on that conflict. British counterterrorism legislation in the early and middle decades of the twentieth century took the form of emergency laws designed to be of a temporary nature. Thus, the Civil Authorities (Special Powers) Act (Northern Ireland) of 1922, which lead to the creation of an emergency regime in the province, was meant to last less than a year but was renewed annually until 1928, when it was extended for five additional years and subsequently made permanent. In the 1980s, these powers were extended to encompass broader threats from international terrorism (Walker, 2002, p. 1).

With the outbreak of significant rioting and violence in Northern Ireland in 1969 and the subsequent deployment of the British army, counterterrorism efforts were, not surprisingly, primarily military in nature and involved military internment without trial and various martial law provisions. After the British government dissolved the Northern Ireland legislature and executive and instituted direct rule from London in 1972, it was decided to implement civilian legal procedures in order to strengthen the perception of legitimacy and legality of British rule. These procedures had to be flexible and robust, however, and it was determined that they could not be based on existing criminal legislation but, rather, needed to be based on emergency laws and a special criminal justice process. This approach formed the basis of the Northern Ireland (Emergency Provisions) Act of 1973 (Walker, 2002, p. 3),

FIGURE 2.3 Northern Ireland mural. This Wikipedia and Wikimedia Commons image is from the user Jove and is freely available at http://commons.wikimedia.org/wiki/File:Mural_bogside1.jpg and has been released into the public domain.

which established a special court system (known as "Diplock courts") without juries to try terrorism cases, allowed for extensive pretrial detention (28 days), and provided broad powers of search and seizure, among other provisions. This emergency law was renewed several times but finally abandoned in August 2000, when it was allowed to expire. By this time the peace process in Northern Ireland had advanced (with the watershed accord commonly known as the Good Friday Agreement having been signed in 1998) and the passage, in 1998, of the Human Rights Act (which incorporated large sections of the European Convention on Human Rights and Fundamental Freedoms into UK law) made it appear that the continuance of emergency legislation would be untenable (Walker, 2002, p. 2).

The need to conduct legal proceedings without fear of intimidation of juries led, in 2007, to the reestablishment of nonjury trials, via the Justice and Security (Northern Ireland) Act. The act empowers the director of public prosecutions for Northern Ireland to certify that indicted offenses should be heard by a judge without a jury if one of the following four conditions exists: (1) the defendant is a member of or associated with a proscribed organization (i.e., terrorist group), (2) any offense being charged was carried out on behalf of a proscribed organization, (3) an attempt has been made to prejudice the investigation or prosecution on behalf or by a proscribed organization, and (4) the offense in question was committed in the context of religious or political hostility (United Nations Office on Drugs and Crime, 2010, p. 69).

With the passage of the Terrorism Act in 2000, the UK had a permanent piece of nonemergency legislation for dealing with terrorism. The act incorporated some elements from previous emergency legislation but also provided stronger guarantees for the rights of suspects and greater allowance for judicial scrutiny. Following the 9/11 attacks, Parliament passed the Anti-Terrorism, Crime and Security Act 2001. This law afforded the executive the power to detain, without charge, non-UK citizens suspected of terrorist activities but who could not be deported to their countries of origin for fear that they would be mistreated—although it did allow for detainees to appeal their status to a special immigration appeals commission, thus ensuring some degree of judicial review. Both of these pieces of legislation include clauses that expire after a set period, but the laws as a whole are not emergency laws of a temporary nature but permanent laws that enhance the normal criminal justice legislation (Walker, 2002, p. 11). Subsequent amendments to these pieces of legislation include the Prevention of Terrorism Act of 2005, which amends the Anti-Terrorism, Crime and Security Act, allowing for, among other things, the use of control orders to limit the freedom of movement of suspects (see the discussion later in the chapter) and the Terrorism Act of 2006, which, among other things, updates the 2000 act of the same name and increases police investigatory powers and creates new terrorism offenses. An additional law, the Counter-Terrorism Act of 2008, gives the police additional powers to investigate individuals, premises, and suspect financial transactions and provides for extended sentences for those convicted of activities with a "terrorist connection." The Terrorism Act of 2006 further strengthened powers by making it an offense to glorify terrorism and even to be in attendance at a place used for terrorist training (even if the person in question did not receive any training or if the training itself was not connected to any specific act of terrorism) (UK Terrorism Act 2006, Part 1, Section 8, n.d.).

This aforementioned package of legislation thus represents, in effect, a transformation in the British approach that began primarily with military executive action against terrorism, then migrated to civilian emergency legislation, and is now based on regular permanent legislation that is primarily special legislation designed to cope with terrorism threats rather than just ordinary criminal legislation (even though this, too, is employed against terrorism suspects when relevant). The UK has not, however, abrogated its right to promulgate emergency legislation. The power to promulgate emergency regulations is enshrined in the Civil Contingencies Act of 2004 and is invested in the Queen (on the advice of her government) or, in the absence of a royal decision due to time constraints, by "senior ministers of the Crown"—meaning the prime minister, the secretary of state (the Home Secretary), and the Chancellor of the Exchequer (the finance minister). These people are empowered to issue emergency regulations if an emergency is occurring that cannot be addressed adequately with existing legislation and there is not enough time to go through proper legislative channels. Such emergency regulations must be designed to safeguard life and health; protect or restore property; the supply of money and the financial system; food, water, energy, or fuel; and the protection of communication and/or transportation networks and provision of health services. Moreover, emergency regulations must be geographically specific, cannot amend basic guarantees of human rights, and must be limited in time (UK Civil Contingencies Act, Part 2, Sections 20–23, n.d.). Despite the existence of these powers, emergency laws are a last-resort option and the preference is to avoid their use. Consequently, central and local governmental authorities have been instructed not to base their planning and response arrangements on the presumption that emergency powers will become available (UK Government, 2005, p. 80). Overall, then, the UK approach has clearly evolved from the days when emergency powers were used as a matter of course to the present situation, in which they are largely held in abeyance.

Canada

Canada provides an example of a country with "dormant" emergency laws. In Canada, emergency legislation takes the form of the War Measures' Act of 1914, which was initially designed to prevent sabotage by enemy aliens during World War I and was also used to put Japanese-Canadians in internment camps during World War II. The law was last used in October 1970 to cope with terrorism on the part of separatists in Quebec. During this crisis, the Canadian army was deployed in Quebec for two months and also sent to guard Ottawa. The War Measures' Act allows the cabinet to issue a proclamation that a state of war, invasion, or insurrection exists and gives the cabinet the power to issue regulations that allow for censorship, preventive detention, control of private property, and control over all forms of transportation. The government has resisted invoking these powers since 1970, and while the Canadian Charter of Rights and Freedoms of 1982 (Canada's bill of rights) would seem to provide significant protections against the powers provided by the War Measures' Act, the charter does note that the freedoms guaranteed by it have limits as established by law. Arguably, then, the Canadian authorities still have recourse to employing emergency legislation (Belanger, 2004).

However, in dealing with the current manifestation of international terrorism, Canada has chosen instead, as noted earlier, to rely on regular criminal legislation and on amendments to existing criminal legislation (the latter because it was recognized that proactive steps were necessary to prevent terrorist acts and the focus could not be exclusively on prosecuting acts that already occurred). Canada's Criminal Code contains a chapter that deals with terrorism and specifically delineates a series of terrorist offenses. The Canadian Parliament passed the Anti-Terrorism Act in 2001, which acts to amend the Criminal Code and further delineate terrorism offenses as well as proactively granting increased powers to the authorities in investigating suspected terrorists, freezing assets and outlawing organizations (some of these provisions were included in clauses that expired in December 2006— more on this later in the chapter). Canadian legislators have, however, balked at allowing the sorts of things routinely practiced in Israel and, to a somewhat lesser extent, in the UK, with respect to following special legal procedures with terrorism suspects. In Canada, terrorism suspects enjoy the same procedural rights as ordinary criminal suspects, and while suspects in terrorism cases are required to answer questions at an investigative hearing even if the answers are self-incriminatory, the evidence derived from them cannot be used in court (UK Foreign Office, 2005, p. 7). Canada has chosen, then, to deal with the terrorism challenge without resorting to emergency legislation and through the use of both ordinary criminal legislation and some special legislative measures.

Australia

As in the case of Canada, and the UK more recently, Australia has chosen to deal with terrorism threats through a combination of ordinary criminal statutes and specialized counterterrorism legislation that act to amend the basic criminal statutes. The two primary pieces of ordinary criminal legislation are the Crimes Act of 1914 and the Australian Criminal Code Act of 1995. The latter was amended several times between 1998 and 2008 in order to incorporate counterterrorism-focused amendments, and these include amendments dealing with bombings and bomb-making (2002), the creation of new terrorism offenses (2002), and the Anti-Terrorism Act, which expands the Crimes Act and Criminal Code Act to strengthen investigative powers, set minimum sentencing standards, make it an offense to associate with terrorists, and so on. (Australian Government, n.d.). As noted above, the Australian authorities have chosen not to employ emergency legislation but, rather, to treat terrorism as a specific variant of criminal activity.

Germany

Given the use of emergency laws as well as extralegal measures under the Nazi regime, it is not surprising that Germany's postwar democratic leadership was not keen on the use of such laws. Nevertheless, the terrorist threat posed by domestic left-wing groups such as the Red Army Faction in the 1970s made it necessary to legislate an emergency law (Notstandsgesetzgebung) as well as three special antiterrorism (nonemergency) laws (Zoller, 2004, p. 472). The current approach to dealing with terrorism threats, however, has been to avoid the use of emergency legislation and to work solely within regular legislative boundaries. The German parliament has both passed amendments to existing law, such as the Strafgesetzbuch (Criminal Code), the Asylgesetz (Asylum Act), and the Ausländergesetz (Foreigners Act), and adopted new counterterrorism laws. As noted earlier, after 9/11 Germany adopted two counterterrorism legislative packages designed to strengthen investigation and surveillance powers, enhance prosecutor powers, limit access to German territory, and improve coordination between intelligence bodies (Congressional Research Service, 2004, p. 4). The Germans thus presently follow a dual approach of legislating changes to existing criminal law as well as special permanent and nonemergency counterterrorism legislation.

France

France, which like the UK was a colonial power, has similarly resorted in the past to military-style activities, emergency legislation, and regular legislation in order to counteract terrorist threats. During France's long and bitter war in Algeria (1954–1962) a full panoply of emergency legislation was employed, and practices such as administrative detention and even torture were widespread. In 1981, the Mitterand government abolished an emergency institution, the state security court (a special tribunal consisting of three civilian judges and two military officers that operated outside the normal judicial system and conducted proceedings in secret with no right of appeal), which was responsible for dealing with all national security cases. At present, France does not employ emergency legislation, although some have argued that its ordinary legislation is quite "emergency-like," in comparison to the laws in the United States (Garapon, 2005).

In the wake of a series of terrorist attacks on French soil from Middle Eastern groups, the French parliament passed a (nonemergency) counterterrorism law (Act of 9 September 1986) that provided for longer terms of imprisonment for those convicted of terrorism offenses, increased investigation powers, and periods of detention but worked within existing judicial procedures. The act stipulated that terrorism cases nationally would be dealt with by the Trial Court of Paris, with normal judges presiding over such cases (Shapiro and Suzan, 2003, p. 77). Since 9/11, additional legislation has reinforced existing criminal law and strengthened powers with respect to terrorism investigations. Criminal law therefore serves as the main legal weapon against terrorism. Terrorism and organized crime are dealt with by combining the charge for the actual crime with the charge for "an individual or collective act whose goal is to seriously disturb public order through intimidation or terror," thus making it possible to hand down more severe penalties (extended to 30 years for felonies and 20 years for misdemeanors) (France, Prime Minister's Office, 2007, p. 53). Additional powers are granted by the Law of 22 July 1996, which allows the intelligence organizations to provide information to examining magistrates that allows them to investigate and recommend prosecution of those providing logistical support for terrorist organizations and those assisting in planning before an actual attack occurs (France, Prime Minister's Office, 2007, p. 53).

French law, specifically the Law of 10 January 1936, also allows the disbanding of associations or groups that promote discrimination, hatred, or violence toward a person or a group of people based on their origin or membership in (or lack of membership in) a particular ethnic group, nation, race, or religion. However, when people undertake such acts not as part of a group, but individually, it is more difficult to prosecute them. The penalty for individual incitement to terrorism or violence is five years in prison and a €45,000 fine. However, the individual incitement offense is not covered in the French Penal Code but, rather, by an 1881 law dealing with freedom of the press, and this law does not allow seizures of evidence or preventive arrest and shortens the statute of limitations to three months. In addition, the law requires that acts of incitement be public and thus excludes incitement and proselytizing carried out on a person-to-person basis (France, Prime Minister's Office, 2007, p. 59). Nevertheless, aspects of civil and administrative law also play a role with respect to immigration issues, supervision of domestic groups, and other matters (Council of Europe, 2006, p. 1).

Despite the fact that terrorism is no longer prosecuted under emergency legislation, French law provides for three separate *crisis regimes* with different implications for state power and civil liberties. The Law of 3 April 1955 allows the declaration of the first type of crisis regime, a *state of emergency*, when there is a serious infringement of public order. Under a state of emergency (which can be declared for an initial period of up to 12 days and can be extended only through legislation), prefects have the power to ban the movement of persons and to implement security zones where individuals' activities can be regulated. The second type of regime can be established under the provisions of Article 36 of the French constitution, which allows for the declaration of a *state of siege*, in which maintenance of order is transferred to the military authorities, and this can be decreed in a situation of armed insurrection. Under a state of siege, military courts assume jurisdiction over any offenses deemed to jeopardize the safety of the Republic, threaten the constitution, or, have a significant impact on public order (Gross and Aolain, 2006, pp. 605–612). Finally, the third type of crisis regime can be established under Article 16 of the constitution, which is designed for extremely serious situations and allows the president of the republic complete power for a limited period (France, Prime Minister's Office, 2007, p. 83). In November 2005, severe rioting across France (with over 1400 vehicles torched in the wake of one night of rioting) led the cabinet to declare a state of emergency, empowering prefects to impose curfews and empowering the police to set up roadblocks, conduct searches of homes, ban public assembly, and put people under house arrest (Gross and Aolain, 2006, pp. 3197–3205).

FIGURE 2.4 November 2005 riots in France. This Wikipedia and Wikimedia Commons image is from the user Mikael Marguerie and is freely available at http://commons.wikimedia.org/wiki/File: Riots_Paris_2007.jpg under the Creative Commons Attribution 2.0 Generic license.

PRECHARGE DETENTION AND OTHER RESTRICTIONS OF FREEDOM OF MOVEMENT OF TERRORISM SUSPECTS

The success of counterterrorism efforts depends in large degree on reducing the number of terrorists that are operational at any given time and, in particular, reducing the numbers of *skilled* operational terrorists in circulation. This can involve a warfighting policy of killing terrorists (discussed later in the chapter), or it can involve a range of measures that restrict the freedom of movement of terrorism suspects. Restricting freedom of movement (a concept that is not restricted to physical mobility but includes the means to transmit orders, instructions, or ideas from one place to another via cellular phone or telephone or via the Internet) is a key counterterrorism tool. Actions that result in the reduction of mobility and accessibility to other members of a terrorist organization can significantly hamper terrorist planning, organization and capacity building, and the ensuing ability to carry out attacks. There are essentially three ways to restrict a person's physical and "electronic mobility": through physical detention at a detention facility, through physical detention at a person's home or other location, or to partially restrict mobility (as is done by some of the countries in the survey) through banning contact between a suspect and specific persons, or banning access to phones and/or the Internet. This form of detention, or limited mobility in the latter case, can either be preventive and designed to disrupt terrorist activities and thus prevent attacks, or it can be a measure of punishment. Accordingly, the focus of this section is on practices relating to arrest, detention, and restriction of mobility.

Israel

In Israel under the Law and Administration Ordinance (1948) and Emergency Authority Law (Arrests) (1979), the minister of defense has the authority to order the detention (known as *Administrative Detention*) of any person for no more than six months if the minister is convinced that the person poses a threat to national security or the security of the public. Moreover, the detention order can be renewed for periods of up to six months [Israel, Emergency Powers Law, Section 2(a–b), 1979]. In the same context, the 1979 law also allows the chief of staff of the Israel Defense Force (IDF) to order a person to be held for up to 48 hours without the possibility of renewal of the period of detention. Under this legislation, detainees must be brought, within 48 hours, before the president of a district court, with the judge having the power to release the detainee, approve the detention order, or alter the detention order. If the detainee is not brought before the court within this time period, he/she must be released [Israel, Emergency Powers Law, Section 4(a), 1979].

With the huge jump in terrorism during the Second Intifadah (uprising 2000–2004), the IDF initially issued an order in April 2002 authorizing its forces to detain suspects in the West Bank and Gaza Strip for 18 days prior to judicial review and access to legal representation, but public pressure and concern that the Supreme Court might intervene and force the military's hand led the IDF to modify its order to 12 days. Within Israel itself, suspects can be detained under emergency legislation for no longer than 48 hours before being brought before a magistrate. Furthermore, the president of the District Court in question is required to hold a follow-up hearing on the detention order within three months. As with trials dealing with terrorism cases, detention hearings (which are held *in camera*) may also be held without the presence of the detainee and his/her counsel if the president of the District Court determines that the information being presented would harm national security if it were to be made available to the detainee or his/her counsel. In all cases of detention hearings, the detainee has the right of appeal to the Supreme Court.

In the West Bank (which is governed under military law) the use of administrative detention is both widespread and easier to implement. According to IDF Order 1226 (1988), a military commander can, for reasons of "area security or the security of the public" issue a detention order for up to three months, which can be renewed consecutively for periods of up to six months. Judicial review of this process is exercised by the requirement that the detainee be brought before a military judge—who has the authority to approve, reject, or alter the terms of the detention—within 96 hours of his/her arrest. The decision of the military judge may be appealed to the military appeals court, with the latter having full authority to alter or cancel the detention order, but individuals detained under this order do not necessarily have the right to attend detention hearings or to receive classified information pertaining to their case. As this is an administrative rather than a judicial order, a senior military commander is authorized to cancel this detention order or to shorten the period of detention (Israel Defense Force, Order No. 1226, Section 9, 1988).

While the administrative detention powers, both within Israel and in the West Bank, are considerable, they are not, as a rule, used as a form of punishment (since punishment for crimes calls for a judicial process—whether via the civilian court system in Israel or the military court system in the West Bank), although human rights groups have argued that they are (B' Tselem, 1997, p. 12). As of April 2009, there were 487 Palestinian administrative detainees and no Israeli detainees. Over the years, the number of Palestinian detainees fluctuated between highs of 1794 and 1007 in 1989 and 2003, respectively, to lows of 18 and 12 in 1999 and 2000, respectively. In addition, since the early 1990s, nine Israeli citizens residing in settlements in the West Bank have been administratively detained for periods of up to six months (B' Tselem, 2009). From an Israeli counterterrorism perspective, administrative detention serves as a quick and efficient alternative to criminal proceedings because it does not require the investment in time necessary to gather sufficient evidence to charge a person and because it allows the authorities to protect intelligence sources and methods. In most cases, people held under administrative detention are eventually either released or tried (either by a civilian court in the case of Israeli citizens and some Palestinians, or by a military tribunal in the case of most Palestinians).

In cases of ordinary criminal investigation, a person can be detained for up to 96 hours before being brought before a magistrate. However, this is not a preventive step but only a precharge detention that provides the police with time to gather evidence prior to the arraignment. This law was amended temporarily in June 2006, pending further legislation. Prior to that, suspects could be held for only 48 hours, which often did not give the police enough time to gather evidence for the arraignment, the result being the frequent release of suspects who should have remained in custody.

Emergency legislation also allows for closures (ringing an area with checkpoints and limiting access to or from that location), curfews, confiscation of mail, closing of newspapers or publishing houses, the banning of public gatherings, and so on. They also allow the authorities to restrict a suspect's access to specific persons or to the Internet. As noted above, these

FIGURE 2.5   Israeli permanent checkpoint. This Wikipedia and Wikimedia Commons image is from the user PalFest and is freely available at http://commons.wikimedia.org/wiki/File:Qalandia-Checkpoint.jpg under the Creative Commons Attribution 2.0 Generic license.

powers are rarely used within Israel proper, but are much more commonplace in the West Bank (and in Gaza when it was under Israeli military rule between 1967 and 2005). Not surprisingly, then, Israel relies a great deal on emergency legislation with respect to the arrest and detention of terrorism suspects as well as other restrictions on them. This is reflective of the desire to protect sources and methods, as the Israelis sense that the country is at war with terrorism and thus cannot treat it as an ordinary criminal matter and require the flexibility afforded by emergency statutes. In all cases, however, the principle of judicial review remains intact and the courts have the final say.

United Kingdom

In the UK, although there was a long history (dating back at least to World War I) of the use of preventive detention within UK territory (with even freer use of executive detention powers in colonial territories), the practice generally ended with the close of World War II. Rather than employing an unlimited executive detention mechanism, the British authorities shifted to a judicial-based system of precharge detention with judicial review. As noted previously, growing violence in Northern Ireland in the early 1970s led to enactment of the Prevention of Terrorism (Temporary Provisions) Act (PTA) in 1974. The PTA allowed for a seven-day precharge detention, with access to an attorney granted to the detainee after 48 hours. This in contrast to the standard criminal procedure in the UK (then and now), whereby a suspect can be detained in a precharge capacity for up to 96 hours, with access granted to legal counsel within the first 36 hours (Blum, 2008, p. 135).

In 2000 Parliament passed the Terrorism Act, which granted permanency to the PTA's seven-day precharge detention provision. In 2003, in response to the rise of the international terrorism threat in the UK and the events of 9/11, the precharge detention period was doubled to 14 days (House of Commons, Home Affairs Committee, 2006, p. 8). After the attacks against the London Underground and a bus on July 7, 2005 in which 52 persons were killed, the government attempted to increase the precharge detention period for terrorist suspects to 90 days, but this was rejected by Parliament. An updated Terrorism Act passed in 2006 did, however, extend the precharge detention period to up to 28 days, but detention can be authorized by a magistrate only to preserve evidence, question the suspect, or decide whether to charge or deport the person (in the case of noncitizens, discussed in a later chapter) and, after 14 days, continued detention requires the approval of a high court judge. This stood in stark contrast to detention under emergency laws in the Israeli case, where the detention is an administrative/executive rather than a judicial act. Moreover, the extended detention period had to be renewed on an annual basis by the Home Secretary. More recently, the government attempted to amend the law so that the period of precharge detention could be raised to 42 days, but while this change narrowly passed the House of Commons, it was rejected outright by the House of Lords (Prince, 2008). In January 2011, the British Government cut the pre-charge detention period back to 14 days.

Under general criminal law, specifically the Police and Criminal Evidence Act of 1984, the police have the authority to detain suspects for questioning for up to 36 hours. At the end of this period the suspect must be charged or brought before a magistrate, who may authorize an additional period of detention with the proviso that the suspect cannot be detained for longer than 96 hours without being arraigned (House of Commons, Home Affairs Committee, 2006, p. 6).

Although the British authorities can no longer detain people under administrative orders, when there is insufficient evidence to arrest a suspect and prosecute him/her in the criminal justice system or that evidence cannot be brought before court for fear of compromising intelligence sources, the authorities have the power to attempt to disrupt a person's actions through the use of "control orders." Under the 2005 update to the Prevention of Terrorism Act there are two types of control orders, derogating and non-derogating. The reason for this distinction has to do with the fact that the UK is a signatory to the European Convention on Human Rights (ECHR; see Sidebar 2-2) and has incorporated its principles into UK legislation in the framework of the Human Rights Act 1998. This requires that the UK follow certain practices stipulated in the ECHR, although that convention does allow for derogation in cases of significant security threats—hence the distinction between derogating and nonderogating control orders.

A derogating control order (which, as its name suggests, derogates from the ECHR) allows for the infringement of the liberty of an individual (such as via house arrest; see Sidebar 2-3), whereas a non-derogating control order allows for the imposition of a set of restrictions on the person's freedom of movement and action, such as prohibiting a person from meeting certain other persons, going to certain locations, and/or accessing the Internet (Police National Legal Database and Andrew Stainforth, 2009, pp. 226–227). Derogating control orders must be confirmed by a vote in each house of Parliament within 40 days in order to continue to be enforced, but thus far, no derogating control order has been sought by the government. Control orders must be authorized by the Home Secretary and approved by a court, and at present only 8 people are subject to control orders. The power to issue control orders under the Prevention of Terrorism Act must be renewed by both houses of Parliament on an annual basis. The British model thus relies largely on criminal procedures for arrest and incarceration but on executive measures with respect to restricting freedom of movement and action via control orders.

Sidebar 2-2: The European Convention on Human Rights

Article 1—Obligation to respect human rights

The High Contracting Parties shall secure to everyone within their jurisdiction the rights and freedoms defined in Section I of this Convention.

Section I—Rights and freedoms

Article 2—Right to life

1   Everyone's right to life shall be protected by law. No one shall be deprived of his life intentionally save in the execution of a sentence of a court following his conviction of a crime for which this penalty is provided by law.

2   Deprivation of life shall not be regarded as inflicted in contravention of this article when it results from the use of force which is no more than absolutely necessary:

   a   in defense of any person from unlawful violence;

   b   in order to effect a lawful arrest or to prevent the escape of a person lawfully detained;

   c   in action lawfully taken for the purpose of quelling a riot or insurrection.

Article 3—Prohibition of torture

No one shall be subjected to torture or to inhuman or degrading treatment or punishment.

Article 4—Prohibition of slavery and forced labor

1   No one shall be held in slavery or servitude.

2   No one shall be required to perform forced or compulsory labor.

3   For the purpose of this article the term "forced or compulsory labor" shall not include:

   a   any work required to be done in the ordinary course of detention imposed according to the provisions of Article 5 of this Convention or during conditional release from such detention;

   b   any service of a military character or, in case of conscientious objectors in countries where they are recognized, service exacted instead of compulsory military service;

   c   any service exacted in case of an emergency or calamity threatening the life or well-being of the community;

   d   any work or service which forms part of normal civic obligations.

Article 5—Right to liberty and security

1 Everyone has the right to liberty and security of person. No one shall be deprived of his liberty save in the following cases and in accordance with a procedure prescribed by law:

   a the lawful detention of a person after conviction by a competent court;

   b the lawful arrest or detention of a person for non-compliance with the lawful order of a court or in order to secure the fulfillment of any obligation prescribed by law;

   c the lawful arrest or detention of a person effected for the purpose of bringing him before the competent legal authority on reasonable suspicion of having committed an offence or when it is reasonably considered necessary to prevent his committing an offence or fleeing after having done so;

   d the detention of a minor by lawful order for the purpose of educational supervision or his lawful detention for the purpose of bringing him before the competent legal authority;

   e the lawful detention of persons for the prevention of the spreading of infectious diseases, of persons of unsound mind, alcoholics or drug addicts or vagrants;

   f the lawful arrest or detention of a person to prevent his effecting an unauthorised entry into the country or of a person against whom action is being taken with a view to deportation or extradition.

2 Everyone who is arrested shall be informed promptly, in a language which he understands, of the reasons for his arrest and of any charge against him.

3 Everyone arrested or detained in accordance with the provisions of paragraph 1.c of this article shall be brought promptly before a judge or other officer authorised by law to exercise judicial power and shall be entitled to trial within a reasonable time or to release pending trial. Release may be conditioned by guarantees to appear for trial.

4 Everyone who is deprived of his liberty by arrest or detention shall be entitled to take proceedings by which the lawfulness of his detention shall be decided speedily by a court and his release ordered if the detention is not lawful.

5 Everyone who has been the victim of arrest or detention in contravention of the provisions of this article shall have an enforceable right to compensation.

Article 6—Right to a fair trial

1 In the determination of his civil rights and obligations or of any criminal charge against him, everyone is entitled to a fair and public hearing within a reasonable time by an independent and impartial tribunal established by law. Judgment shall be pronounced publicly but the press and public may be excluded from all or part of the trial in the interests of morals, public order or national security in a democratic society, where the interests of juveniles or the protection of the private life of the parties so require, or to the extent strictly necessary in the opinion of the court in special circumstances where publicity would prejudice the interests of justice.

2 Everyone charged with a criminal offence shall be presumed innocent until proved guilty according to law.

3 Everyone charged with a criminal offence has the following minimum rights:

   a to be informed promptly, in a language which he understands and in detail, of the nature and cause of the accusation against him;

   b to have adequate time and facilities for the preparation of his defense;

   c to defend himself in person or through legal assistance of his own choosing or, if he has not sufficient means to pay for legal assistance, to be given it free when the interests of justice so require;

   d to examine or have examined witnesses against him and to obtain the attendance and examination of witnesses on his behalf under the same conditions as witnesses against him;

   e to have the free assistance of an interpreter if he cannot understand or speak the language used in court.

Article 7—No punishment without law

1 No one shall be held guilty of any criminal offence on account of any act or omission which did not constitute a criminal offence under national or international law at the time when it was committed. Nor shall a heavier penalty be imposed than the one that was applicable at the time the criminal offence was committed.

2 This article shall not prejudice the trial and punishment of any person for any act or omission which, at the time when it was committed, was criminal according to the general principles of law recognised by civilised nations.

Article 8—Right to respect for private and family life

1 Everyone has the right to respect for his private and family life, his home and his correspondence.

2 There shall be no interference by a public authority with the exercise of this right except such as is in accordance with the law and is necessary in a democratic society in the interests of national security, public safety or the economic well-being of the country, for the prevention of disorder or crime, for the protection of health or morals, or for the protection of the rights and freedoms of others.

Article 9—Freedom of thought, conscience and religion

1 Everyone has the right to freedom of thought, conscience and religion; this right includes freedom to change his religion or belief and freedom, either alone or in community with others and in public or private, to manifest his religion or belief, in worship, teaching, practice and observance.

2 Freedom to manifest one's religion or beliefs shall be subject only to such limitations as are prescribed by law and are necessary in a democratic society in the interests of public safety, for the protection of public order, health or morals, or for the protection of the rights and freedoms of others.

Article 10—Freedom of expression

1 Everyone has the right to freedom of expression. This right shall include freedom to hold opinions and to receive and impart information and ideas without interference by public authority and regardless of frontiers. This article shall not prevent States from requiring the licensing of broadcasting, television or cinema enterprises.

2 The exercise of these freedoms, since it carries with it duties and responsibilities, may be subject to such formalities, conditions, restrictions or penalties as are prescribed by law and are necessary in a democratic society, in the interests of national security, territorial integrity or public safety, for the prevention of disorder or crime, for the protection of health or morals, for the protection of the reputation or rights of others, for preventing the disclosure of information received in confidence, or for maintaining the authority and impartiality of the judiciary.

Article 11—Freedom of assembly and association

1 Everyone has the right to freedom of peaceful assembly and to freedom of association with others, including the right to form and to join trade unions for the protection of his interests.

2   No restrictions shall be placed on the exercise of these rights other than
such as are prescribed by law and are necessary in a democratic society
in the interests of national security or public safety, for the prevention
of disorder or crime, for the protection of health or morals or for the
protection of the rights and freedoms of others. This article shall not
prevent the imposition of lawful restrictions on the exercise of these
rights by members of the armed forces, of the police or of the administration of the State.

Article 12—Right to marry

Men and women of marriageable age have the right to marry and to found
a family, according to the national laws governing the exercise of this right.

Article 13—Right to an effective remedy

Everyone whose rights and freedoms as set forth in this Convention
are violated shall have an effective remedy before a national authority
notwithstanding that the violation has been committed by persons acting
in an official capacity.

Article 14—Prohibition of discrimination

The enjoyment of the rights and freedoms set forth in this Convention shall
be secured without discrimination on any ground such as sex, race, colour,
language, religion, political or other opinion, national or social origin, association with a national minority, property, birth or other status.

Article 15—Derogation in time of emergency

1   In time of war or other public emergency threatening the life of the
nation any High Contracting Party may take measures derogating
from its obligations under this Convention to the extent strictly
required by the exigencies of the situation, provided that such measures are not inconsistent with its other obligations under international law.

2   No derogation from Article 2, except in respect of deaths resulting from
lawful acts of war, or from Articles 3, 4 (paragraph 1) and 7 shall be made
under this provision.

3   Any High Contracting Party availing itself of this right of derogation
shall keep the Secretary General of the Council of Europe fully informed
of the measures which it has taken and the reasons therefor. It shall also
inform the Secretary General of the Council of Europe when such
measures have ceased to operate and the provisions of the Convention
are again being fully executed.

Article 16—Restrictions on political activity of aliens

Nothing in Articles 10, 11 and 14 shall be regarded as preventing the High Contracting Parties from imposing restrictions on the political activity of aliens.

Article 17—Prohibition of abuse of rights

Nothing in this Convention may be interpreted as implying for any State, group or person any right to engage in any activity or perform any act aimed at the destruction of any of the rights and freedoms set forth herein or at their limitation to a greater extent than is provided for in the Convention.

Article 18—Limitation on use of restrictions on rights

The restrictions permitted under this Convention to the said rights and freedoms shall not be applied for any purpose other than those for which they have been prescribed. (Council of Europe, 1953)

Sidebar 2-3: Theoretical UK Case Study to Illustrate the Need for Longer Precharge Detention Periods

This case study has been constructed with the assistance of the Crown Prosecution Service and draws upon issues that have arisen in many real cases. The statistics used are entirely typical of the scale of events that have been seen in terrorist investigations in recent years. The Security Service are told by an agent that a group of men in various parts of the country are planning terrorist attacks on the Houses of Parliament and the British Embassies in Pakistan, Istanbul and Morocco. They have been exploring conventional and homemade explosives as well as CBRN possibilities. It is believed that this will be carried out in 3 months time. The agent is reliable and his information must be acted on for public safety reasons. Surveillance is started on 2 of the men identified and over a period of 2 months they are seen with numerous other people. All of the people seen are unknown to intelligence services and cannot be identified. 5 key addresses were identified and probes put into each over the period. The agent does not know where the dangerous materials are being stored or where they have been obtained from although he believes that some might have been brought in from abroad. The men are believed to be all illegal entrants to this country and are each living on at least 2 false identities. Police arrest 15 people following the execution of Terrorism Act warrants in 4 different areas of the country on day 1. Each arrest requires time-consuming custody procedures; sterile arrest, transportation to the secure suite at Paddington Green, the forensic examination of prisoners and taking of evidential samples. The samples are particularly important as it is thought that the men are not who they purport to be and/or not from the countries they claim to come from. Each has at least one false passport. These procedures have to be completed before any detained person can consult with their legal representatives. On this occasion they took about 8 hours for each person. Some could be conducted simultaneously, but some (like booking in with the Custody Sergeant) had to be done individually. The fingerprints are sent to 5 different countries to see if the men can be properly identified. With 15 people under arrest, a disclosure strategy was required so as to achieve the best evidence from the interviews and test the accounts given. This was done whilst the defendants were being examined and other procedures carried out, and whilst the police were waiting for the solicitors to arrive. Each disclosure package given to the respective legal representative required lengthy consultations with the detainees. 2 firms of solicitors represented all the detained men. Their representatives were not available immediately; the police had to wait 4 hours for one and 5 for another. Each firm only provided 2 representatives. The initial consultations with each client lasted on average 4-5 hours. This time took up some of the time available to the officers to conduct their detailed interviews and enquiries, the clock did not stop running whilst the detainees were taking legal advice.

In addition all 15 men need to be allowed to observe prayer 5 times a day and all say that they need an interpreter. In the first 14 days a total of 165 interviews were conducted. Most of the suspects are saying nothing, but as more evidence is put to them by the 14th day, 2 appear to be getting concerned and might talk. Within the first 4 days of detention, 55 forensic searches were conducted around the country involving residential and non-residential properties and vehicles, again involving an enormous amount of work by officers to speedily assess the relevance of exhibits within the time limits imposed. Each of these required a separate warrant and information received led the police to believe that there could be CBRN material on the premises as well as possibly conventional and homemade explosives. This meant that specialist teams had to be deployed and some of the premises were unsafe to enter until various forms of risk assessments had been done and procedures carried out. There are only a limited number of specialists available to do this work and it was only possible to do one premise at a time. 10 of the premises require this procedure and were in three of the different parts of the country, some about 5 hours drive away from the other. During this period of time a vast amount of exhibits were seized during the searches. This had to be examined, prioritised, sifted for relevance, an assessment made of which individual should be questioned about which exhibit and a decision made on which should be sent to experts in chemical weapons, which on biological, which to FEL and which to the AWE. There were about 4,000 exhibits labelled in the first week with many more outstanding for examination. At least half of the documentary exhibits (about 600) are in Arabic. Most of the available interpreters are being used for the interviews and after trawling the country police manage to locate another 3 who can begin on the documents. There are also several boxes of videos tapes the contents of which the police do not know until they have been viewed. There are no labels on them. A cursory viewing of a handful shows that they are extremist in nature and mostly with Arabic voiceovers or individuals speaking in Arabic on. There is little point in the officers viewing these further as they cannot understand them. A decision had to be taken about where each of the exhibits should go first. It is decided to fingerprint 300 documents first. Half of these are handwritten and will also need to be examined for handwriting analysis. All the identification documents found (at least 100) need to go for expert analysis to see if they are false. 15 of these are French, 10 each are Spanish, Italian and Turkish, but the majority appear to be of Eastern European extraction, maybe Bosnian, and all have to be submitted to their country of origin to check whether they are genuine. There have been over 268 computers seized together with 274 hard drives, 591 floppy discs, 920 CDDVDS and 47 zip discs. The High Tech Crime Unit say that every computer hard drive seized during that period of time takes a minimum of 12 hours to image for the assessment teams at Paddington to then provide to the interviewing officers. The preliminary assessments carried out, due to the time constraints imposed, cannot be considered as thorough and have to be revisited as other factors emerge and different matters become relevant. About a quarter of the computers and hard drives have encrypted material on them and the suspects are refusing to give the keys saying that the computers, even those found in identifiable homes, are nothing to do with them. Assistance is required from a number of agencies here and abroad with regard to this and an assessment has to be made about which computers to prioritise.

It is not clear which of these computers was used the most as the man believed to be the leader and 2 others have been itinerant, using at least 20 of the known addresses over the last 6 month period. The main suspect was of no fixed abode. He had items of personal property at a number of addresses. Some is false; fingerprint and DNA work done in the first 4 weeks enabled police to establish this. During the first two weeks 60 seized mobile telephones, mostly pay as you go, were forensically examined. The sheer volume of material to be gathered from these examinations meant that much of it was not available until the 6th week of investigation. This evidence is crucial as it is needed to corroborate associations and prove movements. DNA analysis is required to discover which telephones have been used by which suspects, again because they have used or visited many addresses. Some 25,000 man hours were spent examining CCTV footage. Some 3674 man hours are used to assess the eaves-dropping material gathered by probes operating 24 hours each day over an 8 week period. There are 850 surveillance and observation point logs that must be assessed for their evidential value. This evidence will be crucial to establish who was present at which meetings and what was said. In the first 4 weeks the police identified 6000 actions in the investigation. 10,000 documents, 2300 statements and 7000 exhibits have been seized or created by week 8 of the investigation. Crucial evidence is still awaited from DNA, other scientific work and from various foreign enquiries coming in gradually over the period of detention. Letters of request for legal assistance in gathering evidence abroad have been written by prosecutors and sent through emergency channels to 17 countries. As the enquiries progress more addresses are being identified, more searches done and more exhibits, computers and false documentation with photographs of the suspects and others are being discovered. In amongst the documents are some bearing the picture of a well known international terrorist being held in custody in another country where it is not easy for the police to obtain access or information. This might be a crucial link with some of the suspects being held and an approach needs to be made through diplomatic channels. Throughout the detention period it is becoming abundantly clear that there were plans to use a dirty bomb in the Houses of Parliament, conventional explosives for an attack on 2 of the Embassies and a possible chemical attack on the third. Each suspect has several identities. We are waiting to hear if the requested countries can establish the true identity of the men. Fingerprints of each man are being found on some documents of a suspicious nature. It is unclear however which role each man took and whether they can be linked to any or all of the planned attacks. The case is largely circumstantial as no chemicals or explosives or anything else of that nature has been found despite the fact that the targeting document (found on the 50th computer to be examined in the 7th week) shows that the attack on Parliament was due to take place 2 days after the arrests. 2 prosecutors are working full time with the Anti Terrorist Branch making applications to extend pre-charge detention, drafting initial and supplementary letters of request and reviewing the evidence as the investigation progresses. Experts from 10 different disciplines are working on exhibits and documents seized as well as scouring addresses and cars for explosive and other traces and ¾ of the police capacity has been involved in various actions including examination of exhibits, computers, interviewing, etc. (House of Commons, Home Affairs Committee, 2006, pp. 57–60)

Canada

As noted previously, suspects accused of terrorism-related crimes in Canada have essentially the same substantive and procedural rights as other criminal suspects. However, in the wake of the terrorist attacks of 9/11, Anti-Terrorism Act (ATA) amendments to the Criminal Code, Sections 83.28 and 83.3, did allow for "investigatory hearings" and pretrial arrest. Investigatory hearings represented a significant intrusion into privacy rights and protections from self-incrimination and involved the summoning of individuals to hearings at which they were required to appear, answer questions, and provide documentation. Pretrial arrest powers allowed law enforcement officials to arrest a person suspected of planning to carry out a terrorist attack without a warrant and without charging them. The person had to be brought before a provincial court judge within 24 hours, if reasonably possible, and the judge had the power to authorize an additional period of detention until charges could be brought against the person or he/she could release the suspect but subject them to travel and other restrictions with a provision for detention of up to one year if the person did not accept these court-ordered restrictions (Canada, Criminal Code, Part II, Section 83.3, 1985). These provisions lapsed when the ATA sunsetted in 2007, and Canada has essentially returned to a model in which terrorism is treated in a manner similar to ordinary criminality.

Australia

Australia's approach is more in keeping with the British model but also includes prominent elements of preventive detention. It appears to be a surprisingly aggressive model given that Australia has, thus far, experienced almost no terrorism on its soil. In Australia the maximum period for pre-charge detention is 24 hours, which can be extended another 24 hours, after which a suspect must be brought before a magistrate. Senior Australian Federal Police (AFP) officials have the power to issue an initial preventive detention order (Australia, Criminal Code, Section 105.8, 1995). This time period can be extended under both commonwealth (federal) and state law. In New South Wales, for example, the Terrorism (Police Powers) Act of 2002 authorizes preventive detention for up to 14 days. During the period of preventive detention, however, the suspect cannot be questioned; consequently, these powers differ from precharge detention, where questioning is allowed [Australia, Anti-Terrorism Act (No. 2), Section 105.42, 2005]. This also makes the Australian approach significantly different from the Israeli approach, where preventive detention detainees can and are interrogated. In addition, the Australian Secret Intelligence Organization (ASIO) is authorized, upon receipt of a warrant from a judge or federal magistrate, to detain people for up to 14 days without charge for the purpose of questioning. This is in contrast to the pre-9/11 period, in which ASIO had no powers to detain (Baldino, 2007). Detainees face a five-year jail term if they refuse to answer questions, including those that might be self-incriminating, and the burden of proof is on the detainee to prove his/her innocence (Australian Secret Intelligence Organization Act, Part III, Division 3, Section 34L, 1979). Although subject to sunset provisions, these measures have been authorized until at least 2016. In general, criminal investigations and prosecution fall under the purview of the states in Australia, but many of the powerful commonwealth provisions noted above are enforced by federal agencies such as the AFP and ASIO.

With respect to the issue of limiting freedom of movement or contact with others, the Criminal Code empowers the attorney-general, upon receipt of a request from a senior member of the AFP, to issue an *interim control order*, and courts also have the power to issue interim control orders. Control orders can be issued only if it is believed that the order will prevent a terrorist act and that the prohibitions and restrictions imposed on a person are reasonable and necessary and can be used to limit the person's access to certain places, require that a person remain at certain premises at certain times, restrict access to the Internet, telecommunications, specific prohibited items, and so on. [Australia, Criminal Code, Section 104.5(3), 1995]. Interim control orders must be reviewed by a court and can be challenged by the detainee subject to the order. Once the interim control order is approved by a court it becomes a *confirmed control order*, which can be extended as long as 12 months. The Australian Anti-Terrorism Act also allows the attorney-general to declare a commonwealth location a *prescribed security zone* for up to 28 days. This power was granted with a sunset provision of 10 years, meaning that it will expire, if the parliament does not make legislative changes, by 2015. All in all, Australia possesses a surprisingly powerful (perhaps even somewhat draconian, given the comparative absence of terrorism) detention regime.

Germany

As with Canada, Germany treats terrorism cases via ordinary criminal procedural rules. In Germany, the police can remand a person into provisional custody but must release him/her from provisional custody at the end of the day following the day of arrest, and consequently, a person could be detained up to 48 hours in the absence of an arrest warrant issued by a judge. Under German law, there is no precise concept of a criminal charge. Investigations are conducted by the state prosecutor, with police assistance, and this results either in an indictment in court or a cessation of the investigation (Russell, 2007, p. 40). In broad terms, the German authorities have the power to detain suspects in one of three situations: under provisional police custody pending a judicial hearing, in the event that a judge has issued an arrest warrant and the suspect is remanded to detention pending a trail, and in the case of detention for public safety reasons (Russell, 2007, pp. 40–41). If a judge issues an arrest warrant (whether for a suspect already in provisional custody or otherwise), the person in question can be detained for a longer period (usually up to six months, although in the case of significant offenses that pose a public danger, detention can be extended to up to one year), but this must be based on the threat that the person poses if released and the strength of the initial evidence on which the investigation is based: a practice known as *remand detention*. German judges usually have very little tolerance for prosecution requests to extend a detention period if they feel that the prosecution has made errors or otherwise failed to build their case in an expedient fashion

(Russell, 2007, p. 43). Preventive detention of terrorism suspects in the German case is thus strictly a legal procedure and does not differ substantively from the process of detaining other types of suspected criminals.

France

As noted above, in France terrorist activities are treated as an aggravated version of traditional offenses. When a suspect is arrested in France, he/she can be detained by the police for an initial period of up to four days without being charged, and in some cases up to six days if there is a serious risk of an imminent terrorist attack or if the authorities are waiting to receive intelligence from overseas—in comparison to the 48-hour maximum for traditional crimes. The placing of a suspect in detention for purposes of questioning is known as garde à vue (GAV—police custody). After this period of detention, suspects must be brought before a liberty and custody judge (juge des libertes et de la detention), who has the authority to extend the detention period for a week. Afterward, the file is passed to an examining magistrate (juge d'instruction) or an examining chamber (chambre d'accusation), which puts the case together. If the magistrate or chamber believe that there are grounds for continued investigation of the suspect, a request will be made by

FIGURE 2.6 French prison block. This Wikipedia and Wikimedia Commons image is from the user Christophe Finot and is freely available at http://commons.wikimedia.org/wiki/File:Autun_prison_cellulaire.jpg under the Creative Commons Attribution-Share Alike 1.0 Generic license.

the examining magistrate or chamber to the liberty and custody judge to extend the period of detention. Once a detainee's case is in the hands of an examining magistrate, the detainee can remain locked up indefinitely as long as the examining magistrate has the approval of the liberty and custody judge (UK Home Secretary, 2004, p. 13). Arrest warrants can be issued by an investigating magistrate or other types of judges (such as the president of the trial court). Once charged, however, suspects in terrorism cases may be held under pretrial detention for up to four years (France, Code of Criminal Procedure, Article 145-2, 2004). The French state thus has significant powers to hold people in custody for a significant time period, thus virtually guaranteeing that the investigation and trial can occur while a person is behind bars and unable either to interfere in the gathering of evidence or to continue to be physically active in planning and executing terrorist acts.

CONDUCT OF INVESTIGATIONS AND JUDICIAL PROCEEDINGS

Democratic governments face a difficult choice in trying to balance civil liberties and privacy rights on the one hand with effective intelligence and law enforcement activity that is able to disrupt the planning and execution of terrorist attacks and/or aid in the prosecution of terrorists on the other. Consequently, the powers of investigation, the admissibility of evidence gathered through investigations, and the judicial proceedings of terrorism cases must represent a balance between respect for privacy and the rights of the accused and the need to gather information in order to prevent attacks. The countries in the survey each make an effort to find this balance but, not surprisingly, approach the issue from different legal and policy vantage points, with some favoring security over privacy and judicial rights, whereas others weight more heavily toward safeguarding those rights, even if this comes at the potential expense of security.

Israel

In Israel, the police have statutory authority to initiate and oversee terrorism investigations. The domestic security service, the Israel Security Agency (ISA, also known by its Hebrew acronym, Shabak, or the first two Hebrew letters of that acronym, Shin Bet) gathers intelligence designed to uncover terrorism networks, prevent attacks, and support terrorism-related investigations. In practice, the ISA conducts most terrorism investigations at the initial stages and provides material to the police when there is a need to conduct evidentiary-based investigation. In the West Bank, the ISA operates in conjunction with the military (which has statutory authority since the West Bank is governed under military law). In Israel proper, the police are responsible for running investigations and gathering evidence for an indictment, but cases are prosecuted by the district attorney's office. The district attorney's office will work with the police investigators and attorneys to determine when enough evidence has been amassed for an indictment and will then take over the case. With respect to non-Israeli citizens in the West Bank, the IDF advocate-general's office oversees the gathering of evidence and building of cases against terrorism suspects. Both the police and the military will use the services of the ISA, and generally speaking, evidence obtained through intelligence measures is admissible in court and can be kept from the accused and his/her attorney.

The military court system, which has jurisdiction in the West Bank and with respect to persons captured in military operations in the Gaza Strip or other areas outside Israel or the West Bank, operates under the auspices of the IDF regional command that has authority over the given area. The West Bank falls under the IDF central command and consequently the major-general who heads that command has legislative authority to issue, amend, and repeal military orders (MOs), and military legislation draws its authority from MOs (Hajjar, 2005, p. 253). The military court system is overseen by the military advocate general (MAG), who assigns judges to specific courts (subject to the agreement of the head of the central command). The MAG also oversees the prosecution, but the prosecutors are administratively distinct from the judges (Hajjar, 2005, p. 254). Military courts are overseen by one judge for comparatively simple charges and a panel of three judges for more complex cases. Three-judge courts have the authority to hand down a sentence of life in prison and even capital punishment (although subsequently, the rare judgment in favor of capital punishment has always been commuted by these courts to life in prison). Sentences can be appealed to a three-judge military court of appeals (Hajjar, 2005, p. 255). The head of the relevant territorial command (as noted above, in the West Bank, this is the central command) has the authority to reduce or commute a sentence handed down by the military court.

In Israel proper, searches of persons and premises must be conducted with a court order from a magistrate court. The judge has the authority to issue a search order if there is reasonable suspicion that the premises are being used, or will be used, for illegal activity, contain illegal items, or a person at those premises is likely to be harmed. Additionally, a police officer has the authority to access and search premises and persons without a court order if he/she believes that a crime is being committed or was recently committed [Israel, Criminal Procedure Law, Section 25(1), 1996]. The police also have the blanket right to stop anyone, request identification, and search his/her person. Under the Prevention of Terrorism Ordinance, arrest and search powers are also vested in military policemen. However, military policemen can conduct searches in civilian zones only if authorized to do so by a magistrate court (which, as noted previously, is a civilian court) or, in certain cases, if authorized by military authorities—although the latter requires that they be accompanied by a civilian police officer [Israel Police, Police Order 14.01.08, Section 4(c), 1995].

After a suspect is arrested, he/she must be granted access to an attorney as soon as is physically possible. The police do have the authority to delay this meeting up to 48 hours if the senior officer in charge of the investigation believes that this meeting will lead to the destruction of evidence, will tip off other suspects who are still at large, or will otherwise lead to the obstruction of justice. However, in terrorism investigations or others that fall within the rubric of what are referred to in Israel as *security offenses*, the senior police officer in charge of the investigation is authorized to delay the suspect's meeting with his/her attorney for up to six days [Israel Police, Police Order 14.01.34, Section 6(d) (1–2), 2008]. When suspects are arrested they must be brought before a magistrate within 24 hours unless there are extenuating circumstances having to do with the nature of the interrogation of the suspect or other facts, in which case suspects must be brought before a judge within 48 hours. However, in terrorism and other security crimes cases, particularly with respect to arrests made in the West Bank that are handled by the military court system, suspects can be held for up to 18 days before being allowed to meet with an attorney (and are usually held *in communicado* during this period of time). In terms of trial proceedings, in order to protect sources and methods, in situations "when national security requires maintenance of secrecy" that cannot be accomplished through other legal means, the Penal Law allows a court to bar the defendant and his/her attorney from the proceedings during specified times provided that the court is able to ensure that this does not prejudice the defense [Israel, Penal Law, Section 128(1)(3), 1977]. Israel thus provides significant executive power, particularly in security crimes cases, to the authorities and they are less constrained, compared to other models, by legal restraints.

United Kingdom

In the UK, the Terrorism Act allows for a number of different types of search powers, including searching residential premises and persons on the premises, searching nonresidential premises, searching cordoned areas, and conducting ad hoc searches (Walker, 2002, pp. 94–98). A search of residential premises requires a warrant, which any police officer ("constable" in British parlance) may apply for to a magistrate. Searching nonresidential premises does not require a warrant but does require that an officer at the rank of superintendent or above request a court's approval—the assumption being that it is necessary to provide for locking down and conducting a mass search in a certain area, as in cases where a bomb-making laboratory may exist (Walker, 2002, pp. 94–95). Interestingly, the restrictions on conducting searches in residential premises do not hold in the case of the third power: that of conducting searches in *cordoned areas*. Under Part IV of the Terrorism Act an officer at the rank of superintendent or above (and in particularly acute emergency situations, a constable of any rank) can designate a particular area a cordoned area for a limited period of time. During that time, constables have the authority to search any person, vehicle, or structure within the cordoned area, order people out of the cordoned area, and restrict access to the cordoned area (UK, Terrorism Act, Sections 33–36, 2000). There are, however, limits to the use of material seized in such a manner, and this is clearly a perogative designed to safeguard public safety in a crisis situation as opposed to being a normal search power. An even more extensive power involves the *power to designate*, which, in cases of extreme emergency, allows an officer at the rank of superintendent or above to effectively issue his/her own warrant to search a premises without that premises having been designated part of a cordoned area. Once again, as with the power to cordon areas, this power is designed with public safety in mind in terms of thwarting a potential attack and is not a standard investigative tool.

The Terrorism Act also affords the police "stop and search" powers whereby, in designated areas, and solely for the prevention of acts of terrorism, for periods of up to 28 days (although this can be renewed), they have blanket authority to stop and search persons and vehicles in those areas without "probable cause" (Walker, 2002, pp. 147–148). Otherwise, the Police and Criminal Evidence Act of 1984 allows police to stop and search people only if they have a probable cause. These powers have come under increasing criticism, and it has been alleged that they have been used disproportionately against ethnic and religious minorities. In January 2010, the European Court of Human Rights (ECHR) ruled that Section 44 of the Terrorism Act 2000 violated Article 8 (which focuses on privacy rights) of the European Convention on Human Rights. The British government is appealing the decision and, in the interim, continues to exercise this power. However, current government policy has restricted such powers to situations in which a senior police officer provides specific authorization on the grounds that a terrorist attack is thought to be imminent and searches may prevent it.

With respect to conducting surveillance, the 2000 Regulation of Investigatory Powers Act (RIPA) allows the Home Secretary, or officials delegated these powers by the secretary, to direct the police or the intelligence agencies

---

Sidebar 2-4: What MI5 can and Cannot Do to Check Up on Individuals

There are strict limitations on what MI5 is allowed to do when investigating an individual. There are laws (covering MI5 and others) which ensure that an individual's right to privacy cannot be overridden without very good cause. In addition, MI5 has its own Act of Parliament that demands it only obtain information in order to carry out its lawful work—in particular, the protection of national security.

MI5 can use what it calls "intrusive techniques" against an investigative target if there is sufficient justification on national security grounds. These techniques might include intercepting telephone communications, interfering with property (for example, planting eavesdropping devices in a person's house or car), "intrusive surveillance" (watching and eavesdropping on private homes or vehicles), or carrying out "directed surveillance" (following and photographing targets and recording where they go, who they meet, and so on).

There must be good justification for using these techniques. In order to intercept telephone communications, interfere with property or conduct "intrusive surveillance" a warrant must be obtained which authorizes precisely what action will be taken. Such warrants are issued by the Secretary of State [Home Secretary] and remain valid until the operation is complete, or for up to six months (whichever is the shorter). The authorizations are reviewed by independent Commissioners to ensure that they comply with the law.

In urgent cases warrants may be signed by a senior official within the Home Office, but only where the Secretary of State has given express permission to the official. These warrants last for between only two and five days (depending on the type of action) unless they are confirmed by the Secretary of State.

"Directed surveillance" is deemed less intrusive (a person being watched in public is a lesser invasion of privacy) and this kind of action can be authorized by officers within MI5. Nevertheless, such authorizations are still subject to independent review by the Commissioners.

The warrant and authorization system, together with the independent review process, is a legal safeguard which ensures that MI5 does not use any intrusive techniques without very good reason. (UK House of Commons Intelligence and Security Committee, 2009, p. 43)

---

to intercept communications and requires that all communication service providers make it possible for their networks to be used for surveillance by the authorities (see Sidebar 2-4). Furthermore, the law requires that communications firms retain communications data for up to one year, and these aforementioned powers were recently expanded to include serious crimes and not just terrorism investigations (Stoddart, 2009, p. 6). Indeed, the remit for surveillance is quite broad. Under RIPA, surveillance can be carried out not only for reasons of national security, public safety, and crime prevention but also to protect public health, serving the economic interests of the UK, and assessing taxes and duties [UK Regulation of Investigatory Powers Act, Section 22(2), n.d.]. Since RIPA came into effect in 2004, the number of intercepted communications has grown significantly and more than 200 agencies, police forces, and prisons now enjoy the authority to intercept communications. In the 2005–2006 period, for example, there were 2407 warrants for interceptions of telephone and mail issued in England and Scotland under RIPA provisions (up from 1466 in 2002) (Stoddart, 2009, pp. 13–14). The mandate for intercepting communications is surprisingly broad. According to RIPA, interception of communications can be undertaken "in the interests of national security, for the purpose of preventing or detecting serious crime, for the purpose of safeguarding the economic well-being of the United Kingdom or . . . of giving effect to the provisions of any international mutual assistance agreement" [UK Investigatory Powers Act, Section 6(2), n.d.]. Information obtained through interception of communications can be used only for intelligence gathering and is not used in court because it must then be made public and could reveal operational methods (Brown, 2006, p. 2). RIFA powers have, however, recently been curtailed significantly.

As far as arrest powers are concerned, Section 41 of the 2000 Terrorism Act affords constables the power to arrest anyone without a warrant who they suspect to be a terrorist. Similarly, warrantless arrests can be carried out in situations of probable cause or on the basis of a warrant issued by a magistrate. Under the Police and Criminal Evidence Act (PACE) of 1984 or the Prevention of Terrorism Act (PTA) of 1974, once a suspect is arrested, he/she has the right to consult a solicitor (attorney) and to have that solicitor present during questioning. However, this right may be amended if there is concern that access to a solicitor will lead to interference with the case, harm to others, tipping off of other suspects in the case, or other forms of obstruction of justice. Under PACE, access to an attorney can be delayed up to 36 hours, whereas under the PTA it can be delayed for up to 48 hours. In practice, however, immediate access to legal representation is rarely denied in the UK (UK Home Office and Northern Ireland Office, 1998). As the discussion above shows, UK search powers are quite extensive, but the right of those arrested to consult legal representation is strongly entrenched, and since the UK does not operate under emergency legislation, access to legal representation cannot be delayed significantly.

Unlike the system in Israel and France, intercept materials cannot, except in very limited circumstances, be admissible in court—something that dovetails with the reluctance on the part of the British Security Service (MI5) to submit such evidence in court, due to the fact that doing so is likely to expose sources and methods. Evidence gathered through surveillance however, is admissible in court. In 2005, the British Home Office floated the idea of designating a number of judges to be *security-cleared judges*, who would review intelligence material and determine admissibility and the manner in which the intelligence would be used as evidence. However, this initiative was blocked by Parliament, which considered it a violation of British common law traditions and an attempt to adopt the continental investigative magistrate system. British judges were also reluctant to accept this idea for fear that it would make them beholden to the government and thus weaken their judicial independence (Foley, 2009, pp. 26–27). To address the dilemma of not providing evidence that might compromise sources and methods to defense attorneys, the British system allows for the appointment of a *special advocate*, whose role it is to defend the interests of the suspect and who has access to materials that neither the accused nor his/her legal representatives can access (Slapper and Kelly, 2009, pp. 3239–3255). Clearly, then, in the realm of conducting searches and investigations, British law provides authorities with significant powers.

Canada

In Canada, counterterrorism-related searches require a warrant from the Federal Court of Canada. In principle, Canadian law operates under the presumption that warrantless searches are unreasonable, and consequently, the onus is on law enforcement authorities to show, in cases where it proved impossible to obtain a warrant, that the search was justified. Changes to the Criminal Code in 2001 gave police and customs officers the power to carry out a warrantless search if the police or customs officer was engaged in the investigation of criminal activity and was authorized to do so by a senior law enforcement officer on the basis of reasonable grounds that such a search was reasonable and proportional to the circumstances (Commission of Inquiry into the Actions of Canadian Officials in Relation to Maher Arar, 2006, p. 42). Evidence obtained during warrantless searches is not automatically excluded from legal proceedings but can only be used under certain conditions, mainly having to do with whether or not the process was reputable. Canada's Public Safety Act (PSA) allows private-sector organizations to collect personal information without a person's knowledge or consent if the information is required by law to be disclosed if the Canadian Security Intelligence Service (CSIS), the Royal Canadian Mounted Police (RCMP), or another authorized government agency requests the information and it relates to national security (Stoddart, 2009, p. 9). Wiretapping requires prior authorization in the form of a judicial warrant, and normally, the warrant must demonstrate that other investigative procedures have been tried and failed (or are unlikely to succeed) or that the matter is too urgent for the application of other investigative means (Commission of Inquiry into the Actions of Canadian Officials in Relation to Maher Arar, 2006, p. 41).

The Canadian approach is thus a compromise between the automatic exclusion of evidence in such cases in the United States and the British common law approach, which allows all relevant evidence to be used in court (Skurka and Pringle, 1999). With respect to arrest and access to legal representation, Section 10 of the Canadian Charter of Rights and Freedoms guarantees the right of arrested or detained suspects to enjoy access to an attorney "without delay." Canada thus follows a very restrictive model that provides strong protections for suspects.

Australia

In Australia, as with Canada, counterterrorism searches require a warrant. If the Australian Security Intelligence Organization (ASIO) obtains evidence that terrorism suspects are planning an imminent attack, it can then apply to the attorney-general for a warrant to employ "special powers," including searches of individuals and premises, access to computer drives, interception of mail, and so on. (Chalk, 2009, p. 21). However, the investigation of a terrorist incident is, first and foremost, the responsibility of the state police service in whose jurisdiction the attack has occurred, although that service will cooperate with ASIO, the Australian Federal Police (AFP), and other agencies.

Law enforcement agencies in Australia have traditionally had the power to stop and search individuals only in the case of a reasonable suspicion that the person has committed or is about to commit a crime and could only conduct stop-and-search operations without a reasonable suspicion in airports or near diplomatic premises. However, with the changes in legislation in Australia post-9/11, the AFP and other law enforcement agencies have had these stop-and-search powers extended to safeguard against terrorism at other locations (Australian Government, 2006, p. 36).

In Australia, ASIO must obtain an *authority to investigate*, which acts as a legal license from the attorney-general to conduct covert surveillance. Since this license requires the provision of information to suggest that a person poses a threat, ASIO's initial investigations often involve regular interviews with community leaders via the agency's Community Contact Program (Chalk, 2009, pp. 20–21). Following an arrest, suspects have the right to more-or-less immediate consultation with a solicitor. This provision holds equally for someone being detained preventively as for someone being arrested under ordinary criminal procedures. In the case of preventive detention, the authorities have the right to monitor the communication between the detainee and his/her legal representative (Australia, Criminal Code, Sections 105.35–105.38, 1995). Australian law thus provides significant search powers and, as noted earlier, allows for preventive detention in terrorism cases, but it also guarantees the immediate right to legal representation, albeit monitored.

Germany

In Germany, Section 103 of the German criminal procedure code (Strafprozessordnung or StPO) allows for searches of premises to be undertaken on the basis of a warrant from a judge or the public prosecution office. If the judge or public prosecutor is not available, law enforcement officers are required to make every effort to call in a local municipal official or two members of the local community to witness the search (German Criminal Procedure Code, Section 105, 1987). The code also prohibits the conducting of searches at night unless there are extenuating circumstances. Additionally, new police powers at the federal and state levels allow police to search persons without probable cause in certain locations, such as trains and train stations, airports and aircraft, and in border regions (Lepsius, 2002, p. 14). As in the case of several other countries in this survey, arrest can occur when law enforcement officers suspect that a crime has been committed, there is a risk that suspects will flee, and on the basis of a judicial warrant.

Once arrested, a suspect has access to legal representation more or less immediately. The German model is thus unique with respect to the procedure for conducting searches and follows an approach similar to those of Canada and Australia with respect to access to legal representation on the part of those arrested.

France

In France, the police have the authority, under an enquête préliminaire (preliminary inquest), to gather evidence via surveillance, witness interviews, and so on, as well as searching premises—although only at the consent of the person whose property is being searched, unless a search without consent is authorized by a judge. In the case of a recently committed offence (enquête de flagrance) the police have the authority to search premises without consent for a period of up to 16 days (Hodgson, 2006, p. 7). Once a suspect is identified, the investigation is then carried out by one of the country's police forces under the supervision of the public prosecutor (procureur), a judicial official who supervises the police investigation. The procureur is thus less a prosecutor than a judge who is responsible not only for supervising the police investigation, but also for safeguarding the rights of those under investigation. In principle, therefore, the procureur closely oversees investigations and ensures that the rights of suspects are scrupulously respected. In practice, however, the public prosecutor is dependent on the police for receipt of information and typically has to handle a large number of cases and is not really able to oversee specific investigations closely. Moreover, most public prosecutors do not see their role as involving close direction of the police but, rather, simply as ensuring that the results of the investigation are arrived at through proper legal procedures (Hodgson, 2006, p. 19).

After the initial stage of a terrorism investigation, if the case is particularly serious or complex, the procureur can decide to transfer it to an investigating (or examining) magistrate (juge d'Instruction) through the formal opening of a judicial inquiry (information). The investigating magistrate then conducts a detailed investigation (instruction) and then decides whether or not to take the case to trial given the seriousness of the charges and the quality of the evidence. If the examining magistrate decides to indict the defendant, the case is then transferred to a cour d'assises, a court consisting of a panel of three judges who act as "assessors" to "assess" the crimes, along with a jury of nine members (Clavier, 1997, pp. 7–8). As noted earlier, unlike the "adversarial" Anglo-Saxon legal system, the French legal system is "inquisitorial" and it is the role of the court to discover the truth (which can also mean exoneration of the suspect) rather than to act as an arena for the prosecution and defense to argue their sides of a case. One of the advantages of this system, certainly in the counterterrorism context, has been the creation of a cadre of counterterrorism judicial magistrates who have built up a considerable amount of expertise and experience in dealing with terrorism investigations. Under this system, the judge can act as the focal point of the investigation and is not only able to centralize authority but also is able to see both the case and the broader picture into which the case fits. Moreover, these judges have often developed strong relationships of trust with the domestic intelligence agencies that allow them access to critical intelligence information (Foley, 2009, pp. 449–450). Judges can also take into account a summary of the intelligence information and admit this as evidence in a trial, although they cannot admit into the proceedings raw intelligence information such as phone intercepts (United Nations Office on Drugs and Crime, 2010, p. 72). Critics of the system, however, argue that investigating magistrates are more concerned with building cases than with being impartial arbitrators and that the relationship between the magistrates and the intelligence agencies biases the investigations (Human Rights Watch, 2008).

One of the problems in marrying the investigatory/evidentiary police work with intelligence is that the former are focused on developing specific cases against specific individuals, whereas the latter are more interested in organizations as a whole: processes and intentions. In many other countries, and this is certainly true of the United States, there is no single authority that can adequately merge the two world views, and this is one of the primary strengths of the French system. The judicial magistrates who specialize in counterterrorism have consequently become "reservoirs" of counterterrorism knowledge and experience. The examining magistrate is thus a powerful figure who has the authority to command the police and the associated domestic intelligence services in their investigations. The magistrates, in effect, are the centers of French counterterrorism investigations and prosecutions and are able to mobilize a wide variety of state resources, including wiretaps, electronic and physical surveillance, raid teams, and the like (Schmitt and Gerecht, 2007, pp. 2–3). The magistrates have the authority to open investigations based on intelligence that they might be conspiring to commit terrorism, can order the arrest of individuals in the groups, can interrogate them, and can indict them (Shapiro and Suzan, 2003, p. 90).

At the same time, the role of the investigating magistrate is gradually being weakened in light of arguments that it represents an unacceptable concentration of power and the procureur are being given some preliminary investigatory powers. At the time of this writing, French President Nicolas Sarkozy has put forward a proposal to further limit the power of investigating judges by allowing the prosecutors to run investigations and giving the investigating magistrates oversight powers (rather than allowing them to run the investigations). Critics of this change argue that this will harm judicial impartiality since the procureur, even though they are judges, are under the authority of the political system via the Ministry of Justice, to which they answer (The Economist, 2009, p. 57). During trials, witnesses can be heard anonymously, and in certain cases the anonymity of investigators is protected (France, Prime Minister's Office, 2007, p. 54). The French system may also view the failure of the accused to answer certain questions as negative and hold that against the accused.

French law provides broad license to law enforcement authorities to conduct searches without consent, including at night, in recently opened terrorism cases. Indeed, investigative magistrates effectively serve as the first line of defense because French law empowers them to open investigations on individuals prior to the carrying out of any suspected criminal acts. This allows the authorities to put pressure on suspected terrorists and disrupt the operation of terrorist organizations (Hodgson, 2006, p. 37). When terrorism suspects are arrested, they may be kept in police custody for up to 96 hours, and police custody can subsequently be extended to six days if there is a danger of an imminent attack in France or overseas or a suspect needs to be held to facilitate international counterterrorism cooperation (Council of Europe, 2006, p. 2). Terrorism suspects are not granted access to legal representation until 72 hours after their arrest. France thus follows a highly centralized model that grants significant powers to judicial authorities to investigate cases and significant leeway for police to keep suspects in custody and, in the initial stages, to deny them access to an attorney.

With respect to electronic surveillance, the 2006 law, Loi Relative á la Lutte Contre le Terroirsme et Portant Dispositions Diverses Relatives á la Sécurité et aux Contrôles Frontaliers, requires Internet service providers, telephone companies, and other communications providers to provide client information to counterterrorism agencies as well as information on calls made by clients (Stoddart, 2009, pp. 6–7). This French law also allies the authorities to collect personal information on all passengers traveling to or from countries outside the EU. All in all, French law allows significant leeway in the conduct of investigations, and its system of investigative magistrates and other judicial officials provides both expertise and significant powers to the state, thus enhancing its ability to conduct investigations.

INTERROGATION OF TERRORISM SUSPECTS

The manner in which terrorism suspects are interrogated and the type of coercion used, if at all, represents a significant moral dilemma. On the one hand, a suspect may have information that is of vital importance to an investigation and may even be able to provide actionable intelligence that might prevent a terrorist attack and the ensuing loss of life, but at the same time, the suspect is entitled to basic legal and human rights, including the right not to be tortured (physically or psychologically). There is also the matter of the question of the effectiveness of coercive interrogation techniques (whether physical or psychological or both) and their utility in extracting the truth, as opposed to inaccurate information provided by the suspect in order to end the coercive interrogation. The moral and operational dilemmas posed by coercive interrogations cannot be solved here. However, it is instructive to contrast the approaches taken by the two countries in our survey most affected by terrorism: Israel and the United Kingdom.

Israel

In Israel the interrogation of terrorism suspects has often involved some degree of physical force as well as putting the detainee in difficult physical surroundings and employing psychological pressures. The Israeli approach traditionally made protecting the Israeli population from terrorist attack through acquisition of intelligence from detainees a significantly higher value than safeguarding the rights of those detainees. This did not mean that there were no restraints on the use of physical means, but the approval for such measures was based on internal vetting processes within the domestic security service, the Israel Security Agency (ISA), approved by a special ministerial committee on ISA interrogations (Israel Supreme Court, 1999). In

1987, however, following widespread complaints of torture, a judicial commission created a set of recommendations (not made public) governing the interrogation process. It was found, however, that in a number of subsequent governmental studies (these, too, were not made public), that the ISA was continuing to violate some of those guidelines. In 1999, the Supreme Court ruled that the ISA could not use physical torture, and the agency claims that it now employs psychological methods exclusively. This assertion, however, is disputed by human rights groups, which claim that detainees are still subject to being put into uncomfortable positions for long periods of time, being held in filthy cells, being shaken violently, being exposed to blasts of cold air, being slapped, and being verbally abused and threatened (United Nations, Office of the High Commissioner for Human Rights, 2001). In its 1999 ruling, the court recognized that the use of physical means might be the only way to save lives in certain situations when a detainee had information about an imminent attack (called *ticking bomb* situations) and when obtaining that information could prevent the death of innocents. The executive's representatives argued before the court that the use of force was permissible in such situations under Section 34 of the Penal Law, which stipulates that "an individual will not be found criminally liable for committing an act that was immediately necessary for the purpose of saving the life, liberty, body or property, either of himself or of another person, from a real danger of serious harm due to conditions prevalent at the time the act was committed, there being no alternative means for avoiding the act" [Israel, Penal Law, Section 34(11), 1977]. The court thus ruled that Section 34 of the Penal Code could be used as a defense in the event of a criminal trial of an ISA investigator. However, it also found that this could not serve as a statutory basis for authorizing such activities in advance. In other words, it was illegal for ISA investigators to employ physical methods (slapping, shaking, putting individuals in uncomfortable positions, etc.) but should such methods be employed in clear ticking-bomb situations and should an ISA investigator subsequently be tried for such actions, that investigator would not be held criminally liable. Needless to say, this legal loophole was not seen by the ISA as providing them with the legal support to obtain crucial intelligence. In the words of then ISA head, Ami Ayalon, "We tell the interrogator: 'Do what you feel is right to save lives and afterward, we will consider whether or not to indict you.' ISA interrogators can no longer rely on such judgments, and I agree with them" (Pedahzur, 2009, p. 103).

United Kingdom

The UK has also had a history of using interrogation techniques deemed illegal under international law. In 1977, the European Court of Human Rights found that the UK had violated the European Convention on Human Rights, to which it was a signatory, through the use of the so-called "five techniques," including forcing detainees to stand against the wall in positions that caused strain, hooding detainees, and subjecting them to noise, deprivation of sleep, and deprivation of food and drink (Gross, 2003, p. 1179). In 1994, however, following an official visit to the UK, the European Commission for the Prevention of Terrorism found that there were no cases of torture and virtually no accounts of brutal measures being used against detainees. The UK has thus moved completely away from a coercive model that involves the, at least theoretical, use of torture.

COUNTERTERRORISM WAR-FIGHTING

The only country in this survey that can be said to employ a primarily war-fighting approach to counterterrorism is Israel. This approach, in the Israeli case, is based on a number of factors, including the scope and nature of the terrorist threat (which is far more significant in terms of time frame, number of attacks, and the toll on human life and the economy than in the case of the other countries surveyed) and the fact that most of the terrorist threat and proactive counterterrorism measures occur either in the territory of foreign states or entities (such as Lebanon or the Gaza Strip) or in areas under military

FIGURE 2.7 Israel air force Apache Longbow. This Wikipedia and Wikimedia Commons image is from the user Galit Luvtzki and is freely available at http://commons.wikimedia.org/wiki/File:Ah-64d.jpg under the Creative Commons Attribution 2.5 Generic license.

occupation (e.g., the West Bank), and consequently, civilian police and legal entities have less sway and the military (which in the Israeli case is a war-fighting agency with no law enforcement role) has overall control.

The single most significant active Israeli counterterrorism policy—as opposed to passive ones such as physical barriers to infiltration—involves large-scale preventive detentions (often leading to subsequent prosecution) and executive actions (known in Israel as *targeted killings*). The targeted killings policy has a number of advantages and disadvantages. At the outset it should be made clear that Israel conducts targeted killing operations only fairly rarely, in comparison to the numbers of people that it arrests. Between 2000 and 2005, Israel targeted fewer than 200 terrorist suspects for assassination while arresting or detaining, at one point or another, some 9000 suspects thought to be involved in some manner in terrorist activities. Only those individuals who cannot easily be arrested will be put on the list (or in very unique cases, such as that of Hamas leader Ahmad Yassin, those that Israel cannot afford to arrest because their arrest would probably lead to heightened terrorism in order to try and obtain their release) and they need to fulfill the criteria of being "arch-terrorists": in other words, senior leaders, planners, bomb-makers, and the like whose removal would significantly undermine the terrorist organization's ability to function—at least temporarily (Morag, 2010, p. 153).

All decisions to assassinate require the approval of the prime minister, acting for the cabinet as a whole, and a case must be made, including provision of a dossier with information implicating the person to be assassinated with terrorist acts and with the potential for contributing to future acts of terrorism. In almost all cases, Israel will prefer to arrest terrorists rather than assassinating them, for two primary reasons, First, as the saying goes "dead men tell no tales." Capturing suspected terrorists usually yields a gold mine of information, whereas assassinated terrorists are clearly useless as intelligence sources. Second, targeted killings often result in international criticism, pressure on Israel, and greater anger and motivation on the Palestinian side for carrying out additional terrorist attacks.

Nevertheless, when coupled with an aggressive policy of detaining suspected terrorists, the policy of assassination has proven highly useful in disrupting terrorist communications, freedom of movement, planning activities, and the like, as well as sowing distrust and fear within terrorist organizations. From time to time, key individuals, such as the notorious Hamas bomb-maker Yihyeh Ayash, have been assassinated, resulting in at least a temporary incapacitation of a critical part of the terrorist organization's apparatus. Moreover, the policy of assassinating terrorists acts to reassure the Israeli public that terrorist leaders and other key individuals are not immune to Israeli "retribution" (even though assassinations are not authorized for purposes of retribution, the policy is often viewed as based on retribution by the Israeli public, as well as by Palestinians and international public opinion). The average Israeli citizen knows that the authorities cannot protect him/her from every terrorist threat; in fact, the authorities frequently remind the public that they cannot provide 100 percent security. At the same time, carrying out an active and aggressive policy of arresting and killing terrorists helps to create a public sense that "something is being done" and this thus helps to reassure the public. As terrorism is more a psychological phenomenon (in terms of creating fear across society) than a physical one (due to the comparatively small number of victims), reassuring the public and making it possible for them to go about their daily lives thus represents an important victory over terrorism. Perhaps the worst feeling that an Israeli citizen can face is one in which he/she senses that Israelis can be killed with impunity, whereas the authorities are powerless to strike back (Morag, 2010, pp. 161–162).

In general, whether Israel, acting in a war-fighting mode, is apprehending or killing suspected terrorists, the essence of the Israeli approach revolves around three key assumptions. First, the number of dangerous terrorists is limited and it is therefore possible to arrest most of them and assassinate those who are not as accessible. This also means that no matter how much rage is produced in Palestinian society as a result of aggressive Israeli counterterrorism policies, this will not translate into more terrorism because very few people possess the resourcefulness and capacity to develop expertise in some area of terrorist operations and thus to become effective terrorists. In other words, more angry people does not equal more truly dangerous terrorists. Being a truly effective terrorist operative is a full-time professional enterprise that requires years of training and development and cannot be the preserve of enraged amateurs desperate to strike out at Israel. Those amateurs can and do fill the ranks of the suicide bombers, but a suicide bomber without an organization behind him or her to supply the wherewithal can only be a *potential* suicide bomber. Second, not every terrorist has to be neutralized for the counterterrorism strategy to be deemed a success. Since terrorists almost always operate as part of a complex organization that involves logistics, internal security, recruitment, leadership, smuggling, bomb-making, and other functions, neutralizing key individuals in one or more of these component areas of the organization can severely hobble the organization, at least temporarily. Third, over time an unrelenting policy of arrests can severely decrease the effectiveness of the organization, and with most of the leadership in jail, lower-level operatives are left demoralized and directionless (Dichter and Byman, 2006, pp. 11–12).

Most terrorist organizations can be viewed as made up of three levels, with the strategic level providing the overall policy guidelines and priorities (as well as inciting the public to violence and glorification of terrorist values); the operational level providing most of the expertise for funding, organizing, and implementing terrorist activities; and the actual perpetrator carrying out the attack. Israeli counterterrorism policy has heavily emphasized the use of intelligence to ascertain who is involved at the operational level as well as to target them for arrest or, in some cases, assassination. The goal in focusing at this level of the organization has been to destroy the capacity of the organization to actually produce terrorist attacks (although, of course, actual perpetrators planning or in the process of carrying out attacks are also targeted).

It should be borne in mind that Israel's policy in this sphere is not predicated on the assumption that effectively combating terrorism requires a one-time operation. Even if the operational level of a terrorist organization is emasculated as a result of a series of successful Israeli arrests or targeted killings, other members of the organization will step in to take their places and fill these functions. The point is that these new planners, bomb-makers, couriers, recruiters, and so on, will require a learning curve, and consequently, their initial operations will be less effective (something that has proven to be true, for example, as Israeli investigators found faulty wiring and poorly made explosives in suicide vests and car bombs used in terrorist operations in the wake of successful arrests or assassinations of key bomb-makers) and this translates into fewer Israelis being killed or maimed in terrorist attacks. Of course, those individuals will eventually develop the necessary expertise to be effective terrorist operatives, but by that time, presumably, Israel will have been able to arrest or assassinate them as well, thus forcing the organization to produce yet another crop of new personnel to step in and fill the breech. Through a methodical and painstaking "shaving off" of these layers each time they surfaced, Israeli counterterrorism officials became convinced that terrorist organizations would become increasingly less effective and that these organizations did not have the capacity to produce forever new crops of personnel for the operational level (Morag, 2010, p. 163).

On the negative side of the ledger, targeted killing often results in collateral damage, despite the use of UAVs and other technical means to try to confirm a terrorist's identity and to assess whether or not civilians have entered the target's immediate area. For example, on July 22, 2002, Israel targeted Hamas leader Salah Shehadeh, a key figure within the organization. Israel believed that by taking him out, Hamas's ability to function would be disrupted for some time. His assassination had been approved by then Prime Minister Ariel Sharon, and the IDF was simply waiting for the opportunity to strike him when the likelihood of producing collateral damage would be smallest. Eventually, intelligence information came in suggesting that

Shehadeh was in a Gaza City apartment building that was not occupied, at the time, by any innocent civilians. An Israeli air force aircraft was deployed, and it dropped a one-ton bomb on the building to ensure that Shehada would be killed. Unfortunately, Israeli intelligence proved faulty and 14 civilians were killed in the attack, including nine children (The Economist, 2002). Following this tragedy, Israel shifted to the use of ordinance with lower yields. On September 6, 2003, Israel received intelligence indicating that the entire senior leadership echelon of Hamas, including the organization's leader, Ahmad Yassin, was meeting in a specific location. This represented a real intelligence coup and could have resulted, in one fell swoop, in total decapitation of the organization. The air force carried out an attack dropping a 250-kilogram bomb (the aircraft could have used a 1000-kilogram bomb) to minimize the risk of collateral damage and not repeat the debacle in Gaza City. The explosion turned out to be too small, however, and the entire Hamas leadership succeeded in escaping, with a few suffering only very minor injuries (CBS News, 2003).

The collateral damage issue thus represents a significant moral conundrum for Israel, something that will be familiar to all students of asymmetric conflict. On the one hand, Israel is a democratic country with a professional intelligence apparatus and armed forces that aspire to do a proficient job combating terrorists without hurting innocents in the process. At the same time, the Israeli military knows that its primary mission is to protect Israeli citizens, so failing to take out key terrorists (who invariably hide among innocent civilians), thus disrupting the terrorist organization, means that more Israelis will die or be injured in terrorist attacks (Morag, 2010, p. 164).

INSTITUTIONAL, ORGANIZATIONAL, AND STRATEGIC ASPECTS OF COUNTERTERRORISM EFFORTS

Counterterrorism policies, laws, and practices are intimately tied to the institutional and organizational framework within which those efforts occur and to the overall strategies employed. Each of the countries in the survey has adopted a different constellation of institutional and organizational tools to cope with terrorism, which are both reflective of their respective histories, governmental systems, and the nature and scope of the threat they have been facing. Just as the over 17,000 state and local law enforcement agencies (not to mention the federal agencies) or the division of the U.S. military into active-duty, reserve and 50 state National Guard forces are a reflection of American history and political arrangements (particularly federalism), the institutions and agencies in the countries surveyed reflect their unique histories and political frameworks.

Israel

Israel has only a small number of agencies actively engaged in various aspects of counterterrorism intelligence gathering, combat, and law enforcement. There is a fairly clear hierarchy and delineation of responsibilities between the primary agencies, and since only a few agencies are engaged in counterterrorism and counterterrorism intelligence collection and analysis, they are usually more than happy to leave activities outside their core mission to other agencies, thus facilitating interagency cooperation. For example, the only agency in Israel that has the authority to enforce laws within Israel proper is the Israel Police, and the police are recognized as the lead agency in all matters relating to coping with run-of-the-mill crime. Some intelligence officers in the Israel Security Agency (ISA, Shin Bet, or Shabak) are granted police powers, but the ISA, as a whole, is not a law enforcement agency [ISA Law, Section 8 (b), 2002]. As none of the other agencies are interested in engaging in traditional law enforcement, they are more than happy to share any information that they come across in the criminal sphere with the police intelligence division, and the police intelligence division, in turn, focuses primarily on criminal intelligence and recognizes that terrorism investigations will be lead by the ISA. Among the intelligence agencies, there is more room for overlap and hence for bureaucratic infighting. Nevertheless, here too each agency has a clearly separate core mission, and the problems usually arrive at the fuzzy edges of each agency's mission but not at the heart of their respective enterprises.

The Israel Defense Force (IDF) is a unitary military force (as opposed to the system of separate military branches practiced in the United States) and is subject to the orders of the IDF chief of staff, who in turn is subject to the authority of the defense minister acting for the cabinet as a whole. In Israel, the cabinet is technically the commander-in-chief rather than the prime minister (or the Israeli president, who serves in what is an almost exclusively ceremonial post). The IDF is the single largest and most powerful governmental agency in Israel, with resources and personnel numbers that dwarf those of all of the other agencies mentioned here. The IDF possesses the country's largest intelligence organization, the Intelligence Branch (IB, also known by its Hebrew acronym, AMAN), and also serves as the primary operational organization tasked with combating terrorism in the West Bank, where the police role is generally limited to policing Israeli citizens living in the settlements and engaging in a few additional limited activities. The IB, the country's lead intelligence agency, is also responsible for providing to the cabinet the national intelligence assessment, which acts as the basis for national security policy decisions. The Israeli equivalent of the U.S. National Security Agency, which serves as Israel's primary SIGINT agency, is also housed within the IDF (known as Unit 8200). The IB includes departments that focus on the various Arab states, Iran, and terrorist organizations and monitors goings-on worldwide. However, in terms of operating HUMINT resources within the West Bank and Gaza Strip, particularly in the area of recruiting Palestinian informants, the ISA serves as the principal agency. The ISA works closely with the IDF in the West Bank and Gaza and the Israel police in Israel proper, but its focus is intelligence gathering and analysis as opposed to law enforcement (although, as noted earlier, some of its personnel possess arrest powers). In the West Bank, for example, ISA personnel, who run assets in a given area and consequently know that area and its population intimately, will typically travel with IDF units (who themselves cycle in and out of particular geographic commands and hence do not necessarily have expert knowledge of specific locales) that are tasked with making arrests, to help pinpoint the suspects to be arrested or the bomb-making facility to be destroyed. Moreover, within the ISA, the desk officer (who plays the role of analyst and researcher) also provides operational guidance and support to the ISA case officers and their military colleagues in the field (Colonel Zohar, 2006, p. 37). The ISA has also been active in recruiting Palestinian inmates in Israeli prisons and detention facilities (Jones, c., 2003, p. 276).

In addition to being an intelligence organization, the ISA also plays an important security role. The ISA is tasked with providing "security guidance" (regulations, training, practice drills) to police, port and airport security officers, ministry security officers (education, industry, transportation, foreign ministry, energy, etc.), and private security firms working in the public sphere, such as Israeli airline security personnel. In other words, the ISA drafts regulations in matters of security and counterterrorism to which a broad range of security agencies, including the police, must adhere. A common agency to enforce procedures at the federal, state, and local levels as well as across much of the private sector is not something that exists in the United States, but it has proven its utility in the Israeli case, as everyone follows the same or similar standard operating procedures and all are "on the same page" with respect to security procedures. Finally, the ISA is also involved in the protection of VIPs (very important persons), certain critical infrastructure sites, and certain communications networks and databases and sets the criteria for the granting of security clearances [ISA Law, Section 7 (b1–b4), 2002].

The Institute for Intelligence and Special Duties (more commonly known by the Hebrew word for "institute"—Mossad), is Israel's premier foreign intelligence-gathering and analysis organization (although the IB also engages in foreign intelligence gathering and analysis, but usually with a focus on military issues). Among other issues, the Mossad also has an active interest in gathering intelligence on terrorist threats emanating from outside Israel's immediate geographic environment.

As noted earlier, the Israel police force (Mishteret Yisrael) is the country's sole law enforcement agency. Aside from times of acute military conflict, when it is put under the overall command of the IDF (see Chapter 5), it is a national organization that operates under the auspices of the cabinet, through the minister for public security, who in turn overseas the senior uniformed police commander, the police commissioner (who holds the rank of police lieutenant-general). The Israel police is also the main terrorism prevention and terrorist interdiction agency within Israel proper and provides law enforcement services to Israeli settlements in the West Bank. The Israel police force is also responsible for crowd control, provision and enforcement of security regulations for businesses, school security, protection of governmental institutions, and protection of installations of national importance, including seaports and airports (Israel Police, n.d., p. 72).

The minister of defense, however, may declare a "limited state of emergency," in which case authority for dealing with a particular incident within Israel proper is technically shifted to the military. This allows the military to enforce orders for people to stay in bomb shelters and shut down schools and places of employment as necessary. Moreover, the police have very limited capabilities in dealing with WMD events and in such cases, the military's Home Front Command (HFC or Pikud Ha'oref) would be authorized to manage the event under the framework of the overall military command (this is discussed in more detail in Chapter 5).

Field personnel in the Israel police are assigned to stations, which in turn form part of the 19 subdistricts that make up the country's six police districts. Individual patrol officers are assigned to a specific station but are under a unified command structure and hence can be moved from location to location as needed. This allows the police commissioner to concentrate forces in certain locations when intelligence indicates probable threats to public safety (as in the case of terrorist threats, riots, large public events, and the like). The senior police command is privy to all of the terrorism-related intelligence information and analysis being generated by the intelligence agencies (primarily the ISA and the IB). That information is then made available to intelligence liaison officers, who are specially designated police officials based in each police district, subdistrict, and station, whose role is then to determine the ways and means of sanitizing and disseminating that information throughout their area of responsibility and down to the level of the patrolman/patrolwoman. In principle at least, every effort is made to provide as much terrorism-related intelligence as possible to the average patrolman/patrolwoman because the senior police commanders recognize that they are not going to prevent terrorist attacks themselves while sitting in their

FIGURE 2.8 Israeli police officers. This Wikipedia and Wikimedia Commons image is from the user Mark Probst and is freely available at http://commons.wikimedia.org/wiki/File:Israel_police_officers.jpg under the Creative Commons Attribution-Share Alike 2.0 Generic license.

air-conditioned offices, and that disruption and mitigation will occur at the level of the officer on the street. The Israel police force includes school security patrols and oversee security officers who work for municipalities and government ministries (Morag, 2010, pp. 154–155).

The Israel police also include a mobile response component for special tasks and as a reserve when additional personnel are needed. The police force has a highly trained antiterrorism SWAT team, known by its Hebrew acronym Yamam, which also carries out operations for the military in the West Bank. Finally, the Israel police also includes a paramilitary component known as the border police or border guard (Mishmar Ha'gvul). The personnel in this force have full police powers but specialize in patrols along some of Israel's borders (the military has responsibility for others) and counterterrorism operations within Israel and also act as a reserve force for additional police activities. In addition, the border police operate in the West Bank, where they are under army command and act as a supplement to military forces conducting counterterrorism operations in Palestinian areas.

Given the fairly streamlined institutional structure described above, it is not surprising that issues relating to information and intelligence sharing in Israel are far less complex than those being dealt with in the United States given the bewildering array of American intelligence, military, law enforcement, and other governmental agencies. This does not mean, however, that Israel is immune to bureaucratic infighting and mutual reticence with regard to information sharing, only that there are fewer opportunities for this, and consequently, on the whole, information and intelligence sharing is both significant and effective. At the senior policymaking level, as Israeli cabinets are typically large (usually more than 20 ministers), most of the work conducted by the cabinet is carried out by smaller cabinet committees (called ministerial committees), which are authorized to make policy decisions for the full cabinet. Cabinet-level committees rarely handle responses but are briefed on an ongoing basis and can make executive decisions as needed.

Israel does not have a clearly articulated counterterrorism strategy, as at least in the Israeli conception of things, a solid strategy might constrain freedom of action and adaptability to new situations. Moreover, it can be argued that Israeli authorities see no need for an overall strategy, as the various components responsible for prevention and response to terrorism threats have considerable experience in doing so and a well-developed operational doctrine. Despite the fact that there is no articulated strategy per se, it can be said that Israeli counterterrorism strategy focuses on five areas: deterrence, intelligence, prevention, arrests/executive action, and public cooperation efforts. Generally speaking, terrorists cannot be deterred from carrying out acts of terrorism by the knowledge that they may die or be incarcerated for long periods of time, particularly in the case of suicide terrorism. Israeli deterrence policy has thus traditionally focused on two areas: the terrorist's family members and morale within the terrorist organization.

Israel had a long-standing policy, which has been challenged through judicial means both in Israel and internationally, of destroying the homes of arrested terrorists or those who died in the course of their attacks. This practice has recently been formally abrogated as a result of a policy reassessment. The logic behind this practice rested on the idea that although a terrorist might have been willing (or was planning to) sacrifice himself or herself, the person would not want family members to suffer undue harm, and destruction of the family's home represents a significant financial loss that would cause the family to suffer. Over the years, however, Palestinian organizations and Arab governments routinely donated money to such families to cover the costs of rebuilding so that the adverse financial impact was minimized. Moreover, there was never any clear proof that this policy actually deterred terrorists from carrying out attacks, although there was some anecdotal evidence to this effect (Guiora, 2006, pp. 10–14). Indeed, during the height of the second intifadah, donations by Arab governments to families of terrorists created a small class of "nouveau riche" Palestinians whose sons or daughters had blown themselves up in Israeli cities. A marginally more effective Israeli policy has been the public exposure, from time to time, of the modus operandi of terrorist organizations. This policy is designed to show the terrorists that their organization has been infiltrated and that the Israeli authorities thus have access to the most sensitive information within the organization. This often leads to the development of a sense of mistrust within the organization and lowers morale. Israel's policy of large-scale arrests as well as of targeted assassinations is also designed, in part, to lower morale by proving to the members of the terrorist organization that the organization is incapable of protecting them and is thus weak in comparison with Israel's counterterrorism capabilities (Morag, 2010, p. 157).

On the intelligence side, Israeli intelligence efforts focus on identifying potential threats at the organizational and individual levels, uncovering sources and methods for financing and provision of other infrastructure-related activities, uncovering sources of arms, and cooperation with like-minded intelligence services—and cooperation between intelligence agencies is generally good. Following the outbreak of the second intifadah, the ISA and IDF developed an effective system for real-time horizontal information sharing which enabled tactical units in the field to exchange intelligence information instead of waiting for intelligence to be pushed up the chain of command and then cross over to the sister agency. This tactical-level information sharing has resulted in significant gains in terms of arrests of suspected terrorists and disruption of terrorist activities. Generally speaking, the IDF–ISA relationship revolves around the IDF providing personnel to go in and make arrests or destroy bomb-making facilities, tunnels, and the like, as well as providing much of the SIGINT (electronic eavesdropping) intelligence gathering, while the ISA focuses more on recruiting and running Palestinian agents (Morag, 2010, pp. 157–158).

As far as preventive efforts are concerned, Israel's counterterrorism strategy is based on the assumption that if terrorists cannot be apprehended, killed, or otherwise have their "work" disrupted before they set off on their mission to attack Israel's population centers, the terrorist will be successful and Israel's counterterrorism policy will be deemed a failure. Hence, preventive activities play a central role, with these efforts naturally being affected by the reality of the short geographic distances between the terrorists' base of operations in the villages and cities of the West Bank or Gaza Strip and Israeli urban centers (which, in the case of the West Bank, are often a 10- to 30-minute drive away), if a terrorist and his/her handlers have already left their town or village on an operation, it will be difficult to apprehend them before they reach an Israeli urban center. Once they've entered an Israeli city, the Israeli security establishment considers them to have completed their mission successfully, particularly if it was a suicide mission. In other words, if a Palestinian terrorist with an explosive vest enters an Israeli city, the terrorist has won and Israel has lost because even if the terrorist does not get to his/her primary target, the person will blow himself/herself up in a manner designed to kill innocent people (Morag, 2010, p. 159).

Consequently, Israeli efforts focus heavily on neutralizing threats within and around Palestinian population centers in the West Bank. This has necessitated an Israeli military presence, usually ad hoc (i.e., per operation) inside Palestinian villages and cities as well as the more or less permanent deployment of troops outside Palestinian population areas and the creation of a system of checkpoints that ring Palestinian towns and lie along major Palestinian highways in the West Bank. The IDF presence in around Palestinian villages and towns and the system of checkpoints has proven extremely burdensome for Palestinians and represents one of the more onerous and obvious signs of Israel's military domination over the Palestinians—thus contributing to Palestinian anger. Nevertheless, this military presence has proven effective in countering terrorist activities. Much of Israel's success in bringing down the frequency and destructiveness of attacks during the second Palestinian intifadah of 2000 to 2005 occurred after Israel had cut off and invaded most of the Palestinian cities in the West Bank during Operation Defensive Shield (March–April 2002). The ability of IDF forces based near Palestinian towns and villages to react rapidly and effectively when real-time intelligence information points to the whereabouts of suspected terrorists or other terrorist activity is infinitely greater than the IDF's capacity to send in units to accomplish the same mission from over the Green Line in Israel itself (Amidror, 2008, p. 18). In general, active military operations in Palestinian villages and cities as well as the widespread use of checkpoints disrupt the movement of terrorist materiel, funds, and personnel from area to area, and drive terrorist organizations deeper underground, thus inhibiting their ability to function.

In addition to intelligence-directed operations targeted at terrorist operational bases, Israel is in the process of completing a controversial barrier of fencing (with sections in or abutting urban areas made up of concrete slabs) within the West Bank designed to make it difficult, if not impossible, for terrorists to infiltrate into Israel. Despite the political, legal, and diplomatic argument over its actual location, the Separation Barrier, as it is commonly known, has proven extremely effective in limiting access to Israel from the West Bank and thus in significantly reducing terrorist attacks.

United Kingdom

The British counterterrorism system is based on a number of agencies. The highest counterterrorism authority in the UK is the Cabinet Office (CO), which is responsible for providing overall policy guidance to the intelligence and law enforcement communities and is under the direct authority of the prime minister. The civil servant who chairs the CO for the prime minister also chairs the Joint Intelligence Committee (JIC), which is responsible for providing the cabinet with intelligence assessments and for tasking the intelligence community in terms of intelligence collection objectives. The JIC provides reports and threat assessments and includes the heads of the three civilian intelligence and security agencies, the head of defense intelligence, and senior officials in the Foreign Office, Ministry of Defense, Home Office, and other departments (British Security Service, n.d.). The Joint Terrorism Analysis Center (JTAC) was established in 2003 with the role of bringing together the intelligence community, police, and other stakeholders in order to share information and disseminate intelligence assessments, and it also sets the terrorism threat levels in the UK. A department within the Home Office, the Office for Security and Counter-Terrorism (OSCT), is responsible for advising ministers and developing policies and security measures to protect the UK from terrorism threats. Operating under the umbrella of the Home Office, the OSCT oversees exercises, protection of VIPs and critical infrastructures, the development of counterterrorism legislation, preparedness for terrorist threats, and coordination between the government and emergency services during terrorist events and counterterrorism operations (UK Home Office, n.d.a).

FIGURE 2.9 Thames House, headquarters of the British Security Service (MI5). This Wikipedia and Wikimedia Commons image is from the user Cnbrb and is freely available at http://commons.wikimedia.org/wiki/File:Thames_house_exterior.jpg under the GNU Free Documentation License, Version 1.2.

The UK has four primary intelligence agencies: the Secret Intelligence Service (SIS, also known as MI6), the British Security Service (BSS, also known as MI5), the Government Communications Headquarters (GCHQ), and the Defense Intelligence Staff (DIS). None of these agencies enjoy law enforcement powers and none are seen as responsible for public safety. Consequently, the police forces in the UK, who, of course, have law enforcement powers and are responsible for the safety of the public, also have important counterterrorism statutory powers. Although all of these intelligence agencies deal with terrorism threats to one degree or another, the BSS is the primary counterterrorism intelligence agency. The Security Service was originally established to combat German espionage during World War I and for much of its history was focused on counterespionage and other missions and dealt only tangentially with terrorism. It also worked independent of the police, and much of the counterterrorism work was carried out by police special branches (see below). In 1984, however, the Home Office, which has oversight over both the Security Service and the police, determined that special branches were henceforth to "assist" the Security Service, and in 1992 the task of countering Irish nationalist terrorism was transferred to the BSS (Beckman, 2007, p. 57). Among other things, the mission of MI5 is to prevent terrorism, prevent foreign espionage and other covert foreign state activity in the UK, disrupt attempts to undermine the political system, prevent the proliferation of material or technology relating to weapons of mass destruction, fight serious crime, and protect critical national infrastructure as well as sensitive government information and assets (British Security Service, 2010).

Its newly adopted regional approach will enhance the ability of the BSS to work with local, specially cleared police intelligence-gathering units (known as special branches) via police counterterrorism units (CTUs) and smaller counterterrorism intelligence units (CITUs). While the BSS was originally a London-based organization, by 2011 it is expected that fully one-fourth of BSS personnel will be deployed to regional headquarters around the UK. This new counterterrorism network will increase intelligence sharing and cooperation between police forces and between the police and the BSS, as well as with local authorities (Prime Minister and Home Secretary, 2009, p. 63). In addition, the country's largest police force, the London-based Metropolitan Police, has a Counter Terrorism Command which works to disrupt terrorist plots and conduct investigations in London and the southeast of England (Police National Legal Database and Andrew Stainforth, 2009, pp. 127–128).

As noted previously, police forces play a critical role in British counterterrorism because of their numbers and presence in the field and their powers of arrest. The BSS, with its limited personnel and lack of intimate familiarity with most localities, is dependent on local special branch units to provide resources so that the BSS can accomplish its mission of counterterrorism intelligence gathering and analysis. There are 52 regional police forces in the UK (43 in England and Wales, eight in Scotland, and one in Northern Ireland), and each is required to include a special branch unit, there being approximately 6000 special branch officers in the country. In addition to functioning as the primary counterterrorism intelligence entities within the territorial police forces, special branch officers are deployed to all major airports and ports. Special branch officers are trained and receive guidance from the BSS and act as the service's eyes and ears in local jurisdictions. In the late 1970s, the Metropolitan Police added the Anti-Terrorist Branch as another tool for tackling terrorism. The drive toward a single counterterrorism police force gained momentum only after the terrorist attacks in London on July, 7 2005 (7/7). The Home Office amalgamated the Anti-Terrorist Branch (ATB or SO13) with the Metropolitan Police Special Branch (MPSB or SO12) to form a new single body for counter-terrorism called the Counter Terrorism Command (also known as SO15), which came into existence as the UK's lead police counterterrorism force on October 2, 2006. The Counter Terrorism Command has a long list of duties; among other things, it is tasked with preventing and disrupting terrorist activity and bringing those engaged in it to justice, supporting the national coordinator of terrorist investigations outside London, gathering and assessing intelligence on terrorism and extremism, partnering with local communities, providing specialist security advice and services, providing a CBRN (Chemical, Biological, Radiological and Nuclear) capability in London, assisting MI5 and MI6, and interfacing with foreign police and intelligence agencies in counterterrorism matters (London Metropolitan Police, 2010). It was subsequently decided that the remnant of the Metropolitan Police Special Branch would perform intelligence gathering while former members of the Anti-Terrorist Branch would do investigative work. In the event of a terrorist attack, as occurred at Glasgow International Airport on June 30, 2007, the local police, who have jurisdictional control, take the initial lead in consultation with the Counter Terrorism Command. As the Glasgow example showed, SO15 has direct jurisdiction only in the greater London area (outside the City of London, which has its own police force) and will only operate in other jurisdictions if invited in by the chief constable of that local jurisdiction.

The special branches outside London focus on intelligence gathering, while, as noted above, the evidentiary-based case work is now carried out by SO15. SO15 currently has four regional offices which assist local SBs with terrorism investigations; in London SO15 is the primary counterterrorism evidentiary and case-based agency; and other SO15s assist special branches in investigating and building cases in local constabularies. The national coordinator for terrorist investigations is a deputy assistant commissioner of the London Metropolitan Police (Scotland Yard) who heads SO15 and is responsible for the police response to terrorism threats. Given the centrality of the police role in counterterrorism in the British context, it is the police, not the security service, that determines when investigations and intelligence-gathering operations need to become arrest operations, based on an assessment of the degree to which terrorists are becoming operational and constitute an actual threat—as opposed to simply planning an operation. Terrorism-financing investigations also involve a major police role, with the primary British entities dealing with such investigations being the BSS, the National Terrorist Financial Investigations Unit (NTFIU, a part of the special branch at the London Metropolitan Police), and the terrorist finance unit of the Serious and Organized Crime Agency (SOCA). Finally, coordination of counterterrorism investigations and surveillance between these national policing entities and the local constabulary is carried out, as noted previously, via counter terrorism units (CTUs) and counter terrorism intelligence units (CTIUs) and occurs under the auspices of a senior Metropolitan Police officer known as the Senior National Coordinator (Bassett, Haldenby, Thraves, and Truss, 2009, p. 13).

The British counterterrorism strategic framework, known as CONTEST (counterterrorism strategy), is based on four work streams: *pursue, prevent, protect*, and *prepare. Pursue* and *prevent* focus on reducing the threat from terrorism. The Terrorism Act and its more recent amendments, as well as other legislation, form part of *pursue*, as do information-sharing and cooperation efforts between law enforcement and intelligence agencies, enhanced technical capacities, overseas operations (military and covert) and efforts in the realm of terrorism prosecution and post-prison supervision. *Prevent* focuses primarily on trying to counter the conditions that may breed home-grown terrorism within the UK, including challenging radical ideologies, increasing the resilience of communities to radicalization, addressing economic and social grievances that extremists attempt to take advantage of, and disrupting the activities of radicalizers, and attempts to steer recruits to radical causes away from extremism. The *protect* and *prepare* streams are designed to reduce the UK's vulnerability to attack. *Protect* focuses on reducing vulnerability to terrorism in the context of critical infrastructures, public places, transportation systems, hazardous materials storage and transport, and border, port, and airport security. *Prepare* focuses on mitigating the impact of terrorist attacks. This includes a range of policies designed to increase local and regional resilience, enhancing response and recovery efforts to terrorist incidents and overall crisis management (Prime Minister and Home Secretary, 2009, pp. 12–15).

Canada

In Canada, responsibility for dealing with terrorism is shared between the country's only civilian intelligence service: the Canadian Security Intelligence Service (CSIS); the federal police force, the Royal Mounted Canadian Police (RCMP); and provincial and municipal police forces. Prior to the creation of CSIS in July 1984, the RCMP's Security Service was responsible for gathering intelligence on terrorist threats, but abuses carried out by the RCMP Security Service in the 1970s during the struggle with violent separatists in Quebec led the Canadian government to conclude that law enforcement and security work needed to be separated (Rosen, p., 2000, pp. 5–6). CSIS is only allowed to gather information on people suspected of engaging in espionage, sabotage, foreign-influenced activities detrimental to Canadian interests, subversion, or political violence and terrorism (Canadian Security Intelligence Service, 2005). CSIS, which is under the authority of the minister of public safety and emergency preparedness, interfaces with the rest of the intelligence and law enforcement community via the Integrated Threat Assessment Center (ITAC). Established in 2004, ITAC produces threat assessments for all levels of government, from first-responders to critical infrastructure entities in the private sector. ITAC includes representatives from CSIS, the RCMP, and other law enforcement bodies, and a range of other governmental agencies, including the correctional service, Transport Canada, and other agencies. Interestingly, Canada does not have a foreign intelligence service, and foreign intelligence is not technically within the CSIS remit— although in attempting to cope with terrorism threats, through the authority provided by Section 12 of the Canadian Security Intelligence Service Act, CSIS also gathers intelligence overseas if it is of relevance to protecting Canadian security (Canadian Security Intelligence Service, 2008, p. 24). CSIS is mandated to conduct investigations and gather intelligence, among other areas, in the field of political violence and terrorism as well as conducting security assessments. The Canadian intelligence community also includes the Intelligence Division of the Department of National Defense (DND) and the Communications Security Establishment (CSE), Canada's primary SIGINT agency. Both of these entities are under the responsibility of the minister of national defense. The Canadian military also possesses a counterterrorism unit tasked with dealing with hostage situations or other counterterrorist operational actions.

The RCMP, as Canada's national police force, acts as the primary law enforcement counterpart to the CSIS and under the Security Offences Act has lead jurisdiction in investigating offenses that affect the security of Canada. The RCMP, which, like the CSIS, is overseen by the ministry of public safety and emergency preparedness, also leads integrated national security enforcement teams (INSETs). INSETs share and analyze information on persons

FIGURE 2.10 Graduates of the Royal Canadian Mounted Police Academy. This Wikipedia and Wikimedia Commons image is from the user Brian Dell and is freely available at http://commons. wikimedia.org/wiki/File:Graduating_Mounties.jpg and has been released into the public domain.

deemed a threat to national security and work to apprehend targets. The RCMP provides a unique model, as it serves as a national police force with jurisdiction across Canada, but is also contracted as a provincial police force (with the exception of the provinces of Quebec and Ontario). The RCMP also contracts with over 200 municipalities in Canada to provide law enforcement at the local level (more on this in Chapter 3). Consequently, the RCMP plays a role at all jurisdictional levels in Canada, and RCMP officers have law enforcement powers in all parts of Canada.

In terms of overall strategy, Canada's national security policy is based on three pillars: protecting Canadians and Canadian territory, ensuring that Canada is not a base for threats to its allies, and contributing to international security. As is evident from these pillars, Canada attaches great importance to trying to prevent its territory from serving as a staging point for attacks against the United States. This is not surprising given Canada's dependence on trade with the United States and its keen interest in keeping its southern border open for unimpeded trade with its massive neighbor.

Australia

The Australian counterterrorism community includes all the major intelligence and law enforcement agencies. On the intelligence side, this includes the Australian Security Intelligence Organization (ASIO), which reports to the attorney-general. Modeled on Britain's BSS, ASIO is a domestic intelligence and security service without arrest powers. Its missions in the counter-terrorism context include intelligence collection and provision of protective security in key critical infrastructure sectors. Although ASIO also has non-terrorism-related missions such as counterespionage, its focus has shifted increasingly to counterterrorism. In 1998, some 35 to 40 percent of the agency's resources were devoted to the counterterrorism mission, but this had risen to 70 percent by 2004 (Flood, 2004, p. 77). Over time, ASIO has also taken on an increasingly law enforcement–oriented role, as it has been called upon to gather evidence for criminal prosecutions, and it has been argued that ASIO could eventually move into a role more akin to that of the U.S. Federal Bureau of Investigations (Pearlman, 2009). ASIO is in the process of setting up a multiagency counterterrorism control center to manage counterterrorism priorities, identify intelligence requirements, and oversee the process of collection and distribution of counterterrorism intelligence products (Australian Department of the Prime Minister and Cabinet, 2010, p. 28). Other intelligence services with a significant counterterrorism mission include Australia's external intelligence service, the Australian Secret Intelligence Service (ASIS, which is under the purview of the Foreign Ministry), the primary military intelligence-gathering and assessment agency, the Defense Intelligence Organization (DIO); and Australia's premier electronic intelligence agency, the Defense Signals Directorate (DSD). In the wake of the September 11, 2001 attacks in the United States, Australia established a liaison forum to link the intelligence community with the law enforcement community and to help facilitate the transfer of intelligence information to investigations (Chalk, 2009, p. 30). This entity is known as the Joint Counter-Terrorism Intelligence Coordination Unit (JCTICU). Interestingly, Australia also has an ambassador for counterterrorism, who is responsible for building international contacts and leading negotiations on counterterrorism agreements with other countries.

On the law enforcement side, the Australian Federal Police (AFP) serves as a national police force and also as the chief link between ASIO and the rest of the intelligence community as well as the municipal and state police forces (Australia's Northern Territory is, in effect, now a state, although it still maintains the territory designation in its name). In the counterterrorism and information-sharing context, the AFP cooperates with state police forces via joint counterterrorism teams (JCTTs), of which, as of 2006, there were 12, with one in each capital city. The AFP is also active globally as the sole international representative of Australian law enforcement and has contributed, along with ASIO personnel and state law enforcement agencies, to investigating the 2002 Bali bombings (which killed 98 Australians) and possesses a rapid-response force (i.e., scene forensics, intelligence support, financial investigation assets, etc.) that can be deployed outside the country when needed (Australian Federal Police, 2006, pp. 25–26). The AFP also maintains a counterterrorist first response (CTFR) capability at 11 major Australian airports and protects government offices, embassies, and certain critical infrastructure sites.

In terms of strategic counterterrorism policy, the commonwealth (federal) government is responsible for determining overall national counter-terrorism policy as well as for overseeing some of the key counterterrorism bodies, including ASIO and the AFP. Canberra also supports state counter-terrorism response, determines the national counterterrorism alert level, and has the authority to declare a *national terrorist situation* (Australian Govern-ment, 2005). The commonwealth government will, in consultation with state governments, decide whether to declare a national terrorist situation based on the nature of the incident, whether it is multijurisdictional, the effect on civil aviation, the maritime sector or other critical infrastructures, and other factors. If such a status is declared, the commonwealth government takes overall responsibility for policy and strategy throughout the country. The factors that may lead to the declaration of a national terrorist situation have to do with (1) the scale and nature of the incident, including whether or not CBRN materials are involved; (2) whether or not commonwealth interests are involved; (3) the scope and significance of the threat; (4) the impact on civil aviation or maritime operations; (5) the involvement of critical infra-structure; and (6) the involvement of foreign or international interests (Western Australia Police Service, n.d.). The commonwealth government also sets the national alert levels (low, medium, high, and extreme). The national alert level is not necessarily linked to ASIO threat assessment levels with respect to particular persons, places, sectors, and so on, and conse-quently, different alert measures will be put in place for those areas, sectors, or persons in question, irrespective of the national alert level (Australian Government, 2005, p. 9).

The state governments are responsible for maintaining counter-terrorism plans and have primary operational responsibility in responding to terrorist incidents within their respective jurisdictions, as spelled out in the Agreement on Australia's National Counter-Terrorism Arrangements of December 2002 between the commonwealth and state governments. Interestingly, in keeping with the strong federative character of the Australian Commonwealth, Section 10 of the Agreement allows either the federal or state governments to withdraw from the agreement, although it is unclear what effect this will have on terrorism investigations and counterterrorism policy (Agreement on Australia's National Counter-Terrorism Arrangements, 2002).

The commonwealth government's role is to provide particular capabilities that state/territory governments do not possess, such as the intelligence organizations, the military, border security, and some emergency management capabilities. Australia has a range of intergovernmental committees responsible for coordinating counterterrorism policy and response, including at the commonwealth level, the National Security Committee of the Cabinet (NSC), the National Crisis Committee (NCC), the Australian Government Counter-Terrorism Policy Committee (AGCTPC), and the Australian Government Counter-Terrorism Committee (AGCTC). At the state level, there are state crisis committees (S/TCCs) that support state/territory policies and coordinate information and strategic communications. Finally, there are interjurisdictional committees that bring together commonwealth and state officials, such as the Council of Australian Governments (COAG), which is the primary intergovernmental forum, as well as the National Counter-Terrorism Committee (NCTC) and a range of security, police, emergency management, health, and transport security committees (Australian Government, 2005). Finally, each state maintains its own committees, plans, and mechanisms to mitigate and respond to acts of terrorism. New South Wales, for example, maintains policy committees at the level of the state cabinet and a counterterrorism disaster recovery directorate and a state emergency management committee to coordinate and implement policy in the area of law enforcement and emergency response (New South Wales Government Counter Terrorism Plan, 2008, pp. 6–7).

Germany

The key elements of Germany's counterterrorism policy include arresting terrorists and their supporters and breaking up their infrastructure; assisting countries in danger of becoming failed states; addressing the social, economic, and cultural roots of terrorism; and combating the proliferation of weapons of mass destruction. This policy is balanced by the high priority given to the protection of the basic rights of suspects (both citizens and noncitizens) (Miko and Froehlich, 2004, p. 2).

Among the countries surveyed, Germany has arguably the most diffuse intelligence and law enforcement structure, based on its profoundly federalist political and administrative structure. Moreover, given its history, Germany has maintained the principle of Trennungsgebot (the separation of police and intelligence powers) as part of a broader policy of the prevention of the aggrandizement of power in the hands of any single institution. Both the federalist principle and that of the separation of policing and intelligence are reflected in the role and powers of the federal domestic intelligence service, the Bundesamtes für Verfassungsschutz (Federal Office for the Protection of the Constitution—BfV). The BfV is primarily an intelligence-coordinating body and has limited intelligence-collection powers. The primary role of the BfV is to gather and analyze information on groups and individuals that threaten the existence or security of the Federal Republic of Germany or its free and democratic order. It is also tasked with counterespionage intelligence gathering, the gathering and analysis of intelligence on threats to German interests overseas, and the gathering and analysis of intelligence against individuals and groups that threaten "international understanding, in particular peaceful co-existence" (German Federal Ministry of the Interior, 2005, p. 11). In addition, each of Germany's 16 Länder (states) has a regional intelligence organization known as Landesämter für Verfassungsschutz (LfV), and these have the same status as the BfV, so that federal intelligence coordination does not supersede that carried out by the states, and in many cases, the LfVs have greater intelligence-collection powers than those of the BfV (Warnes, 2009, pp. 93, 102). Moreover, if the BfV wants to conduct intelligence-collection operations, it almost always first secures the cooperation of the relevant LfVs in whose jurisdiction the operation will be carried out.

The BfV and the 16 LfVs enter information into a federal intelligence database, the Nachrichtendienstliche Informationssystem (Intelligence

FIGURE 2.11   Logo of the German Federal Office for the Protection of the Constitution (BfV). This Wikipedia and Wikimedia Commons image is from the user Fornax and is freely available at http://commons.wikimedia.org/wiki/File:Bundesamt_f%C3%BCr_Verfassungsschutz_Logo.svg and has been released into the public domain.

Service Information System—NADIS). However, due to the principle of separating intelligence gathering from policing, the primary law enforcement agencies, the Bundeskriminalamt (federal criminal police—BKA) and the Landeskriminalamt (state criminal police—LKA), cannot access this database and have no access to the NADIS system (Warnes, 2009, p. 100). Both the BKA and the LKA are primarily investigatory agencies, similar in that respect to the U.S. Federal Bureau of Investigation. This system comprised of a federal domestic intelligence service and 16 state counterparts, each independent of the other, has been recognized as grossly inefficient, yet despite calls from senior politicians and officials to merge the LfVs with the BfV, state governments have effectively blocked tighter centralization on the grounds that this would impinge on their autonomy. A law passed in 2009 provides the BkA with the authority to investigate terrorist threats that either involve more than one state or in the event that the state's LkA requests BkA support (Gujer, 2010, p. 72).

The other major intelligence agencies that have a counterterrorism role are the Bundesnachrichtendienst (Federal Intelligence Service—BND), which focuses on foreign intelligence and the Militaerischer Abschirm-dienst (Military Security Service—MAD), which focuses on the security of military forces. As far as policing is concerned, the federal government, via the Ministry of the Interior, operates two primary police forces, in addition to the investigatory-based BKA, there is the Bundespolizei (Federal Police—BPOL). Just over half of BPOL's 40,000 personnel are involved in border protection, railway policing, or aviation security, with the remainder serving as rapid reaction units, intelligence analysts, and members of special units such as the counterterrorism force, the GSG-9, or air marshals (German Federal Police, 2005, p. 3). Each German state (Land) has its own policing agencies in addition to its respective investigatory-based LKA. Recognizing the difficulties in coordination between federal and state authorities, between states and between federal and state law enforcement bodies and their respective intelligence counterparts, the German government established a Joint Counterterrorism Center (Gemeinsames Terrorabwehr Centrum or GTAZ) in 2004 that brings together 39 agencies to facilitate the exchange of information—although this is not an operational body and has no command authority (Gujer, 2010, p. 73).

France

In France, the primary counterterrorism policymaking body is the Internal Security Council (Conseil de Sécurité Intérieure), which is chaired by the country's president. The council sets the priorities and then a number of committees, such as the Interministerial Intelligence Committee, which is chaired by the prime minister, and the Interministerial Counter-Terrorism Committee, chaired by the minister of the interior, then coordinate activities and focus on setting goals at their level. The primary body that coordinates the operational response between and among the various agencies that deal with terrorism is the Unité de Coordination de la Lutte Anti-Terroriste (the Anti-Terrorism Coordination Unit—UCLAT), which is located within the Ministry of the Interior. At the operational level, the agencies that have a counterterrorism role reside primarily in the Ministry of the Interior, the Ministry of Defense, and the Ministry of the Economy, Finance, and Industry. The Ministry of the Interior oversees the civilian police forces and, consequently, the civilian law enforcement counterterrorism agencies. Most of the services that deal with counterterrorism are attached to the Direction Générale de la Police Nationale (National Police Directorate or DGPN). The DGPN also includes a dedicated force that focuses on counterterrorism judicial investigations, known as the Sous-Direction Anti-Terroriste (Anti-Terrorist Sub-Directorate or SDAT). In addition, the DGPN can call upon the resources of the Police aux Frontières (Border Police or PAF) and the Gendarmerie Nationale. The Gendarmerie is a military police force that is part of the Ministry of Defense (but, generally speaking, is under the operational control of the Ministry of the Interior), which polices much of the French countryside (which makes up approximately 90 percent of the land area and some 10 percent of the population).

The former counterterrorism domestic intelligence and policing agency, the Direction de la Surveillance du Territoire (Territorial Surveillance Directorate or DST) and the former multipurpose, non-law enforcement, domestic intelligence service responsible for monitoring extremist groups and observing economic and social trends, the *Direction Centrale des Renseignements Généraux* (Central Directorate of General Intelligence or DCRG) were merged in July of 2008 to form the *Direction Centrale du Renseignement Intérieur* (Central Directorate of Domestic Intelligence or DCRI). Since the DCRI is a police agency as well as an intelligence agency, it has the advantage of having judicial jurisdiction and law enforcement powers similar to those of the U.S. Federal Bureau of Investigation. DCRI officers normally gather intelligence in support of a decision to open a judicial inquiry and are then put under the authority of an investigating magistrate in order to gather evidence (Hodgson, 2006, p. 38). As the principal counterterrorism law enforcement and intelligence agency, the DCRI provides information to central government officials and regional prefects to help them set threat levels.

Finally, France's external intelligence service, the *Direction Générale de Sécurité Extérieure* (Directorate for External Security or DGSE) and the *Direction du Renseignement Militaire* (Directorate of Military Intelligence or DRM), both of which are part of the Ministry of Defense, as well as the *Direction*

Générale des Douanes et Droits Indirects (General Directorate of Customs Information and Investigations or DGDDI), which operates under the auspices of the Ministry of Finance, also play counterterrorism roles. The *Secrétariat Général de la Défense Nationale* (National Defense General Secretariat or SGDN) is the primary interministerial coordinating body for security matters. It reports to the prime minister, liaisons with the president's office, and helps coordinate committee work at the cabinet level as well as with working groups within the ministries.

CONCLUSION

The countries discussed in this chapter can be arrayed along the domestic counterterrorism continuum noted at the beginning of the chapter: from Israel, with its primary focus on counterterrorism as a war-fighting activity, to Canada, Australia, the UK, France, and Germany which view domestic counterterrorism almost exclusively as a law enforcement activity. With respect to the manner in which these countries define terrorism (which, of course, forms the basis for their legal action against suspected terrorists), all the countries surveyed in this chapter opt for some form of flexibility in designating terrorist organizations although they all focus on defining terrorism through the specific actions of members of terrorist organizations— with the caveat that those illegal actions are undertaken for the purpose of intimidation and/or destruction in order to achieve some sort of political, social, or religious agenda. In returning to Lord Carlile's typology of four counterterrorism legislative approaches, the countries surveyed essentially fall into only two of Lord Carlile's models, with Germany providing an additional approach. Israel, the UK, Canada, and Australia have all opted for a broad definition of terrorism with either specific counterterrorism legislation and amendments to existing criminal legislation, as in the case of Israel and the UK, or simply utilizing amendments to existing criminal legislation, as in the case of Canada and Australia. France, following a slightly different model, has a very specific definition of terrorist crimes in its criminal law but does not have special laws to deal with terrorism (although it does institute special procedures in dealing with terrorism offenses). Finally, Germany does not utilize a specific definition of terrorism and it uses existing criminal legislation in prosecuting terrorism crimes, but it does utilize special procedures in terrorism investigations and allows for harsher penalties in terrorism cases.

There is, naturally, a correlation between the overall approach taken toward combating terrorism by the countries surveyed in this chapter and the use of emergency legislation (with countries adopting a more military approach employing emergency legislation more extensively than those following a law enforcement approach). Consequently, Israel still relies quite heavily on emergency legislation, whereas since the demilitarization of counterterrorism operations in Northern Ireland in the mid-1970s, the UK and the other countries surveyed here have little recourse to emergency legislation (although it exists on the books for extreme situations of insurrection and the breakdown of social order).

With respect to pretrial detention or other restrictions on the freedom of movement of terrorism suspects, the countries in the survey differed quite significantly in policies and laws. Not surprisingly, Israel made the heaviest use of preventive detention as a way of preventing terrorism suspects from planning or carrying out attacks while evidence was being gathered against them for a judicial process. The UK and France, both countries that used preventive detention heavily in the past, prefer to employ a judicial-based system of precharge detention for terrorism suspects that allows for significant periods of detention, whereas Canada, Germany, and Australia essentially follow a standard criminal model of pretrial detention that allows minimal time for the police and public prosecutors to bring charges against suspects. In terms of investigations and surveillance, Israel and France arguably have the greatest leeway in opening investigations and carrying out surveillance of various kinds and can generally use evidence gathered via intelligence operations in court. Investigative and surveillance powers are more restricted in the British case and arguably more so in the Australian, Canadian, and German cases. Ultimately, countries with a history of a significant terrorism threat and a population that is accustomed to significant governmental powers (as with Israel, France, and arguably, the United Kingdom) generally have stronger tools and means to cope with terrorism than do countries such as Canada and Australia that have a tradition of greater mistrust of government (reflected, in part, through their federalist systems of government). Germany stands alone as an example of a state with comparatively weak tools for carrying out counterterrorism policies, due to the historically based German fear of governmental power.

From an institutional perspective, all of the countries surveyed in this chapter use a combination of intelligence and law enforcement agencies (although Israel also uses the military extensively) in dealing with terrorism. Each differs, of course, with respect to the nature of these agencies and their makeup and with respect to the balance between law enforcement and intelligence agencies. The French have opted for an approach similar to that of the United States through the creation of a domestic law enforcement agency that also has significant intelligence functions, whereas the UK, Israel, Canada, and Australia have opted for the model of separate domestic intelligence services that are not law enforcement agencies. The line between intelligence and law enforcement is perhaps starkest in Germany, where these functions and agencies are very clearly separated. Regardless of whether a given country separates law enforcement from intelligence institutionally, there is no question that the law enforcement function is critical to counterterrorism and to homeland security in general, and it is this function that is the topic of Chapter 3.

ISSUES TO CONSIDER

- What are the various legal strategies employed by the countries surveyed in this chapter with respect to dealing with terrorism?
- What are the pros and cons of employing a war-fighting approach to counterterrorism versus those involved in utilizing a law enforcement approach?
- How do different preventive detention regimes across the countries surveyed in this chapter work?
- How do the countries surveyed in this chapter utilize domestic intelligence services, and how do they utilize law enforcement agencies with respect to the overall counterterrorism mission?

LAW ENFORCEMENT INSTITUTIONS AND STRATEGIES

Policing organizations in the countries surveyed were mentioned briefly in Chapter 2. In the present chapter we focus in greater detail on policing agencies and on community policing efforts. It is recognized across all the countries surveyed in this book that law enforcement serves as a critical "force multiplier" in counterterrorism efforts, and traditional policing practices, such as community policing and intelligence-led policing, can sometimes be just as useful in tracking down terrorists as in tracking down traditional criminals. Moreover, law enforcement agencies play a critical role in disaster response and in response to public health emergencies. Consequently, looking at a few examples of law enforcement and community policing approaches and institutions among a group of countries in our survey will be instructive. In this chapter we look at a slightly different mix of countries because they possess diverse models of policing and provide a good overview of different types of approaches.

By way of introduction, there are three models of policing and policing organizations in democracies: Napoelonic, centralized, and decentralized. In *Napoleonic systems*, such as those of France, Italy, and Spain, military and civilian policing systems exist side by side (with the military policing system designed primarily to police civilians and both systems designed to counter-balance each other). Frequently, the civilian police also maintain a judicial arm (or judicial function) that deals with investigations and prosecutions. In *centralized police forces*, there is one civilian police agency with a single command structure deployed throughout the country, and in *decentralized police forces* there are a multiplicity of police agencies with no unified police command

Comparative Homeland Security: Global Lessons, First Edition. Nadav Morag.
© 2011 John Wiley & Sons, Inc. Published 2011 by John Wiley & Sons, Inc.

FIGURE 3.1 Nineteenth-century British police constables. This Wikipedia and Wikimedia Commons image is from the user Linda Spashett and is freely available at http://commons.wikimedia.org/wiki/File:Westgate_062a.jpg and has been released in the public domain.

structure. Decentralized police forces can be primarily local, with one or more of these local agencies also having national-level functions (as in the British model) or they can exist alongside federal agencies (as in Germany, Canada, Australia, and the United States). Roughly corresponding to these models are two different policing philosophies related to the ultimate objective of policing. The first, which generally corresponds to the Napoleonic model, views policing as designed primarily to defend the state and the social order. The second, which generally corresponds to the decentralized model, focuses much more on ensuring quality of life and the security of the citizenry (Brogden and Nijhar, 2005, p. 109). The centralized model lies somewhere in the middle in terms of having elements of both goals—which is not to suggest that the decentralized model is devoid of interest in ensuring the security of the state and the broader society or that the Napoleonic model is not concerned with quality of life and local security—just that the emphasis differs across the various models. The existence of particular models of policing in different countries, as with respect to counterterrorism policy, is a function of the history, laws, and institutional environments of these countries and also relates to the nature of the traditional role of policing and the nature of the criminal threat.

POLICING AGENCIES

Israel

As befitting a centralized state, Israel has a single national police force, Mishteret Yisrael (Israel Police) under the command of a commissioner

FIGURE 3.2   Israeli border policeman. This Wikipedia and Wikimedia Commons image is from the user James Emery and is freely available at http://commons.wikimedia.org/wiki/File:Israeli_Border_Guard_Police.jpg under the Creative Commons Attribution 2.0 Generic license.

who reports to the minister for public security. The force is organized into six districts (corresponding to Ministry of the Interior districts) and 13 subdistricts with 67 local police stations. In addition to law enforcement roles, the police develop legal cases, and police personnel with law degrees are responsible for prosecuting misdemeanors in magistrate courts.

Police personnel can be deployed anywhere in the country, thus allowing for a rapid response in the event of terrorist attacks or public disturbances. As noted in Chapter 2, the Israel police force includes a multipurpose paramilitary component known as *Mishmar Ha'Gvul* (the border guard or border police, and also known by the Hebrew acronym *Magav*), whose personnel wear military-style uniforms and carry assault rifles. Although, in principle, their role focuses more on the provision of security in many of the border areas and counterterrorism and public order duties, they have the same police powers as those of other police personnel and can be used for normal policing. Unlike most of the rest of the police force, the border police are often deployed in the West Bank to conduct patrols and counterterrorism duty, and when they do so, they serve under the military chain of command. Consequently, while the border police is not a military policing agency such as the French GN (see below), parts of it can be put under military command in certain types of deployments. The Israel police also possesses a rapid-response and public order unit known as the Special Patrol Unit (known most commonly by its Hebrew acronym, Yassam) and a SWAT counterterrorism and hostage rescue unit known as the Special Police Unit (also known most commonly by its Hebrew acronym, Yamam). Even though Israel has traditionally followed the centralized policing model, this has not prevented several Israeli mayors from lobbying for the creation of parallel local/community police forces under joint municipal and national police authority to deal with quality-of-life issues, neighborhood disputes, school security, enforcement of municipal bylaws, traffic safety and directing of traffic, and so on. In fact, traffic laws were amended in 1997 to allow municipal inspectors to direct traffic and enforce traffic and parking laws, contingent upon the approval of the police district commander. Several cities already run joint teams consisting of police officers and municipal inspectors who patrol and respond to calls. However, this arrangement, known as *integrated policing*, has not yet been legislated and consequently varies from area to area based on local traditions and "chemistry" (or the lack thereof) between the local governmental leadership and the local police commanders (Yehezkeli, 2001, p. 72).

As far as community policing efforts are concerned, the Israel police created the Community Policing Unit (CPU) in 1994 in order to coordinate cooperation between the police station personnel and the mayor's office in a given city (Haberfeld and Herzog, 2000, p. 63). The goals of the CPU are to develop a policing strategy appropriate to the each geographic region, developing models for community partnership, and publicizing the work of the police and the community through outreach to the media. The CPU trains police officers and community representatives and provides local police stations with advice as to who to contact in the community to develop working relationships, disseminates models of community policing in other localities as templates for other areas starting their own community policing efforts, and initiating research evaluating the effectiveness of various forms of community policing (Meyr, 1999, p. 105). Many of the cities participating in this program have established community policing centers (CPCs), which act as ministations within neighborhoods (somewhat reminiscent of the Japanese Koban, discussed below, although devoid of the all-encompassing role played by this institution in Japan). CPCs are used throughout Israel but especially in the Israeli–Arab sector, where they serve as a way of improving ties between the police and this ethnic minority, many of whose members are sometimes at odds with the government [12 Israeli Arabs were killed by the police during violent riots in Arab–Israeli areas in October 2000, corresponding to the earliest stages of the second Palestinian Intifadah (uprising)]. One study of community policing in Israel found that police supervisors and commanders were significantly more enthusiastic about community policing than the average patrol officer and that both police commanders and patrol officers were less involved in community policing over time. Overall, the study found that most police officers and commanders still view the professional model of policing (in which the police, not the community, plays the central role) as the most credible (Weisburd, Amir, and Shalev, 2001, pp. 55–56). The department responsible for community policing at the national headquarters of the police, which initially reported to the Israel police commissioner, has since been put under the auspices of the civil guard department—something that some authorities on community policing consider to be the organizational "kiss of death" to community policing in the country (Yehezkeli, 2001, p. 88).

Israel also enjoys a long and significant tradition of volunteerism in the realm of security, law enforcement, and emergency services. Community volunteers can serve as part of the civil guard (Ha'mishmar Ha'ezrahi), of which there are over 70,000 members. Civil guard personnel range from uniformed officers who have been trained and enjoy full police powers to neighborhood watch patrols (although in Israel, unlike the United States, these are typically armed). The types of civil guard services also include security patrols; school protection; rapid-response units; activities in the tourism, public transportation, and environmental protection sectors; bomb disposal; intelligence; and investigations. The civil guard operates on the basis of Section 49 of the Israel National Police Act of 1974 and the 1988 amendment to this law. The civil guard's duties are defined as assisting in the prevention of terrorist acts, organizing neighborhoods for emergency operations in the event of a terrorist attack, and assisting civil defense efforts during a mass conscription of reservists and in the event of war (Geva, 1995, p. 3).

In sum, Israel still follows a centralized model of policing, with one national police agency. Over time, there has been a slight chipping away of this model as mayors and heads of local and regional councils (responsible for towns and rural areas, respectively) have brought some lower-level law enforcement functions under their control. Ultimately however, given the security challenges facing Israel, it is unlikely that the flexibility afforded by a single national police force will be given up in the interests of regionalizing policing.

United Kingdom

In the UK, the policing model is largely decentralized, although all 52 territorial police forces are under the control of their agency's chief constable (or commissioner for the London Metropolitan Police and City of London Police), the local policing authority (two-thirds of whose members are elected councilors and one-third magistrates), and the Home Office (McBride and Collins, 2002, pp. 15–16). However, the Home Office's role is primarily in the field of influencing legislation and providing funding, and each of the territorial police forces has a large degree of autonomy. The police forces themselves do not operate under the auspices of local governments but, rather, draw their authority from the Crown, although they do usually have jurisdictional boundaries that conform to local government boundaries. The largest police force by far is the London Metropolitan Police (Scotland Yard).

FIGURE 3.3   London Metropolitan Police officers. This Wikipedia and Wikimedia Commons image is from the user Patrick Scales and is freely available at http://commons.wikimedia.org/wiki/File:Very_friendly_MPS_officers_in_London.jpg and has been released into the public domain.

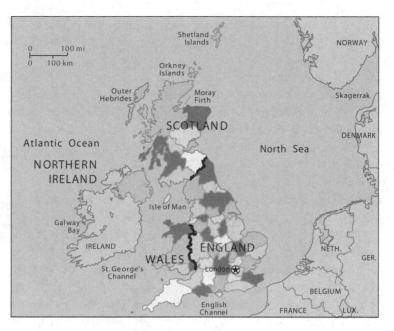

FIGURE 3.4 UK police divisions.

There are also a number of nonterritorial national policing agencies that deal with particular issues and, because of the increasing importance of these, the autonomy enjoyed by the territorial police forces (known as constabularies) is weakening (Lawday, 2001, p. 6). The Serious and Organized Crime Agency (SOCA) and the Counter Terrorism Command (CTC, also known as SO15) are examples of such national policing agencies. SOCA is a national investigatory agency focused on fighting organized crime and provides support to the territorial police forces (Serious and Organized Crime Agency, 2008, p. 7) and the Counter Terrorism Command, as noted in Chapter 2, investigates and builds cases against suspected terrorists. The CTC is part of the London Metropolitan Police (although its remit is national) and the London "Met" is also responsible for protecting the royal family, the government, and foreign dignitaries. The British Transport Police (BTP) is another example of a national policing agency, which in this case is responsible for policing the railway system, including the London Underground. Additional national policing units include the National Terrorism Finance Investigation Unit (NTFIU), the National Counter Terrorism and Security Office (NaCTSO), the Police International Counter Terrorist Unit (PICTU), and the National Extremism Tactical Coordination Unit (NETCU). There are also small specialized police forces with a national remit that protect the nuclear energy infrastructure (UK Atomic Energy Authority Constabulary) and Ministry of Defense installations (Ministry of Defense Police), as well as military police agencies. In addition to these police forces, there are four senior police officers that play a national role in coordinating constabularies in particular issue areas, including the national coordinator for special branch, the national coordinator of ports policing, the national coordinator for terrorist investigations, and the national coordinator for domestic extremism. Finally, the UK also has "deputized" security officers at airports, ports, and waterway authorities as well as wardens at national and local parks who are not police officers in the strict sense of the word, but are sworn in as constables and enjoy many police powers (McBride and Collins, 2002, pp. 17–18).

Police forces in the UK are required by law to support each other. This support is coordinated by the Police National Information and Coordination Center (PNICC), which coordinates mutual aid between police forces during incidents and keeps the central government apprised of developments. Mutual aid between the territorial police forces is coordinated by the Police National Information Center (PNIC), which is the central point for arranging the deployment of police support units (PSUs) and special services (such as armed police) from one constabulary to another. The PNIC is based at New Scotland Yard but is technically a unit of the Association of Chief Police Officers (Bassett, Haldenby, Thraves, and Truss, 2009, p. 13).

Overall, the general trend in UK policing seems to be toward greater centralization and the amalgamation of forces. For example, prior to 1967, there were 300 police forces in England and Wales and in reforms that year the number was cut to 43. The argument for greater amalgamation and centralization rests on the contention that modern police forces, which require significant resources to cope with a wide range of crimes as well as terrorism threats, need to be larger in order to take advantage of economies of scale (BBC News, 2005).

Community policing has a long history in the British context, stretching back to the establishment of the London Metropolitan Police by Sir Robert Peel in 1829. Peel considered the police to be "citizens in uniform," in contrast to the militarized police system popular on the European continent. Peel based his approach on a much older tradition (some 800 years old) of the *tythingman*, who was a member of the community elected by that community and empowered to collect taxes and maintain the peace. English common law granted the tythingman, and indeed any member of the community, arrest powers on the principle that any member of the community had the right to take any other member of the community to court (Brogden and Nijhar, 2005, p. 25).

At present, community policing efforts are based on the use of police community support officers (PCSOs), who are tasked with community interface, reassuring the public, supporting vulnerable populations, and dealing with antisocial behavior, quality-of-life issues, and low-level crime (see Sidebar 3-1)(National Policing Improvement Agency, 2008, p. 10).

Sidebar 3-1: PCSO Standard and Discretionary Powers

British Police Community Support Officers (PCSOs) enjoy both "Standard" and "Discretionary" law enforcement powers. Standard Powers include:

- Issuing fixed penalty notices for littering, dog control orders and cycling on a footpath.
- Requiring that individuals provide their name and address where they have reason to believe a person has committed a relevant offence or relevant licensing offence, for antisocial behavior, road traffic offences or possession of drugs.
- Confiscating alcohol from persons in designated places and from those under the age of eighteen.
- Seizing tobacco from those under sixteen years of age.
- Seizing drugs.
- Entering and searching a premises to save life or prevent serious damage to property.
- Seizing vehicles used to cause alarm or those that are abandoned.
- Controlling traffic and carrying out road checks.
- Enforcing cordoned areas and conducting stops and searches in authorized areas under the *Terrorism Act 2000*.

PCSO Discretionary Powers include:

- Issue fixed penalty notices for disorder, truancy, excluded pupil in public place, dog fouling, graffiti and flyposting (5 separate powers).
- Detain person (for 30mins if awaiting an officer or can accompany to police station if requested) who does not give name and address when required to do so.
- Enforce byelaws.
- Deal with begging.
- Enforce certain licensing offences.
- Search detained people for dangerous items.
- Use reasonable force in relation to a detained person or to prevent a detained person making off.
- Disperse groups and remove under 16s to place of residence.
- Remove children contravening bans imposed by a curfew notice to place of residence.
- Remove truants to designated premises.
- Search for alcohol and tobacco.

- Enforce park trading offences.
- Enter licensed premises (limited).
- Stop vehicles for testing.
- Direct traffic for the purposes of escorting abnormal loads. (UK Home Office, 2008, p. 18)

PCSOs deal with matters such as vandalism, crowd control, support for victims of crimes, and community regeneration, and they enjoy limited policing powers to detain individuals until a constable arrives, to direct traffic, and to issue fines for antisocial behavior. The Police Reform Act of 2002 grants PCSOs the power to give traffic fines and fines for "dog fouling" and littering as well as to confiscate alcohol from people consuming it in public places, to seize tobacco from underage persons, and to carry out traffic checks and seize vehicles. There are currently some 16,000 PCSOs in England and Wales, and 13,500 sergeants and constables have been assigned to neighborhood policing duties across those areas of the UK (UK Home Office, 2008, p. 15). Section 38 of the Police Reform Act 2002 allows a chief constable to assign up to 53 policing powers (not including enforcement of local authority bylaws) to PCSOs, and there is a standard list of 20 powers that, at a minimum, constitute the powers of PCOs across England and Wales (the policing system differs in Northern Ireland, and criminal laws and the judicial system differ in Scotland).

The British approach to community policing involves employing what is referred to as the *extended police family* and includes police officers, PCSOs, and wardens (which may be variously defined as local government officials, private security officers, housing association officials, and others). Most of the UK police agencies also make use of special constables, part-time volunteers trained and equipped by the police who enjoy full police powers. UK police agencies can also call upon police support units (PSUs), which are trained and equipped to deal with public disturbances.

Unlike PCSOs, special constables are police volunteers who enjoy full police powers and carry out many of the same duties as those of full police officers. However, unlike regular police officers, who can exercise police powers anywhere in the country, special constables can generally exercise police powers only within their own police area and those immediately bordering their area (Almandras, 2008, p. 2). As part of special constable recruitment, some UK policing authorities have established employer-supported policing (ESP) programs, which involve a partnership between a number of constabularies and local employers in which employees are allowed to train as special constables and are given paid time off by their employers to carry out their duties and receive training (see Sidebar 3-2).

In exchange, the special constables working under this arrangement patrol local areas or premises relevant to their employer's industry with six industry-specific programs in operation, including ShopWatch for retail areas, BoroughBeat for local government, CampusWatch for universities, HospitalWatch for hospitals, and BusBeat for the bus system.

Sidebar 3-2: The Met Employer-Supported Policing ESP

ESP began in the retail sector in London's Oxford Street in November 2004, following a pilot in Camden. As part of 'ShopWatch', retail employees, with the backing of their managers, are invited to train as Specials. Their workplace allows them paid time off to carry out their duties. There are now 15 live 'ShopWatch zones' where one or more staff members are trained Specials. The Met has more than 500 Specials who are part of the ESP programme, including police staff who are also supported for the same one day a fortnight. Neil Barrett, head of Employer Supported Policing for the Metropolitan Police and the National Advisor on ESP, says: "We want to make this as big as we can in London. There are still a lot of sectors that are untapped. We can also help other forces follow the same path and use the same high-quality national schemes. They don't have to start with a blank page, as we did four years ago." Hampshire, South Wales, West Midlands, Sussex and Suffolk are just a few of the 20 forces that are considering the scheme. Merseyside recently launched ShopWatch in Liverpool in the run-up to its year as European Capital of Culture in 2008.

ESP Developments

The main ESP scheme covers staff from organisations in a wide range of industries. They patrol local areas or premises relevant to their employers' industry. Within the overall programme, there are now six industry-specific schemes: including ShopWatch for retail staff; BoroughBeat for council staff; CampusWatch for students and staff; HospitalWatch for hospital staff; ArtBeat for those with expertise in the art world and BusBeat for bus drivers and depot supervisors. As many as a quarter of Specials in London are recruited through ESP, including civilian police staff who joined in this way. "It's important to get rock-solid support from businesses before you think about recruiting staff," says Neil. "It's a great way to grow your strength and use the skills of people who work in a wide range of industries and have a body of relevant knowledge, as well as access to intelligence. These people know their industry well."

As part of its community policing efforts, Crime and Disorder Reduction Partnerships (CDRPs) have been set up in England and corresponding Community Safety Partnerships (CSPs) have been created in Wales. There are currently over 370 CDRPs and CSPs. CDRPs and CSPs bring together police, fire, local governmental authorities and community organizations to devise strategies and obtain buy-in for crime reduction measures ("The Met," 2008, p. 9).

Overall, then, the British model is largely one of partial decentralization, with regional constabularies augmented by national policing agencies, national-level coordinating officers, and specialty units belonging to the London Metropolitan Police that provide countrywide services.

Canada

Canada operates on the basis of a semidecentralized model with three categories of police agencies, corresponding to the federal, provincial, and municipal levels of government. The Royal Canadian Mounted Police (RCMP) is a federal policing agency focused, in the first instance, on enforcing federal laws (most laws relating to criminal justice matters are, however, provincial). The RCMP, in its federal law enforcement capacity, focuses on the enforcement of federal laws in such issue areas as smuggling, drug crimes, commercial and securities fraud, organized crime, and immigration violations and also fills an international liaison and training function and provides protective services to federal officials and foreign dignitaries. At the same time, the RCMP also provides contract police services to eight Canadian provinces and the country's two territories (thus enforcing provincial and territorial law, as well as federal law, in those jurisdictions). When operating as provincial police, the RCMP functions under the jurisdiction of the provincial attorney-general (even though its command structure is national). The RCMP does not operate as the provincial police in the provinces of Ontario or Quebec, as these provinces have their own police agencies. These two provincial policing agencies also provide municipal policing services in areas where municipalities are not required to maintain their own police force. Finally, at the municipal level, there are over 150 police agencies, and many more towns and cities contract with the RCMP to provide municipal policing services. Canada also has over 50 tribal (indigenous Canadians are known as "First Nations") policing agencies. In parallel with the RCMP, non-RCMP local and provincial police also have the authority to enforce federal statutes (Reichel, 1999, p. 180). The Canadian model thus provides a significant degree of flexibility because the type of law being enforced and the law enforcement entity enforcing that law are not necessarily from the same level of government. Consequently, Canada possesses federal, provincial/territorial, and municipal laws, and all of these laws can be enforced by the RCMP (where it possesses authority to conduct provincial and municipal policing) as well as by the provincial police within one of the two provinces where they exist and the municipal police of a given city or town.

Community policing in Canada has its origins in the fact that Canada, like the United States, was a frontier society as late as 130 years ago. This meant that when RCMP officers attempted to enforce the law in the vast frontier

FIGURE 3.5 Vancouver police officers. This Wikipedia and Wikimedia Commons image is from the user Bobanny and is freely available at http://commons.wikimedia.org/wiki/File:Bikecops. jpg and has been released into the public domain.

regions, they were entirely dependent on community support for food, horses, and shelter and thus had to work with the community rather than impose the law on the community. Moreover, the self-reliance necessary for frontier life meant that neighbors were expected to help each other, and the police could call upon the local community for help in periods of crisis (Lindsay, 2003, p. 44).

Germany

Germany acts as a good example, again befitting its federalized governing structure, of a decentralized policing model (although not nearly as decentralized as the United States). The German constitution (Grundgesetz) grants states significant powers in virtually all realms of governance, and this naturally includes law enforcement. Accordingly, each of the 16 German states (Länder) has its own police force under the authority of that state's Ministry of the Interior. While state police forces enforce laws and state criminal courts pass judgment, state criminal law must comply with federal legislation, and states enforce federal laws as well as their own. Each state has three categories of police forces: municipal police (Schutzpolizei), who serve as first responders and handle basic aspects of law enforcement and investigations; criminal police (Kriminalpolizei), plainclothes detectives

FIGURE 3.6 German federal police (Bundespolizei) patrol car. This Wikipedia and Wikimedia Commons image is from the user Mattes and is freely available at http://commons. wikimedia.org/wiki/File:BPOL-BMW-blau-silber.JPG and has been released into the public domain.

who handle criminal investigations, surveillance, and similar duties; and readiness or standby police (Bereitschaftpolizei), who are officers-in-training and live in barracks (and consequently can be called upon as a rapid-response force to deal with civil disturbances or disasters). At the federal level, the federal police (Bundespolizei, formerly known as the federal border police: Bundesgrenzschutz), under the federal ministry of the Interior, carry out airport, port, and railway security and respond to major disturbances or disasters. The federal government can also call upon a police investigative service (similar to the state criminal police and referred to in Chapter 2) known as the Federal Criminal Investigation Office (Bundeskriminalamt or BKA), which provides support to the states in matters of criminal investigations, forensics, and the provision of information. The federal police can also be used to reinforce the state police if requested to do so by the state government.

With respect to issues of community policing, the German response to rising crime levels in the 1980s was, among other things, to form "prevention councils", and contact police officers [Kontachbereichtsbeamter (KoB)]. Each officer is responsible for policing a particular area, and duties include interface with the public, crime prevention, traffic control, and accident prevention (Dammer and Fairchild, 2006, p. 127).

France

Broadly speaking, the police in France, in keeping with the Napoleonic model of policing, act on behalf of the state, whereas in the British model of policing, the police are said to act on behalf of the public, a principle known as *policing by consent* (Hodgson, 2006, p. 11). In France, policing, like many other things, is largely centralized. There are two primary police organizations, the Police Nationale (PN) and the Gendarmerie Nationale (GN). The PN is a civilian agency overseen by the Ministry of the Interior while the GN is a military policing organization which, until 2002, was under the authority of the Ministry of Defense but is now also under the authority of the Ministry of the Interior. The Ministry of the Interior is responsible, among other things, for maintaining public order and operates via préfets (prefects), who are responsible for maintenance of public order in their own départments (regions) and can request that the minister of the interior order police forces in their respective regions to deploy in order to deal with public order

FIGURE 3.7 French national police officers in riot gear. This Wikipedia and Wikimedia Commons image is from the user David Monniaux and is freely available at http://commons. wikimedia.org/wiki/File:CRS_tenue_maintien_ordre_p1200484.jpg under the GNU Free Documentation License, Version 1.2.

problems (Lawday, 2001, p. 22). The PN is primarily responsible for urban areas and the GN for small towns and rural areas. GN personnel live in barracks and, in addition to patrolling rural areas, perform specialized duties such as protecting public officials and public events, airport and port security, and overseas missions. More recently, and only in some cities, municipal police forces, Police Municipale (PM), have been created, but these are usually small and have fairly limited duties. PM agencies are autonomous of central government and under the control of their respective mayors (although their selection processes and training are similar to those of the PN) and their authority varies depending on the municipal policy. In some cities they carry arms and fight crime, and in others they are primarily responsible for the maintenance of order (Reichel, 1999, p. 177).

The French also make a distinction between *administrative policing* and *judicial policing*, the former category dealing with provision of security and maintenance of public order and the latter dealing with investigation of serious forms of criminality. Most PN and GN officers can fulfill both roles (although judicial policing is usually carried out by police personnel from the officer and NCO ranks and is overseen by the Ministry of Justice). Moreover, prosecutors and some government officials (such as mayors) can also act as judicial police in investigating serious offenses and ordering the detention of suspects.

In France, community policing has only fairly recently been seen as an important policing objective. Traditionally, the PN and GN viewed community policing as peripheral, and after the creation of municipal police forces in the early 1980s, these police services acted as the closest thing to community policing agencies and represented a kind of auxiliary police that allowed the national police agencies to deal with what were viewed as the central policing missions. Of the two national agencies, it was actually the military-based GN that was usually in closer contact with citizens, due to its deployment in most suburbs, small towns, and the countryside and the fact that the local *gendarme* often processed an entire case, from the complaint to the investigation to the interface with the prosecutor (Brogden and Nijhar, 2005, p. 111).

The current French community policing policy, known as *policing de proximité* (proximity policing), was introduced in 1999. This policy involves a cooperative venture between the police and local private and public agencies and establishes greater social service functions for the police. It is based on five operational principles: (1) increased visibility of police patrols and provision of autonomy to local police stations to deploy their personnel based on local needs, (2) increased responsibility for officers at all levels, (3) the expectation that officers be able to perform a broad range of policing and community–interface tasks, (4) permanent contacts and cooperation with nonpolicing agencies, and (5) better "customer service" with respect to serving the public (Dupont, 2002, p. 4). There is, however, evidence to suggest that French police agencies still primarily follow the professional model of policing. The police usually adopt new policies without properly notifying the public and rarely consult with the public. As one study notes, "police attendance at community meetings generally takes the form of announcing steps being taken, explaining how the police will enforce certain laws or ordinances, and taking note of complaints. Public involvement is minimal and is limited to receiving information. The members of the community usually have no control or oversight in that process" (Jones and Wiseman, 2006, p. 4).

Police reform efforts that created the proximity police approach also led to an increase in recruitment of special constables (police auxiliaries), who are young people recruited for five year periods. These special constables wear police uniforms with their own insignia and do not have legal powers to conduct investigations or make arrests. These positions have allowed larger numbers of ethnic minority youths to be part of the police force, thus making the police more diverse and allowing it to interact more closely with different communities (Dupont, 2002, p. 5).

One of the primary problems in instituting community policing strategies in France had to do with the fact that the natural partners for police in any community policing effort—the regional and municipal governmental agencies—were themselves part of centralized structures (given the high degree of centralization in France). Since these agencies were part of national-level entities, they felt less need to cooperate with the police at the local level if it did not serve their institutional interests more broadly. The central government's answer to this problem was to form a new administrative strategy that would integrate the relevant governmental and nongovernmental actors at the local level. In the context of this new strategy, police, judges, prosecutors, corrections officials, educators, health and social services administrators, municipalities, housing authorities, public transport operators, and so on, often sign a local safety contract (Contrat Local de Sécurité—CLS). The CLS outlines priorities for the maintenance of security and the roles that each signatory must play (Brogden and Nijhar, 2005, p. 114). This thus represents a devolution of sorts of centralized state power to the local level that is somewhat unusual given the centralized nature of French governance. Thus far, over 700 CLS agreements have been signed in France. The process begins with the identification of the relevant entities that will be participating in the CLS by the prefectural authorities. Once this is completed, an outside consulting firm or group of academics is commissioned to ascertain the primary problems, including issues such as crime rates, school attendance, urban decay, and fear of crime. Once the problems have been identified, the parties are invited to negotiate a set of common objectives and the role that each entity will play, and this forms the basis of the CLS (Dupont, 2002, p. 6).

Japan

Japan is an interesting case, due primarily to the nature of its community policing strategy. The country has one centralized police agency, the National Police Agency (NPA), which is the central supervisory agency for the Japanese police system and which oversees seven regional police bureaus that liaise with the 47 prefectural police agencies and large local agencies (e.g., the Tokyo Metropolitan Police, which is the country's largest police force). These prefectural and local police agencies do the actual work of policing in Japan (although the NPA does engage in intelligence gathering). Typically, high-ranking officers in prefectural and local police forces are employees of the NPA (and are rotated from police force to police force), whereas the lower-level officers and rank and file are employed by their

FIGURE 3.8 Japanese police Koban in Kyoto. This Wikipedia and Wikimedia Commons image is from the user Chris Gladis and is freely available at http://commons.wikimedia.org/wiki/File: Kyoto_Koban.jpg under the Creative Commons Attribution 2.0 Generic license.

particular police force. Consequently, while Japan has a centralized police structure and force, that centralization is expressed primarily in matters of resources, planning, and the senior command, while there is a considerable degree of decentralization and localization with respect to actual policing functions (Reichel, 1999, p. 168).

The Japanese have long viewed community policing as an integral part of law enforcement. At the heart of Japan's community policing structure are the urban neighborhood police posts (Koban) and the village police posts (Chuzaisho). Officers assigned to these posts serve as first-responders when crimes or accidents occur, provide instruction to the local community on prevention of crime and accidents, and provide administrative and social services (i.e., provide directions to visitors, handle lost and found articles, provide counseling services for residents, etc.). Japanese police are seen as multipurpose public servants and it is not unusual for them to act as counselors for troubled individuals, couples with marital problems, people with money problems, and parents with rebellious children (Dammer and Fairchild, 2006, p. 120). Japanese police interface regularly with other governmental organizations and the private sector to decrease the likelihood of criminal activity. Thus, they will lobby for the construction of pedestrian overpasses, request that the sanitation department pick up trash and abandoned items, and ask business owners not to serve children during school hours (Bayley, 1991, p. 85). Essentially then, a Koban/Chuzaisho policeman (there are few such policewomen) is more like a postman than a firefighter responding to calls and spends most of his time on a daily round consisting of low-key activities.

In criminal matters, Koban officers are tasked with documenting crime scenes and tracking down witnesses, but they rarely deal with arrests. Decisions to arrest are usually made at the central police station, where the primary processing of suspects also occurs. Compared to other developed countries, low crime rates in Japan seem to suggest that this community policing model is successful (although there is some dispute regarding the accuracy of Japanese crime data, as, for example, domestic violence crimes are rarely dealt with as a police matter). There are, however, criticisms of the model. Some suggest that due to the less desirable nature of these duties, the Koban/Chuzaisho are repositories for incompetent veteran officers and young, inexperienced officers (Brogden and Nijhar, 2005, p. 91). Moreover, Japanese society is largely collectivist and conformist, and the public is expected to cooperate with the police in reporting on irregular behavior. Overall, Japan's system of community policing is seen as a model to be considered, if not emulated, and Japan manages to have a centralized police force that is also able to focus locally and be responsive to local needs.

European Union

The EU has a number of unique policing powers and responsibilities, one of which is its own arrest warrant (see Sidebar 3-3).

Sidebar 3-3: Policing in Europe: The European Arrest Warrant

A European Arrest Warrant, valid throughout the European Union has replaced extradition procedures between Member States of the enlarged Europe. Such a warrant may be issued by a national issuing judicial authority if the person whose return is sought is accused of an offence for which the maximum period of the penalty is at least a year in prison, or if he or she has been sentenced to a prison term of at least four months. A decision by the judicial authority of a member state to require the arrest and return of a person should therefore be executed as quickly and as easily as possible in the other Member States of the European Union. The European Arrest Warrant means faster and simpler surrender procedures and no more political involvement. It also means that Member States can no longer refuse to surrender to another Member State their own citizens who have committed a serious crime, or who are suspected of having committed such a crime in another EU country, on the ground that they are nationals. Simplifying and improving the surrendering procedure between EU Member States was made possible by a high level of mutual trust and cooperation between countries who share the same highly demanding conception of the rule of law.

The Member States of the European Union were required to introduce legislation to bring the European Arrest Warrant (EAW) into force by 1 January 2004. On 13 June 2002, the EU Council of Ministers adopted a *Framework Decision on the European Arrest Warrant* and the surrender procedures between Member States of the European Union.

An EAW may be issued by a national court if the person whose return is sought is accused of an offence for which the penalty is at least over a year in prison or if he or she has been sentenced to a prison term of at least four months. Its purpose is to replace lengthy extradition procedures with a new and efficient way of bringing back suspected criminals who have absconded abroad and for people convicted of a serious crime who have fled the country, in order to forcibly transfer them from one Member State to another for conducting a criminal prosecution or executing a custodial sentence or detention order. The EAW enables such people to be returned within a reasonable time for their trial to be completed or for them to be put in prison to serve their sentence.

The EAW is based on the principle of mutual recognition of judicial decisions. This means that a decision by the judicial authority of a member state to require the arrest and return of a person should be recognized and executed as quickly and as easily as possible in the other Member States.

What Exactly is a European Arrest Warrant?

The European arrest warrant introduces some novelties compared to the former extradition procedures:

Faster procedures. The State in which the person is arrested has to return him/her to the State where the EAW was issued within a maximum period of 90 days of the arrest. If the person gives its consent to the surrender, the decision shall be taken within 10 days.

Simpler procedures. The dual criminality principle—which means that both the country requesting extradition and the country that should arrest and return the alleged criminal, recognize and accept that what he or she is alleged to have done, is a crime—is abolished for 32 serious categories of offences. These 'include participation in a criminal organization, terrorism, trafficking in human beings, sexual exploitation of children and child pornography, illicit trafficking in arms, ammunition and explosives, corruption, fraud including fraud pertaining to the financial interest of the European Union, money laundering and counterfeiting of money including the euro. European arrest warrants issued in respect of crimes or alleged crimes on this list have to be executed by the arresting state irrespective of whether or not the definition of the offence is the same, providing that the offence is serious enough and punished by at least 3 years' imprisonment in the Member State that has issued the warrant. For offences which are not in the list or beyond the 3 years' threshold, the dual criminality principle still applies.

No political involvement. In extradition procedures, the final decision on whether to surrender the person or not, is a political decision. The EAW procedure abolished the political stage of extradition. This means that the execution of these warrants is simply a judicial process under the supervision of the national judicial authority which is, inter alia, responsible for ensuring the respect of fundamental rights.

Surrender of nationals. EU countries can no longer refuse to surrender their own nationals. The EAW is based on the principle that EU citizens shall be responsible for their acts before national courts across the EU. This means that it will not be possible in principle for a Member State to refuse to surrender one of its citizens who has committed a crime in another EU state on the ground that he or she is a national. On the other hand, it will be possible for a Member State, while surrendering this person, to ask for their return on its territory to serve its sentence in order to facilitate future reintegration.

Guarantees. The EAW ensures a good balance between efficiency and strict guarantees that the arrested person's fundamental rights are respected. In implementing the framework decision on the EAW, Member States and national courts have to respect the provisions of the European Convention on Human Rights and to ensure that it is respected. Anyone arrested under an EAW may have a lawyer, and if necessary an interpreter, as provided by the law of the country where he or she has been arrested. If judgment was given in his absence against anyone later arrested under an EAW, he has to be retried in the country requiring his return (see infra).

Grounds for refusal. The surrender of the person can be refused on several grounds (see Arts. 3 and 4 of the Framework Decision) among which:

- the *"ne bis in idem"* or double jeopardy principle. This means that the person will not be returned to the country that issued the arrest warrant if he or she has already been tried for the same offence.
- amnesty: A Member State can refuse to return a person if an amnesty covers the offence in its national legislation.
- statutory limitation: A Member State can refuse to return a person if the offence is statute barred according to its law (which means that the time limit has been passed and that it is too late under that country's law to prosecute the person).
- the age of the person: A Member State can also refuse to return a person who is a minor and has not reached the age of criminal responsibility under its national laws.
- It is also possible for a Member State to execute directly the sentence decided in another Member State instead of surrendering the person to that Member State.

Life sentence. Where someone arrested under an EAW may be sentenced to life imprisonment, the state executing the EAW may insist, as a condition of executing the arrest warrant, that if sentenced to life, the accused person will have a right to have its personal situation periodically reconsidered.

Death penalty. There is no mention of the death penalty as the death penalty has been abolished in the European Union.

Relations with third countries. The European Arrest Warrant only applies within the territory of the EU (15 Member States, and 25 as from 1 May 2004). Relations with third countries are still governed by extradition rules. If a person has been surrendered to another EU country according to the EAW and is afterward demanded by a third country, the Member State which authorized the surrender in the first place shall be consulted.

The European arrest warrant implements a decision taken in October 1999 by the European Council—the heads of state or heads of government of all 15 EU countries—at Tampere, Finland, to improve judicial co-operation in the European Union and, in particular, to abolish formal extradition procedures for persons "who are fleeing from justice after having been finally sentenced."

Its effectiveness depends on EU Member States trusting each other's legal systems and accepting and recognizing the decisions of each other's courts. Its objective—which is agreed by all EU states—is to ensure that criminals cannot escape justice anywhere in the EU.

Had Efforts Already Been Made to Simplify Extradition Procedures Between Member States?

Yes. Extradition is currently ruled by the European Convention on extradition from 1957 which was negotiated in the framework of the Council of Europe. In the context of the European Union, two Conventions were elaborated in the 1990s:

1 the Convention of 10 March 1995 on simplified extradition procedure between the Member States of the European Union (1995 EU Extradition Convention),

2 the Convention of 27 September 1996 relating to extradition between the Member States of the European Union (1996 EU Extradition Convention) and

3 Extradition is also ruled by some provisions of the Convention of 19 June 1990 implementing the Schengen Agreement (Schengen Implementation Convention).

The aim of both EU Extradition Conventions and the Schengen Implementation Convention was to supplement and facilitate earlier extradition conventions, such as the 1957 European Convention on Extradition (Council of Europe), that were applicable between the Member States.

The 1995 and 1996 EU Extradition Conventions have not entered into force as not all Member States have ratified them yet. A Member State has, however, the possibility to make a declaration that it will apply the conventions in relation to other Member States that have made the same kind of declaration. Both EU Extradition Conventions are therefore applicable between almost all Member States.

The 1995 EU Extradition Convention requires agreement (consent) from the person to be extradited. If consent has been given the simplified procedure applies. The Schengen Implementation Convention also contains rules on consented extradition. The Framework Decision on the EAW has incorporated these rules into a single legal framework on the surrender procedures.

The 1996 EU Extradition Convention contains several innovations, such as the abolition of two important impediments to extradition; political offences and own nationals. The 1996 Convention provides thus that no offence may be regarded by the requested Member State as a political offence, as an offence connected with a political offence or an offence inspired by political motives and that extradition may not be refused on the ground that the person claimed is a national of the requested Member State. However, there is a possibility for the Member States to make a reservation in relation to the articles on political offences and own nationals, which has been widely used. This possibility has disappeared in the Framework Decision on the EAW. As mentioned above EU countries will no longer refuse to surrender their own nationals.

Who Does What?

The European Commission shares the right to initiate proposals with Member States in the area of justice and home affairs. Concerned with the overall strategy, the Commission is responsible for revising policies and actions and implementing them. The Commission issued a proposal for a European arrest warrant on 2001 September 19, in order to replace extradition procedures by a more efficient instrument based on mutual recognition (see above). The Commission has also evaluated the implementation of the Framework decision in a report on the 23.2.2005.

European judicial cooperation in criminal matters is also dealt with by **international organizations** like the Council of Europe and the United Nations, as well as being included in work related to the enlargement of the EU and to the transatlantic dialogue with the United States and Canada.

How Has the EAW Been Implemented by Member States?

Despite some initial delays, the EAW is now operational in most of the cases planned and its impact is positive, in terms of de-politicization, efficiency, and speed in the procedure for surrendering people who are sought, while fundamental rights are respected throughout.

The **effectiveness** of the EAW can be gauged provisionally from the 2,603 warrants issued, the 653 persons arrested and the 104 persons surrendered up to September 2004. The surrender of nationals, a major innovation in the Framework Decision, is now a fact, though most Member States have chosen to apply the condition that in the case of their nationals the sentence should be executed on their territory.

Since the Framework Decision came into operation, the average time taken to execute a warrant is provisionally estimated to have fallen from more than nine months to 43 days. This does not include these frequent cases where the person consents to surrender, for which the average time taken is 13 days.

The improvements due to the arrest warrant also benefit the persons concerned, who in practice now consent to their surrender in more than half the cases reported. The Framework Decision is more precise than previous provisions, as regards for instance the ne bis in idem principle, the right to the assistance of a lawyer or the right to the deduction from the term of the sentence of the period of detention served. Furthermore, as a result of the speed with which it is executed, the arrest warrant contributes to better observance of the "reasonable time limit" principle.

This overall success should not make one lose sight of the effort that is still required by some Member States to comply fully with the Framework Decision. The report highlights the difficulties that remain in that respect. A few Member States considered that, with regard to their nationals, they should reintroduce a systematic check on double criminality or convert their sentences. Noticeable in some Member States is the introduction of supplementary grounds for refusal, which are contrary to the Framework Decision, such as political reasons, reasons of national security or those involving examination of the merits of a case. Moreover, there are cases in certain Member States where the decision-making powers conferred on executive bodies are not in line with the Framework Decision. Lastly, by ruling out the warrant's application to acts that occurred before a given date, a few Member States did not comply either with the Framework Decision. The extradition requests which they continue to present therefore risk being rejected by the other Member States.

The Commission considers this first evaluation as provisional and accordingly reserves the right to present proposals for amending the Framework Decision in the light of further experience. (European Union, 2005)

HARNESSING THE COMMUNITY AS A SOURCE OF INFORMATION FOR LAW ENFORCEMENT AND PUBLIC SAFETY

Most countries realize that the public can be a critical source of information and a true force multiplier in helping law enforcement, intelligence bodies, and public health agencies cope with the threat of terrorism or pandemics. Consequently, there are a number of examples of methodologies in place for engaging the public in suspicious activity reporting and reporting on potential public health emergencies.

Israel

The Israel Police provide regular guidance on how to spot potential terrorist activity. The Israeli public is continually encouraged to contact the police emergency number (100) whenever they see suspicious objects or behavior, and people regularly call the police to notify them of suspicious objects (which, in Israel, is really anything that exists in the public domain—backpacks, purses, automobiles, etc.—which does not have a clearly identifiable owner). It is a rather common sight in Israel to see traffic backed up as police bomb squad personnel obliterate someone's briefcase or backpack forgotten on a park bench or at a bus stop in the wake of a tip by a private citizen. In most cases, the object in question is harmless, but there have been cases of bombs discovered in this manner. It should be noted that while the degree of this phenomenon is sometimes exaggerated, there is generally a culture in Israel of personal accountability toward things in the public sphere, and people are quite willing to take responsibility for issues that in other societies would be seen to be the exclusive purview of governmental authorities. Hence, the participation of individual citizens in the evacuation of the injured to hospitals (as noted in Chapter 8) or the reporting of suspicious objects to the police.

United Kingdom

In the UK, if there is an imminent threat to life or property, the public is encouraged to contact the police emergency number (999) or to contact the police via a dedicated counterterrorism hotline. For longer-term issues, citizens also have the option to contact MI5 directly via their web site, a hotline, or by mail. The MI5 web site makes it clear that the public can help prevent terrorism through alertness and reinforces this by noting that:

> Terrorists have to live somewhere, and they need to plan and prepare for attacks. They buy and store materials, fund their activities, move around, prepare equipment and weapons and possibly undergo training. They may have people helping them—and these people might come and go at strange times of the day and night. They may make unusual financial transactions or use false documents to hide their real identities. They may be behaving differently to how you've known them to behave in the past. Members of the public may spot such activities, and if reported in time, a planned terrorist attack may be stopped before it happens.
> —British Security Service, 2009

CONCLUSION

The countries surveyed in this chapter have a wide range of approaches toward police structures and roles as well as community policing initiatives. Israel, being a small and highly centralized country that requires a police force that can be deployed rapidly in different region has opted for the centralized model of policing with some localized community policing efforts and reliance on a comparatively large pool of volunteers. The UK's police forces developed organically over nearly two centuries and has gradually been moving from a localized and dispersed model toward a more centralized set of structures. The London Metropolitan Police (Scotland Yard), in addition to being, by far, the largest force in the country, is also the nation's police force with respect to specific policing functions in the realm of terrorism and serious crime and the British have gradually been amalgamating their police forces—although it is highly doubtful that they will ever create a fully centralized national police force. In the area of community policing, the British also have a series of institutions and policies in place. Community policing perhaps comes more naturally to the British, as the London "Bobbie" or country constable was traditionally viewed as a "beat cop" who was embedded within the community in his/her respective patrol area. Canada, having a federal structure, follows a decentralized policing model, with the important caveat that Canada has a federal police agency, the RCMP, which, in practice, enforces laws in most of the provinces and local jurisdictions as a contracted law enforcement agency. Hence, the RCMP is a centralized force, but, it usually finds itself enforcing provincial and local laws on contract as a de facto local policing agency. Germany, having a federal system of government like Canada but with much stronger state powers, relies primarily on its state police agencies to enforce laws and has a comparatively weak pair of federal investigatory and policing agencies. France, being a centralized state, follows a centralized model of policing but also employs a military policing agency (see Chapter 5), making it different, in this respect, from the other countries surveyed here. Finally, Japan has a national police agency, but since most policing is carried out by the 47 prefectural police agencies, Japan has a de facto decentralized policing system, and the Japanese, with their Koban/Chuzaisho system, are true pioneers in the area of community policing. One of the benefits of effective community policing institutions and policies is that it allows the authorities to interface positively with immigrant communities and minority communities and hence be in a position to spot radicalization trends and try to mitigate the type of overall discontent and anger at the authorities that often precedes the radicalization process. In Chapter 4 we focus on radicalization and integration issues in the European context.

ISSUES TO CONSIDER

- How do the various countries addressed in this chapter organize their police forces?
- Why do some countries opt for centralized police forces whereas others opt for decentralized forces?
- How do the various countries surveyed in this chapter approach the issue of community policing?
- What is relatively unique about the Japanese system of community policing?

CHAPTER 4

INTEGRATION AND COUNTER-RADICALIZATION

In this chapter we focus on the position of Muslim minorities in a number of European countries, the attitudes of these countries toward national identity and the integration of immigrants, the status of Muslim populations (in terms of educational levels and geographic dispersion), and the nature of Muslim group identity. We also briefly survey the role of Muslim communal organizations and Imams, describe elements of the radicalization process, and provide some examples of European governmental responses to radicalization. While extremism and acts of terrorism based on radical ideologies are by no means limited to radical variants of Islam, at this juncture in history, and in the wake of major terrorist attacks in New York and the Pentagon (2001), Bali (2002), Madrid (2004), London (2005), and several failed plots of major potential ramifications (such as the 2006 airlines plot), extreme and violent interpretations of Islam thus represent the primary common threat across the countries surveyed in this book. Western Europe has been chosen as the focus of this chapter because Muslim communities are very prominent in many western European countries and the links between western Europe and the Islamic world are long-standing and intensive. Moreover, western Europe, in part due to its geographic proximity to the Muslim world and the presence of significant numbers of Muslim immigrants, has had to cope with radical Islam in a more intensive manner, and western European countries have served for longer as targets of radical Islam and, sometimes, as logistical and recruitment centers for such radical activities. Unlike previous chapters, this chapter is organized topically rather than country by country.

Comparative Homeland Security: Global Lessons, First Edition. Nadav Morag.
© 2011 John Wiley & Sons, Inc. Published 2011 by John Wiley & Sons, Inc.

FIGURE 4.1 Birmingham Central Mosque. This Wikipedia and Wikimedia Commons image is from the user George Daley and is freely available at http://commons.wikimedia.org/wiki/File: Flickr_-_BB_B_-_Birmingham_Central_Mosque.jpg under the Creative Commons Attribution-Share Alike 2.0 Generic license.

BACKGROUND

Over the last five decades, Muslims have become a significant minority population throughout most of western Europe. The development of Muslim communities in Europe can be seen to have followed a more-or-less universal path of modern migration. During the first phase, the early "pioneers" set up the first points of contact for their compatriots and were followed by a second phase in which a wave of generally unskilled and uneducated male workers from the countryside of their respective countries migrated in search of work (in this case with the initial intention of working in Europe for a few years and then returning with their earnings to their countries of origin). The men often lived together in crowded conditions to conserve funds, sent regular remittances home, and saw little reason to learn the language or develop an understanding of the culture of their host countries, as their presence there was perceived of as temporary. In the third phase, family members were brought over from the home countries and religious and other institutions were set up to serve the needs of these growing communities. This phase required learning (at least partially) the local language and developing some understanding of the majority society. Finally, the fourth phase involved the raising of children born in the host country and the beginnings of the incorporation of this new generation into the broader society (Lewis, 2003, pp. 67–68).

At present, the largest concentration of European Muslims is in France, which has an estimated 5 million Muslim inhabitants (the number constitutes approximately one-third of the Muslim population within the entire European Union). Germany has the second largest Muslim population (an estimated 3.3 million), followed by Britain (with some 1.6 million) and Italy and the Netherlands (with approximately 1 million Muslim inhabitants each). Although there have been a few cases of conversion (less than 1 percent of all Muslims in Europe are converts), the vast majority of Muslims living in Europe are either immigrants or the children or grandchildren of immigrants. In fact, approximately 50 percent of Muslims in western Europe were born there and the overall Muslim birthrate is approximately three times that of non-Muslims, thus ensuring that Muslims continue to be the fastest-growing sector of the European population. By 2015, Europe's Muslim population is expected to double, and by 2050, given the high Muslim growth rate coupled with the negative non-Muslim demographic growth rate, conservative projections suggest that Muslims will make up one-fifth of Europe's population (Savage, 2004, p. 28).

MODELS OF NATIONAL IDENTITY AS THE SOURCE OF IMMIGRATION POLICY

European states initially viewed immigration from the Muslim world as a temporary phenomenon needed to fuel western Europe's postwar economic boom. In fact, the attitude toward immigration in general varied in line with the overall approach toward national identity that was customary in each country. Broadly speaking, there are three models on which national identity is based: ethnic, civic, and pluralist. The *ethnic national model* (of which Germany used to be the prime example) views the role of the nation-state as representing the members of a specific ethnic-national community. Countries that follow this model traditionally raised significant barriers to immigration and naturalization for those who were not members of the ethnic community represented by the state (while, at least in past years, facilitating the immigration of co-ethnics from outside the country's borders). According to the ethnic national approach, a person's identity is determined by the identity of his/her parents and grandparents and not by the circumstances of his/her birth (i.e., where they were born).

The *civic national model* (exemplified in Europe by France) is based on the creation of a common identity among the inhabitants of a particular country (based, in the French case, on the values of the Republic and with little regard for ethnic identities) and the expectation that immigrants (who are more accepted than under the previous model) will assimilate in the general culture. Identity, under this model, is therefore determined by the location of one's birth and not by ethnic, religious, or other identities.

Finally, the *pluralist national model* (also known as *multiculturalism*, and traditionally practiced in many countries in Europe, including the UK and the Netherlands) provided for relatively liberal immigration policies while also recognizing the right to (and, proponents argued, the beneficial nature of) allow immigrants and other minority groups to maintain their cultural, linguistic, and religious differences. This model essentially combines the civilc model with the ethnic national model in that it splits identity between one's formal status and overarching identity (determined by the circumstances of one's birth) and one's communal identity (based largely on ethnicity).

Not surprisingly, then, the real-world immigration, naturalization, and integration policies of European countries followed the dominant model of national identity in each country. In Germany, workers recruited from the Muslim world (particularly Turkey) in the 1950s and 1960s to help fuel the extraordinary recovery and growth of the West German economy, known as the *economic miracle* (Wirtschaftswünder), were expected to remain in the country for a limited period (usually two years) and then be replaced by a new crop of migrant workers. These were known as *guest workers* (Gastarbeiteren) and even when the German government suspended, for financial reasons, the rotation of workers in the early 1960s, it still viewed this community as a temporary one that would eventually be largely repatriated (Ireland, 2004, p. 30). With the perspective of hindsight, the German assumption that poor Turkish and Kurdish guest workers would happily return to their home country after a few years appears extremely naive, as these migrant workers chose to stay and to move their families to Germany. By the 1980s, most Muslim immigrants were asylum seekers fleeing conflict and persecution in Turkey and the former Yugoslavia. Despite the changing makeup of immigrants and the fact that very few were leaving Germany, it took several decades for the government to finally recognize that Germany was a country of immigration, allowing, with the promulgation of the new citizenship law (Staatsangehörigheitsrecht) in January 2000, for children born in the country to foreign parents to gain citizenship. This has not, however, as noted below, meant that Germany has been able to integrate its immigrant communities successfully, as the old model has proven difficult to completely abandon.

FIGURE 4.2   Caricature criticizing French clergy opposed to the French revolutionary constitution by Jean-Baptiste Gobel (1792). This Wikipedia and Wikimedia Commons image is from the Library of Congress (Digital ID cph.3b43646) and is freely available at http://commons.wikimedia.org/wiki/File:Jeann-Baptiste_Gobel_caricature_1792.jpg and has been released into the public domain.

France, given its civic model of national identity, followed a different approach toward immigration, naturalization, and integration. Naturalization and integration processes were designed to make immigrants French and the Republic related to all of its citizens, at least in theory, on a common basis as French citizens with no official acknowledgment of the existence of ethnic minorities based on language, culture, and religion. Religious differences, in particular, were not acknowledged given France's official state secularism (known as Laïcité), which developed from the rejection of the Catholic Church as part of the old order overthrown in the French Revolution.

Religion was thus seen as something that should be confined to the private sphere by what some have termed an official *combative secularism* (Kuru, n.d., p. 9). Official efforts to homogenize immigrants were, however, implemented unevenly (particularly with respect to the integration of immigrants within the veteran population and the provision of equal opportunity to immigrants), and consequently, the French were never entirely true to their model of a state that is officially disinterested in its inhabitants' ethnic and religious affiliations. In addition to the fact that the republic did not entirely live up to its ideals with respect to the equal treatment of all, French society

itself proved to be even less magnanimous. Muslim immigrants, in particular, were generally not accepted as being French (regardless of the fact that
most were granted French citizenship) and even second-generation French
Muslims were still viewed by many as "migrant workers" (Open Society
Institute, 2007c, p. 7).

The pluralist, or multicultural, model is much more common in Europe
than the other two models (as well as existing in non-European countries such
as Canada). According to this model, the host society's acceptance of diversity
and the maintenance of separate identities among immigrant groups and
between immigrants and the veteran population will not only enrich the
country but will endear the host society and state to the immigrants because
their cultures will be respected, valued, and accepted as part of the makeup of
the larger society. Although this model sounds as though it strikes a perfect
compromise between the need to create a common society and the respect for
diversity, the reality has been significantly different. Immigrants were, on the
one hand, made dependent on state assistance and not required to become
proficient in the host society's language rather than being forced to "sink or
swim" in the larger society because it was seen as necessary to maintain their
communities as separate (in order to safeguard the culture and social bonds).
At the same time the authorities were reluctant to address the addiction to
social welfare, the lack of local language skills, unemployment, and the
associated problems of criminality and other antisocial behavior within
immigrant communities for fear of being accused of being discriminatory
toward the immigrants. As Ruud Koopmans et al. note:

> Particularly in the Netherlands, the combination of these factors has led to
> a vicious circle in which state policies have reinforced the image of
> migrants as a problematic, disadvantaged category in need of constant
> state assistance. To the majority population, migrants thus appear as a
> group deserving help, respect, tolerance, and solidarity, but not the kind
> of people that anyone would want to employ or would want one's child
> to be in school with. As a result, in spite of the liberal rhetoric, the much
> better legal position, and the much higher level of tolerance for cultural
> diversity in the public sphere and in political debate, levels of ethnic
> residential segregation, segregation in the school system, as well as levels
> of unemployment and social security dependence (relative to the majority
> population) are higher in the Netherlands than in Germany.
> —Koopmans, Statham, Giugni, and Passy, 2005, p. 15.

The problems produced by the pluralist model have been so acute that both
in the UK and in the Netherlands they have led to a retreat from multiculturalism toward a civic-national model. In the Netherlands this has led to
the development of a concept called Inburgering and immigration laws

FIGURE 4.3   Dutch and Turkish flags hang side by side in a multiethnic neighborhood in Eindhoven, the Netherlands. This Wikipedia and Wikimedia Commons image is from the user kees@eindhoven and is freely available at http://commons.wikimedia.org/wiki/File:Turkish_Dutch_ Flag_Football.jpg under the Creative Commons Attribution 2.0 Generic license.

based on this. Although this concept cannot be translated adequately into English, it basically means learning the Dutch language and adapting to and accepting Dutch values and mentality—clearly an assimilationist process that effectively represents a pronounced movement away from multiculturalism.

THE DEGREE OF INTEGRATION OF MUSLIM POPULATIONS IN EUROPE

There are a number of indicators that suggest the degree of integration (or lack thereof) with respect to Muslim minorities in Europe. Typically, such indicators include educational achievements, employment levels, geographic dispersion, and political representation. We shall explore each briefly.

Educational Achievements

Educational achievements such as number of years of schooling, enrollment in and completion of programs in higher education, and the like are an important indicators of integration because employment opportunities and levels of income generally correspond to higher educational achievement.

In Germany, official statistics refer only to foreign students and not to Muslims specifically. Nevertheless, immigrant children (of which Muslims form the bulk) generally do poorly in school. Close to one-fourth of immigrant students do not complete their studies and achieve a diploma (at any school: from trade schools to academic prep schools in the multitier German school system) as opposed to approximately 7 percent of native German students. Similarly, whereas close to one-fourth of German students qualify for higher education, less than 8 percent of foreign students do so. Interestingly, German-born students of immigrant parents tended to do worse in school than those who had immigrated to Germany, suggesting that second-generation immigrants might actually be less able to integrate into the broader society (Open Society Institute, 2007a, pp. 26–27).

In France, official statistics do not include religious and/or communal affiliation and consequently it is difficult to obtain accurate statistics relating to the French Muslim community. Nevertheless, some studies have shown that the percentage of students failing to complete primary school is twice as high among Muslims (16 percent) than in the general population (Open Society Institute, 2007c, p. 32). In the Netherlands only 10 percent of Turks and Moroccans (who constitute the bulk of the Muslim inhabitants) completed secondary or higher education. Approximately 40 percent of Turks and 45 percent of Moroccans have either no formal education or are educated only to the primary school level (Open Society Institute, 2007b, p. 18).

In the United Kingdom, adult Muslims are the least likely group to have secondary education qualifications. Muslim students, particularly of Bangladeshi and Pakistani origin, fall significantly below the national average in educational attainment. At the same time, the UK arguably has the largest population (by proportion) of Muslim university students anywhere in Europe, and it is estimated that there are some 90,000 Muslim students at higher education institutions in Europe (Reed, 2005, p. 8).

Employment Levels

In Germany, Turkish immigrants, with an unemployment rate of 23 percent, make up nearly one-third of unemployed immigrants, who, as a group, have nearly twice the unemployment rate of native Germans. In France, immigrants and their children (most of whom are Muslim) have higher

FIGURE 4.4 Turkish Day procession in Berlin. This Wikipedia and Wikimedia Commons image is from the user Danyalov and is freely available at http://commons.wikimedia.org/wiki/File: Turkisch-day-in-Berlin.jpg and has been released into the public domain.

unemployment rates than people of ethnic French origin. Among young people, for example, the unemployment rate is 30 percent as opposed to 20 percent among the native French population (Open Society Institute, 2007a, p. 42). In the Netherlands, Muslim unemployment rates are high, with 27 percent of Moroccans and 21 percent of Turks unemployed (as opposed to 9 percent of the ethnic Dutch population). Not surprisingly, given their overall lower level of education in comparison to the native population, half to 60 percent of Muslim immigrants have lower-level employment, compared to 28 percent of native Dutch (Open Society Institute, 2007b, p. 22).

In the United Kingdom, the Office of National Statistics in 2004 found that Muslims were the most likely group to be unemployed or otherwise economically inactive (31 percent of Muslim men compared with 16 percent of Christian men). For Muslim women, seven in 10 were found to be economically inactive compared with four in 10 women among the Christian population (UK Office of National Statistics, 2009).

Geographic Dispersion

In Germany, Muslims live mainly in "ethnic districts" within urban areas and tend to remain in these areas even when their socioeconomic status improves. In fact, there is some evidence that physical segregation is actually increasing in Germany. In France Muslim immigrants typically settled in industrial areas with cheap housing and have largely remained in these areas. Thus, 35 percent of the population in the greater Paris region (Île-de-France), 20 percent of the population in Marseilles and its hinterland (Provence-Côte-d'Azur), and 15 percent of the population in the Lyons region (Rhône-Alpes) are Muslim (Open Society Institute, 2007c, p. 14).

In the Netherlands, most Muslim immigrants are concentrated in the four main cities: Amsterdam, Rotterdam, Utrecht and The Hague and make up approximately one-fourth of the population of these cities. Most Muslim immigrants live in neighborhoods where immigrants make up half of the population, and studies have found that some 70 percent of Turks and 60 percent of Moroccans associate primarily with members of their own ethnic community. These immigrant neighborhoods, referred to by Dutch authorities as *concentration neighborhoods*, are, not surprisingly given the low levels of educational achievement and employment, also areas characterized by decay and crime (Open Society Institute, 2007b, p. 24).

In the United Kingdom, most Muslim immigrants are urban with, 38 percent living in the greater London region. Generally speaking, Muslim communities in the UK cluster by area of ethnic origin. Thus, east London has a large Bangladeshi community, Arab immigrants tend to concentrate in the southeast of London, and Pakistani communities can be found in London and throughout the English midlands and farther north in the Yorkshire region (Change Institute, Communities and Local Government: London, 2009, pp. 19–20).

Education of Muslim Students

One of the most important ways in which government and society in any country have an impact on the individual is through the educational system. This is not difficult to fathom given the fact that education is mandatory (at least in the developed world) and the role of state education systems, in addition to imparting knowledge and analytical tools, is to convey socially accepted norms of behavior and thinking (including systems of belief). With respect to the position of Muslim minorities in Europe, the primary educational concerns have traditionally revolved around three issues: the hijab (headscarf worn by religious Muslim girls and women), Islamic education in the public schools, and the role of private Islamic schools. There has been quite a bit of controversy and press coverage revolving around the issue of whether girls in public school (as well as female teachers) should be allowed to wear the hijab. The approaches to the issue vary across Europe and are, to a certain extent, reflective of the model of national identity being followed in the country in question. France stands at one end of this continuum, having legislation that does not allow "ostentatious" displays of religious identity in the public schools—although, in practice, there is a considerable amount of ambiguity over the issue, and school principals play an important role in determining whether or not teachers and school administrators at a given school will turn a blind eye to the wearing of the hijab (Fetzer and Soper, 2005, p. 5). The ostensible barring of the hijab from schools has created some controversy, and some in the Muslim community have seen this as affirmation of their belief that the French system is anti-Islamic. At the same time, a commission created by then president Jacques Chirac to explore the issue found that many secular Muslim girls and their parents wholeheartedly approved the ban for fear that otherwise the girls would be subject to peer pressure, insults, and threats because they did not wear the hijab (Shore, 2006, p. 79). Germany, with its tendency to view its Muslim population traditionally as somehow temporary (which would imply leniency with respect to the religious practices of that minority) but tempered by its strongly federal system, left the issue up to the states, four of which enacted a ban on the hijab (Shore, 2006). Britain, at the other end of the continuum, and in keeping with its multicultural approach, did not ban the hijab in schools.

The issue of Islamic education in the public schools has been an important one for many Muslim parents—who feared that the absence of such education would lead their children to stray from the religion and value system that they considered important to their identity. There is nothing particularly unique about this, as virtually all parents desire to have their children educated in the creeds and values that they themselves hold dear. Nevertheless, European countries were slow (or in some cases refused) to allow for Islamic education in state schools. France, being true to its republican traditions and official creed of Laïcité, does not allow for religious instruction in its public schools. In Germany, as with many other policy areas, this matter is left to the states. Several German states accordingly, and in keeping with the traditional view that the large Turkish minority was on a temporary footing, instituted Islamic education classes in Turkish in the public schools. The instructional materials and teachers were appointed by the Turkish Ministry of Religious Affairs (Germany's decision to allow Turkey to provide religious services to Turks in Germany is discussed in greater detail below) although in some cases the materials were modified by school authorities and local Islamic associations. By the late 1990s, however, the diversification of the Muslim population in many states, coupled with the deterioration in Turkish language skills among second- and third-generation Turkish immigrants, led many of the states to switch to instruction in German. Nevertheless, instruction in Islam is far from being available in all German states, due to the fact that Islam is not recognized as a religious corporation and thus not given the right, under the German constitution, to receive tax funding and conduct religious education in the schools. The lack of such instruction in many places is one of the most hotly contested educational issues in the country. The absence of official recognition of Islam at the national level has, not surprisingly, meant that different Islamist movements operating out of the mosques have been able to more-or-less monopolize Islamic instruction (Tol, 2008, p. 14). In response to this, a number of German universities have attempted to offer Islamic theological training and there are also plans to offer such programs to high school teachers (Laehnemann, n.d., p. 2).

The Netherlands has long had a traditional of publically funded private denominational schools (Bijzondere Scholen) existing alongside a secular state school system (Openbare Scholen). Consequently, creating a series of state-backed Muslim denominational schools did not present the same sort of problem as it did for other European countries. By 2006, there were 46 Muslim primary schools and two secondary schools in the country and there are now two private Muslim universities, both of which aspire to become state universities (Open Society Institute, 2007b, p. 20). Despite this, there has been growing concern that these schools might be used for instilling radical ideas, and the government's inspectorate for education has been involved in monitoring these schools and recommending actions designed to enhance the governance of these schools and attempt to ensure that radical and/or antidemocratic messages not be taught in these schools. The advantage, of course, in educating Dutch Muslim children in Islam at publically funded schools as opposed to in the mosques or elsewhere is that the state can monitor and guide this instruction. Separate Islamic schooling has also been a contested issue.

The number of Islamic schools varies across Europe. In Germany, only two Islamic primary schools have been officially recognized (they receive some state funding), while in the UK there are some 100 Islamic schools, most of which are privately funded. Not surprisingly, France, with its secular principles, does not provide state funding to Islamic schools. Although the issue of state support for religious schools can be seen as part of the long-standing debate in Europe and elsewhere over whether or not the state should be thoroughly secular, Koopmans et al. (2005) argue that Islamic schools differ from Catholic or other religious schools in that, unlike the latter which promote a liberalized version of the faith, Islamic schools put religion at the forefront and promote a set of values that are not in keeping with those of the larger society and are less respectful of state authority. No doubt, this issue will continue to be a bone of contention between Muslim communities in Europe and their respective governments.

Group Identity Within Muslim Communities

Several public opinion studies of Muslim communities in Europe have been conducted since 9/11. A 2007 British study found that the younger generation of Muslims identifies itself as more religious than their parents (including a preference for Islamic schools, Islamic dress, and the Shari'a) and feel that they have less in common with non-Muslims than their parents have (Mirza, Senthilkumaran, and Ja'far, 2007, p. 5).

Interestingly, this study showed that despite concerns about discrimination, 84 percent of Muslim respondents believed that British society treated them fairly and over one-fourth believed that British multiculturalism policies were exaggerated in their attempts to placate Muslims. On the issue of popular attitudes toward communal organizations, the study found that only 6 percent of respondents felt that the country's primary Muslim communal organization, the Muslim Council of Britain (MCB), represented them, and just over half felt that no Muslim organization represented their views (Mirza, Senthilkumaran, and Ja'far, 2007, p. 6).

A study of German Muslims commissioned by the Federal Ministry of the Interior found that the overwhelming majority of the population, both young and old, described themselves as devout or very devout (87.3 percent), although daily and weekly mosque attendance was much lower (31.4 and 28.5 percent, respectively) (Brettfled and Wetzels, 2006, p. 12). On the issue of

FIGURE 4.5   Religious Muslim woman wearing the hijab and the niqab, UK.   This Wikipedia and Wikimedia Commons image is from the user Tasja and is freely available at http://commons.wikimedia.org/wiki/File:Burqa_England.jpg under the Creative Commons Attribution 2.0 Generic license.

attachment to the host country, only 5.1 percent of schoolchildren felt closely attached to Germany (as opposed to 12.2 percent among the overall Muslim population). The majority of Muslims surveyed felt that Christian society was immoral (55.9 percent) and that Islam was the only true religion (65.6 percent) (Brettfled and Wetzels, 2006, p. 14). The study further found a high percentage of Islamic orthodoxy, traditionalism, and fundamentalism (all ideologies that involve a significant degree of nonintegration into the larger society and in the case of fundamentalism, general hostility toward the larger society). Among the general Muslim population in the sample, 21.9 percent were found to hold orthodox religious views, 21 percent to hold traditionalist/conservative views and 39.6 percent to hold fundamentalist views. Among school-children, as with the findings in the UK, there was a visible increase in orthodox, traditionalist and fundamentalist views: 23.4, 23.2, and 42.6 percent, respectively (Brettfled and Wetzels, 2006, p. 16). Finally, nearly half (46.7 percent) of the Muslim respondents in the sample agreed with the statement that religion was more important than democracy, and a large proportion (65.3 percent) favored government control over the press in order to preserve morals and order.

Communal and Religious Leadership and Their Regulation

All Muslim countries regulate, in one form or another, the teaching and practicing of Islam. This is because, as with the old adage "war is too important to be left in the hands of generals," in this case *religion is too important to be left in the hands of imams and ulemma*—religious scholars (or at least those who are not appointed and controlled by the state). Naturally, Western countries, with their largely secular ethos and view of religion as an individual choice that reflects one's basic rights, have tended to take a much more laissez-faire approach to regulating religion. This approach has been reflected in the independence of mosques and religious seminaries and the absence of control over the imams, the majority of whom were brought in from overseas without significant knowledge of the host country and with the objective of teaching the strain of Islam followed in the country of origin. Not surprisingly, this has often led to the preaching of messages in the mosques that are hostile to Western values and sometimes incite to violence, as in the case of the Finsbury Park Mosque in London, where "shoe bomber" Richard Reid and others were radicalized.

This fear of radicalized messages coming from imams with little or no understanding of, or respect for, the host society has led most European countries to conclude that the state must have some control over the way that Islam is preached. France, for example, has begun to push for imams to be either French-born or at least educated in the country and in the French language. There are also a number of Muslim religious seminaries operating in France designed to train imams. In the Netherlands, the Islamic University of Rotterdam has trained over 20 practicing imams. In 2004 the Dutch parliament passed a law, which took effect in 2008, that allows only those imams who have studied at one of the two Dutch seminaries to work in mosques, thus effectively banning the "importation" of imams. Officials in the Netherlands put much stock in the imam as a principal figure in the socialization of youth and are convinced that imams should demonstrate loyalty to Dutch values (i.e., gender equality, non-discrimination based on sexual orientation, secularism, etc.). It is not clear, however, whether this policy is likely to be effective, because it will require a new conception and role for the imam, who traditionally focused on the strengthening of religious ties and following of religious traditions among believers. Certainly in Sunni Islam, it is the responsibility of the believer to follow the dictates and practices of the faith; the imam's role is to educate, not to lead (Boender, 2008, p. 22). Nevertheless, even these governmental initiatives designed to promote the training and certification of imams with more tolerant views have come under suspicion, because many of these institutions are funded by foreign governments or associations—such as the Algerian government in the case of some of the French seminaries and a Turkish Islamic movement in the case of an institution in Rotterdam (Sciolino, 2004).

In the United Kingdom, a Mosques and Imams National Advisory Board (MINAB) was proposed by a Home Office task force on extremism in the wake of the July 7, 2005 London suicide bombings. This organization has since been created but functions as a communal organization rather than a governmental entity, and its primary role is to advise mosques on governance issues and the suitability and improvement in the performance of imams (Mosque and Imams National Advisory Board, n.d.). However, according to a report by the Quilliam Foundation, most UK-based imams are "disengaged, isolated and highly socially conservative" who are "physically in Britain, but psychologically in Pakistan or Bangladesh" (Dyke, 2009, pp. 4–5). In a survey conducted in September 2008, the Quilliam Foundation found that some 97 percent of imams in the 254 mosques surveyed came from outside Britain, and 44 percent of these mosques do not preach wholly or partially in English (Dyke, 2009, p. 8). While foreign-born imams are required to prove some level of competence in English in order to receive a visa to the UK, a "Britishness" test for foreign-born imams was reportedly canceled in 2005 after protests from the Muslim community.

In addition to the religious leadership of Muslim communities, there is a growing pool of Muslim communal organizations in Europe. European governments take a variety of approaches toward these organizations in keeping, at least to some degree, with their overall model of national identity and integration. These approaches vary from viewing such organizations as representatives of the Muslim community as a whole to viewing these organizations as unofficial associations that do not enjoy a special status with governmental authorities. The French follow the latter approach. While in 2002 the French authorities did create an official Muslim association, the French Council for the Muslim Religion (Conseil Français du Culte Musulman), this organization is meant to represent the religion of Islam rather than the Muslim community and deals with issues such as the construction of mosques, observance of Muslim holidays, provision of Halal food, and the like.

In Germany, the domestic intelligence service, the Office for the Protection of the Constitution (Bundesamt für Verfassungsschutz or BfV), has been tasked with evaluating whether Muslim organizations are moderate or extreme. Governmental organizations often look to the BfV to provide guidance before they decide whether or not to fund or otherwise interact with Muslim organizations. Unlike Judaism and a wide range of Christian denominations, Islam does not have the status of a public corporation (Körperschaft des Öffentlichen Rechts) and thus there is no central Islamic organization authorized to receive funds gathered by the government from members of recognized religious denominations (a "religion tax") or teach Islamic education in the schools (something that would be problematic in any case given the decentralized nature of Islam and the divisions within the Islamic community). Historically, Germany has "outsourced" the management of Islam (in keeping with the traditional approach that viewed non-natives as foreigners) and has long relied on a Turkish quasi-governmental agency, the Turkish–Islamic Union for Religious Affairs (Diyanet İş leri Türk-Islam Birliği or DİTİB) (International Crisis Group, 2007, p. 2). The DİTİB handled a variety of different administrative matters for the German authorities, including the vetting of imams (who gave sermons approved by Ankara), permits for mosque construction, and the vetting and provision of teachers for Islamic education in the public schools. The German government relied on DİTİB services both because it allowed the government to avoid making unpopular decisions with respect to coming to grips with the reality of the permanence of the Muslim minority and because of the assumption that the Turkish authorities had a vested interest in mitigating Islamic radicalization so that Turkish communities in Germany did not become centers for subversion against the still largely secular nature of the Turkish Republic.

In 2006, the Federal Ministry of the Interior created an initiative called the German Islam Conference which brings in segments of the Muslim community into a structured debate with federal and local governmental representatives as part of a governmental outreach effort. The German

FIGURE 4.6 Turkish Islamic Union mosque in Mosbach, Germany. This Wikipedia and Wikimedia Commons image is from the user AlterVista and is freely available at http://de.wikipedia.org/w/index.php?title=Datei:Mosbach_Moschee.JPG&filetimestamp= 20080322140223 under the GNU Free Documentation license.

authorities purposefully chose people from all walks of life within the community and exclude communal leaders in order to remove some of the vested interests involved in dialogue with communal leaders. The conference conducts a series of working groups dealing with issues such as the reconciliation of different religious points of view, the role of religion in the schools, economic programs, integration efforts, the role of the media, and so on (Paris, 2007, p. 8).

In the United Kingdom, the policy designed to prevent Islamic extremism (and in keeping with that country's multicultural approach) focused, in part, on backing Muslim community associations and allowing them, to some degree, to represent the Muslim community, even when such groups advocated Salafist ideas but stopped short of advocating violence. This sort of approach was summed up by then London Metropolitan Police (Scotland Yard) Commissioner Ian Blair when he suggested adopting a "traffic light system" toward Muslim associations, in which the government would speak to any nonviolent Muslim communal association (green) and would never speak to terrorists (red) but would leave the door open to talking to certain types of activists whose views were highly controversial and walked a thin line between nonadvocacy and advocacy of violence (yellow) (Caldwell, 2006). While this multiculturalist policy largely characterized the approach taken by the government of Tony Blair, the government of his successor, Gordon Brown, was divided on the issue, with some ministers advocating backing away from most of the existing (and suspect) Muslim communal associations that had heretofore been the British government's main partners. Thus, in early 2009, then communities secretary Hazel Blears suspended relations with the Muslim Council of Britain (MCB), the largest of the Muslim communal associations, after a senior MCB official allegedly endorsed violence against foreign ships deployed to prevent arms smuggling to the Gaza Strip (The Economist, 2009). Interestingly, some argue that British Muslim communal leaders are unrepresentative because they are too acceptable and comfortable in the eyes of the non-Muslim majority. As British Muslim law enforcement consultant Mehmood Naqshbandi argued in a study for the UK Defense Academy:

> Leaders of Muslim organizations with a high public profile are unrepresentative. Muslim communities are desperately self-conscious about their spokesmen and women, only putting forward in public those people that they believe meet the expectations of the non-Muslim world—successful businessmen, career politicians, shaved, scrubbed, suited professionals, who rarely know where the local mosque even is. There is a yawning gulf between active, concerned Muslims and community engagement efforts through government institutions who only deal with 'community leaders'. Muslim community leaders among local politicians and parliamentarians are entirely made up from people who have in all essentials, left the Muslim community far behind. . . . No one who meets the norms of a respected practical Muslim is currently able also to gain popular electoral support. Leaders of Muslim institutions represent the opinions of older generation Muslims, preoccupied by status and wealth. They get very short shrift from the youth.
>
> —Naqshbandi, 2006

THE RADICALIZATION PROCESS

Generally speaking, first-generation Muslim immigrants in the West in the 1950s, 1960s, and 1970s (as opposed to some temporary guest workers) tended to desire to integrate within their new societies. The phenomenon of radicalization, although spurred on in some cases by first-generation immigrant imams who often do not even speak the language of their host country, is most common among Muslim youth in Europe. These youth (usually second- or third-generation immigrants, although some are first generation, particularly in countries without an established Muslim population) have often lost their links with the old country (sometimes including the ability to speak the native tongue) and thus are forced to engage in the creation of a new, universalized (and idealized) Islamic "culture." Indeed, it has been argued that the very concept of a universal Islamic culture, is a product of life in the West. According to this view, there is no "Islamic culture" in the Islamic world, but rather, a range of cultures corresponding with the different countries, ethnicities, and traditions—all of which include an Islamic component (Roy, 2004, p. 151). Young Muslims from a wide variety of national backgrounds in the West, therefore, having largely lost their native cultures (i.e., Algerian, Pakistani, Palestinian, Egyptian, etc.), band together and adopt a substitute concept of an Islamic culture which they view as a purer form of Islam shorn of familial, tribal, local, or national customs. Accordingly, the new generation of Western-born and educated Muslims do not want to be Pakistanis, Turks, or Moroccans; they want to be Muslims first (Roy, 2004, pp. 23–25).

As Roy (2004) explains, the process of Muslim immigration to Europe changes the nature of religiosity because it entails: " (1) the dilution of the pristine culture, where religion was embedded in a given culture and society; (2) the absence of legitimate religious authorities who could define the norms of Islam, coupled with a crisis of the transmission of knowledge; and (3) the impossibility of any form of legal, social or cultural coercion" (p. 151). It should be no surprise that the interpretation of Islam adopted by those who wrap themselves in the mantle of an "Islamic identity" is often a Salafist one (which is focused on trying to create what is believed to be the original Islamic community and lifestyle of the Prophet Muhammad and his immediate successors). Indeed, this new concept of a universal Islamic nation, culture, and identity lends itself well to the global Jihadi movements, whose members outside the West also tend to be estranged from their national roots and often forgo their own national struggles in favor of the larger global Jihad, with which they identify much more readily.

Even among first-generation Muslim immigrants in contemporary Europe who have joined extremist organizations, only a minority were fanatics in their country of origin before coming to Europe, and most of these young men (many of them middle class) became radicalized in Europe in the wake of a personal "culture shock" and rejection of the perceived secularism, materialism and social and sexual promiscuity of Europe (Savage, 2004, p. 33). One of the things that characterizes this "globalized" Islamic revivalism is its anti-intellectual stance and its highly personalized nature. To the adherents of this new, universal "Islamic identity," faith is something that needs to be felt and enjoyed at the personal, individual level and not to be analyzed, interpreted, and made into dogma, schools of jurisprudence, and the like. Emotion is more important than logic; sensing something is more important than understanding it. This approach, too, lends itself to the simplistic, literalist, and unquestioning attitude of the Jihadists.

An additional element that ties together radicalized Western Muslims is the collective anger in the face of real or imagined discrimination. Socioeconomic discrimination is, of course, endemic in the West as it is elsewhere, but the combination of a sense of alienation from Western values and a sense of economic marginalization and hopelessness can often form a highly combustible mix. This is not to say, however, that all "homegrown" Jihadist terrorists in the West come from disadvantaged backgrounds and that they were unable to learn the local language and otherwise integrate, acquire an acceptable education, and so on. Many of them, such as the assassin of Dutch filmmaker Theo Van Gogh and the 2005 London bombers, were well integrated into their host societies (Laurence and Vaisse, 2006). Effective terrorism, one must bear in mind, requires effective and capable terrorists at the middle and upper ranks of their organizations, and the smarter, more educated, and better integrated such a person is, the more dangerous he or she will be. Radicalization, it should be remembered, is not necessarily the result of one's real-life experience and the racial slights one has had to endure or the barriers that have been put in one's way. Radicalization could also be the product of imagined slights and barriers and failures attributed to society that are really the product of one's own inadequacies.

A study for the UK government's Department for Communities and Local Government identifies four essential indicators of radicalization: the individual's perception of the degree of their acceptance into the larger society, the individual's perception of the degree of opportunity afforded his/her ethnic/religious community, the degree to which one feels integrated into the larger society, and the extent to which the individual identifies with general societal values (Choudhury, 2007, p. 5). Solving problems of integration among Muslims is of course necessary to avoid the creation of Muslim ghettos (which already exist in much of Europe, as in the case of some of the Parisian suburbs or banlieues), which may, and frequently have, become centers for lawlessness and at least partial "no-go zones" for the authorities. Moreover, from a counterterrorism perspective, such neighborhoods could potentially serve not only as breeding grounds for Jihadists, but also as logistical centers with bomb-making facilities and the like.

As far as recruitment of young Western Muslims into terrorist cells is concerned, the Jihadist movements typically try to create a sense of comradeship, a kind of "brotherhood in Islam." Jihadist cells attached to a particular sheikh will attempt to create strong social bonds between members while isolating them from the broader society. The lack of any recognized religious hierarchy within the Jihadist movement contributes to the decentralized, cellular structure of Jihadist groups, in which anyone with some religious

FIGURE 4.7 Russell Square Underground station, one of the locations of the suicide bombings in London on July 7, 2005. This Wikipedia and Wikimedia Commons image is from the user Francis Tyers and is freely available at http://upload.wikimedia.org/wikipedia/commons/b/b3/ Russell_square_ambulances.jpg under the GNU Free Documentation License, Version 1.2.

knowledge becomes a "sheikh" and can lead a group. Once formed, these groups target three categories of potential recruits: (1) first-generation Muslims (in countries whose Islamic community is of very recent origin, (2) second- and third-generation Muslims in countries with a longer-standing Muslim presence, and (3) converts to Islam (Stemmann, 2006). Among all three groups, petty criminals have often been targeted both within and outside the prison systems.

The recruitment process itself involves a gradually increasing isolation of the recruit from family and friends, frequent meetings and outings with "brothers" designed to bond the recruit to other group members, and intensive indoctrination focusing on the real and imagined "sins" of the West and the rewards of Paradise. Eventually, in some cases, the recruit is typically sent abroad (before 9/11 this was to camps in Afghanistan) to undergo training. At each stage of the process, the recruit is made privy to increasingly sensitive information and achieves an enhanced status within the organization. The entire process typically takes approximately a year and a half (Jenkins, 2006, pp. 86–89). A New York City Police Department report issued in 2007 identified four stages in the radicalization process: pre-radicalization, self-identification, indoctrination, and jihadization. According to the authors of the report, the pre-radicalization stage involves the point of origin of the recruits, which is usually ordinary and unremarkable with little, if any, criminal history. The self-identification stage involves the initial exploration of Salafist and other radical ideas due to internal or external factors (e.g., international issues, a sense of economic or social isolation, discrimination, frustrated professional mobility) and the gradual association with like-minded individuals. The indoctrination phase involves the progressive intensification of radical ideology, usually facilitated by a spiritual guide, and the progressive isolation of the individual from family, previous friends, and the larger society. Finally, the jihadization phase involves the decision to participate in the Jihad and the commencement of operational planning for a terrorist attack or participation in an overseas conflict (Silber and Bhatt, 2007, pp. 6–7).

EUROPEAN GOVERNMENTAL RESPONSES TO RADICALIZATION

Rabasa et al (2010) note that, in contrast to the practice in Muslim countries (where radicalization is viewed through a theological lens), European countries are uncomfortable with addressing theology and tend to view radicalization primarily as an economic and social issue (p. 122). European governments have adopted different strategies to try and cope with the phenomenon of Islamic radicalization, with the authorities in the Netherlands employing a particularly interesting approach. In a 2004 report, the Netherlands general intelligence and security service (Algemene Inlichtingen- en Veiligheidsdienst—AVID) argued that radicalization was a threat to the democratic political order even when radicals did not overtly support the use of violence and noted that the European Court of Human Rights ruled, in a 2003 decision, that governments "may act against groups seeking to subvert or undermine the democratic legal order, in particular if there is a real risk that their activities may have effective results" (Netherlands General Intelligence and Security Service, 2004, pp. 34, 37). AVID recognizes that operating against groups that espouse antidemocratic ideas requires a difficult balance between allowing for the freedoms guaranteed under Dutch law and at the same time mitigating the dissemination of ideas that are a threat to democracy and Dutch values. Moreover, they note that many of the radicals are aware of the legal boundaries and make an effort not to cross these publicly (Netherlands General Intelligence and Security Service, 2004, p. 50). In keeping with their view that the growth of an ideology that is hostile to democracy and Dutch culture is a threat as great as that of actual terrorist activity, the Netherlands Ministry of the Interior and Kingdom Relations identified the creation of a "Salafist culture" as something that needed to be understood, presumably in order to evaluate the degree of the threat of this lifestyle and belief system to the Dutch system (National Coordinator for Counter-Terrorism, 2008, p. 63). In terms of counter-radicalization, AVID recommends a series of measures that include cooperation with moderate Muslims leaders, promotion of moderate Muslim narratives, development of youth identity (with an emphasis on educating youth with respect to their responsibility to society at large), promotion of positive role models, and encouragement of the emancipation of women and the promotion of their communal roles (Netherlands General Intelligence and Security Service, 2004, pp. 50–51).

Local governmental authorities in the Netherlands have been pivotal in addressing the issue of Islamic radicalization. Local authorities are responsible for town planning and the preservation of law and order and consequently are tasked with providing approval for the construction of mosques and Muslim cemeteries as well as Islamic schools. These powers allow them to interface with the local Muslim community and help ensure the propagation of moderate messages. Local authorities are also responsible for provision of welfare, employment, education, culture, and recreation services. In each of these capacities, the local authorities are able to conduct outreach with and provide services to the Muslim community (Open Society Institute, 2007b, p. 47). In the wake of the murder of filmmaker Theo van Gogh by a radicalized Dutch Muslim in November 2004, the municipality of Amsterdam developed a program known as "We the People of Amsterdam" (Wij Amsterdammers), designed to encourage democratic norms, combat discrimination, and counteract radicalization. This program is based on three sets of policies:

1. *Resilience*—strengthening the capacity of Muslims to resist radical ideologies.
2. *Alternative supply*—providing moderate religious messages and answers that can compete with radical ideologies.
3. *Breeding ground*—mitigating external factors that create a basis for radicalization (e.g., the perception of injustice, alienation, discrimination, limited economic opportunities) (Mellis, n.d., p. 4).

As part of the overall "We the People of Amsterdam" program, the municipal authorities set up an "information house" (Informatiehuishouding) unit within the Department of Public Order, Safety and Security, which was designed to implement antiradicalization policies and act as a radicalization "early-warning system" for the municipality on a case-by-case basis. The Dutch authorities make a distinction between *thinking* and *acting* in their approach to dealing with radicalization, whereby as long as a person is at the stage of ideological radicalization, the responsibility for dealing with this lies with the municipality, but once there are indications of actual preparatory

FIGURE 4.8 Municipality of Amsterdam. This Wikipedia and Wikimedia Commons image is from the user Ijanderson977 and is freely available at http://commons.wikimedia.org/wiki/File: Dam_Square_03_977.PNG and has been released into the public domain.

actions (even if these are legal, such as recording a video testament, purchasing potentially explosive substances, or photographing potential targets), the issue becomes a police matter. The Information House unit was tasked with developing expertise (and maintaining up-to-date information) in order to recognize the different forms of radicalization and the stages of radicalization. The unit was also responsible for developing and maintaining networks that included national and local counterterrorism officials, operators of municipal hotlines and municipal crime prevention and social service networks, neighborhood organizations, and city district administrators. These networks were augmented with ties to representatives of Muslim communities, social workers, and teachers. Once a case was brought before the unit's staff (via their network of contacts or others who come forward with information), their first priority was to determine the degree of radicalization of the person in question. After that initial evaluation was completed, a case management team (CMT), consisting of the information house unit and other municipal officials, met biweekly to discuss the various cases and decide on courses of action. The CMT was then either provided guidance to the person who reported the case (these were usually youth workers, teachers, police or parole officers, or local community leaders) in terms of trying to take steps (both ideological and material) to stop the radicalization process or it selected someone (generally from their contacts in the Muslim community) to fulfill this function of intervention

(Mellis, n.d., p. 7). Due to privacy concerns, the municipality closed Information House in 2009.

As of this writing, the Netherlands government, in cooperation with the Association of Netherlands Municipalities (Vereniging van Nederlandse Gemeenten—VNG), is developing a system to share information and best practices on radicalization prevention and counter-radicalization between municipalities and other interested parties. The system will work through provision of access to a new entity, Nuansa (the Knowledge and Advice Center on Polarization and Radicalization), which will serve as a clearinghouse for information and best practices. In addition, the Ministry of the Interior will cooperate with the VNG in order to encourage municipalities to adopt successful models (whether domestic or foreign), share practical experiences through sponsorship of regional and national meetings for sectoral partners (police, social workers, teachers, etc.), and host training sessions. Moreover, the Ministry of the Interior will award an annual local government prize to the best municipal initiative on countering radicalization (Netherlands House of Representatives, 2009, pp. 5–6).

The Dutch government is focusing on three overarching policy goals: (1) increasing resistance to radicalization and fostering greater identification with mainstream society among those populations vulnerable to radicalization, (2) increasing the capacities of local authorities to work with young people to combat radicalization, and (3) isolating and curbing radicalization at an early stage. Within this framework, specific initiatives will be developed for resistance to radicalization training and peer education, creation of platforms for dialogue on religion and its role in society as well as integration in Dutch society, spreading knowledge about democracy, the rule of law and responsible citizenship, countering radical messages, conflict management, deployment of positive mentors and role models, employment services, combating negative messages regarding Muslims in the media, and training of a range of actors that interface with the Muslim community (Netherlands House of Representatives, 2009, pp. 7–13).

The Dutch have taken a strident approach toward dealing with advocates of Salafist principles, including those that do not openly espouse violence. The Dutch parliament has noted that "the Dutch Constitution guarantees freedom to religious movements, including orthodox Islamic ones. The limit to that freedom is reached when a religious movement proclaims a radical political message that is incompatible with democracy and the rule of law. This is the case with political Salafism, a movement that actively seeks to reform society along ultra-orthodox lines . . ." (Netherlands House of Representatives, 2009, p. 14). This is in contrast with other countries, such as Germany and the UK, that do not openly attempt to block religiopolitical movements that do not advocate the use of violence.

In Germany, most integration efforts are carried out at the local and state levels and most policies in the areas of education, naturalization, integration, and religious affairs are carried out at the state level. Moreover, the federal government's ability to set policy in these areas is further constrained by the fact that the German parliament's upper house, the Bundesrat, is controlled by the states and can intervene to block legislation initiated at the federal level, as was done in 2002 when the federal government attempted to overhaul immigration policies.

In the United Kingdom, the government's Counter-Terrorism Strategy (CONTEST), mentioned briefly in Chapter 2, is based on four "strands" or categories of policies, known as *pursue, prevent, protect,* and *prepare.* The *prevent* strand is tasked with mitigating the terrorist threat and involves a set of five counter-radicalization objectives. The first is to challenge the ideology behind extremist ideas and provide alternative Muslim voices that can put forward a more moderate interpretation of Islam (see Sidebar 4-1 and 4-2). The second objective is to disrupt the activities of those who promote radicalization. Third, the strategy calls for the provision of social and economic support for

---

Sidebar 4-1: UK Home Office Proposal for Promoting Mainstream Islam

1 Bring about the development and provision of subsidised training, upskilling and qualifications for home-grown Islamic faith leaders. Training to focus on pastoral, community leadership and management skills. Action in hand, by Learning and Skills Council and Home Office (with FCO involvement). Subsequent roll-out of LSC-subsidised courses.

2 Raise the standards required from ministers of religion including Imams seeking admission and extension of stay. Package to include immediate English language requirement. Religious qualification requirements and civic engagement tests to follow after consultation, in stages during 2004/5.

3 Assist mainstream organisations to promote the many UK-based courses on Arabic and theology, taking away the need for Muslim youth to travel to seminaries in the Islamic world, many of which preach extremist doctrines. Encourage mainstream organisations to put their material on the web.

4 Seek opportunities through Government engagement and recognition, to promote awareness of moderate scholars with followings amongst young Muslims, such as Imam Hamza Yusuf and Imam Suhaib Webb.

5 Strengthen moderate Muslim media organisations (radio stations and publications, such as MCB Direct, e.g. by giving them stories and interviews). (UK Home Office, n.d.b.)

Sidebar 4-2: Case Study—Luton Ambassadors Project

The 'Ambassadors for Islam' project works with a group of young Muslims to build understanding and equip them with the theological arguments to counter extremist ideologies, dispel misapprehensions and develop their role as citizens, leaders and positive role models, so that they can become 'ambassadors' for mainstream Islam and assert their British identity.

The project was commissioned by the Luton Borough Council through the Islamic Cultural Society and is based at the Central Mosque in Luton... The project initially ran from October 2007 to March 2008 and has been extended until July.

Tweny-four young men have taken part in classes. They are taught by a British-born Islamic scholar with seminars from various visiting Muslim and non-Muslim speakers. Visitors to the classes have included a senior rabbi from London, local police officers and church leaders, all of whom have given very positive feedback.

The project is working towards forging stronger working partnerships with the Chaplaincy of the University of Bedfordshire, local colleges, the charity Crime Concern and other local and national faith/non-faith organizations. (UK Government, 2008, p. 20)

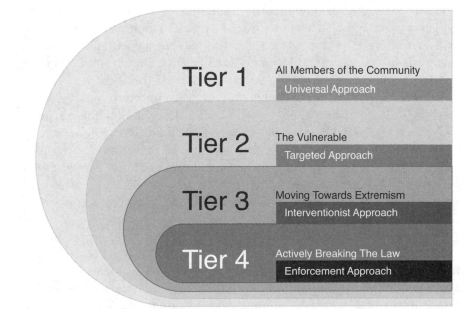

FIGURE 4.9 Tiered model of intervention to address the *prevent* strategy.

individuals who are vulnerable to recruitment by extremists or have already been recruited. The fourth objective is to increase the resilience of communities in coping with radicalization, and the fifth and final objective is to attempt to address grievances that are being exploited by radicalizers. These goals are to be achieved, at a communal level, through two cross-cutting "work streams": the gathering, analysis, and dissemination of information about radicalization within Muslim communities and the use of strategic communications to counter-radicalization (UK Government, 2009, p. 83).

While the British counter-radicalization strategy calls for top-down efforts at the national level, it also recognizes that local authorities and local communities will have to play a significant role in developing and implementing this policy. In addition to the local police authorities and local government, *prevent* strategy calls for partnerships with social, cultural, sports, and leisure services, community representatives, schools and universities, prison and probation authorities, public health agencies, and other governmental and nongovernmental entities (UK Government, 2009, p. 9). The strategy also calls for local authorities and police to determine where radicalization is occurring in their jurisdictions (in some cases through mapping out ethnic and religious communities and socioeconomic factors), which groups may be most vulnerable to radicalization and which sectors of the local community need to be supported in resisting radical messages and overtures (UK Government, 2008, p. 12).

As noted above, the *prevent* strategy supports engagement between local governmental authorities and Muslim communal organizations for the purposes of countering radicalization (see Sidebar 4-3). The strategy recognizes, however, that it is important that local authorities engage with the appropriate organizations and not work inadvertently with organizations that are pursuing a radical agenda or have radical tendencies. Accordingly, the strategy sets out criteria for engagement with Muslim communal organizations and indicates both that organizations that closely follow these criteria should be engaged with more intensively than those who follow it only loosely and that engagement should constantly be reevaluated in light of possible ideological changes in the communal organization being engaged (See Sidebar 4-4 for another appoach). An updated version of *prevent* released in July 2011 put even greater emphasis on combatting the ideology of extremism (UK Government 2011).

In France, official policy is not to favor any particular interpretation of Islam but rather to insist on adherence to the law and thus to prosecute any incitement to hatred, discrimination, and/or violence. The French authorities have defined five target groups for different types of messages regarding the struggle against radicalization: the broader population, persons and organizations engaged in the fight against terrorism and radicalization, populations

Sidebar 4-3: *Prevent* **Criteria for Engagement with Muslim Communal Associations**

(a) The organization actively condemns and works to tackle violent extremism. Factors to consider as part of this criterion include whether the organization:

- Publicly rejects and condemns violent extremism and terrorist acts, clearly and consistently;

- Can show evidence of steps taken to tackle violent extremism and support for violent extremism;

- Can point to preventing violent extremism events it has supported, spoken at or attended;

- Can show that its actions are consistent with its public statements; and

- Can show that its affiliated members or groups to which it is affiliated meet these criteria.

(b) The organization defends and upholds shared values including:

- Respect for the rule of law;

- Freedom of speech;

- Equality of opportunity;

- Respect for others; and

- Responsibility towards others. (UK Government, 2008, p. 60)

In an official correspondence between the British Cabinet Secretary and the Home Secretary in April 2004, an action plan was laid out with nine action items. This plan called for:

- Improving understanding of the extent and causes of extremism among young Muslims—primarily via focus groups and surveys.

- Combating the recruitment of young British Muslims by terrorist organizations—via identification of a "terrorist career path" and developing a comprehensive intervention strategy involving community organizations, religious organizations and local and central government.

- Combating Islamophobia—via a change in official terminology, strategic communications efforts and supporting mainstream Muslim communication channels.

- Increasing dialogue with young Muslims and building leadership capacity—via creating a cadre of young Muslim leaders to represent the community in the media and overseas and to lead Muslim youth and student organizations.

- Reaching out to disadvantaged and socially-excluded populations (including prison inmates).

- Judicious use of anti-terrorism powers—designed to ensure that the Muslim community does not perceive arrests and prosecutions as directed at the community via evidence-based, intelligence-led proportionate use of powers.

- Responding to Muslim concerns by dialogue with Muslim communities and provision of information regarding policies designed to protect Muslims against discrimination and hate crimes.

- Promoting mainstream interpretations of Islam via raising the requirements for foreign Imams, providing UK-based courses on Arabic and theology (thus removing the need for Muslim youth to travel to religious seminaries in the Muslim world), promoting moderate religious scholars and strengthening moderate Muslim media organizations.

- Remedying the exclusion of Muslims from public life via enhancing Muslim representation in public life. (Turnbull, 2004)

that serve as a potential source of recruitment for terrorists, France's foreign partners, and the national and international media. The Central Directorate of General Intelligence (DCRG), one of France's domestic intelligence organizations, has been tasked with monitoring the approximately 1700 mosques in France in order to disrupt violent networks. This work is carried out by a unit of the DCRG called Millieux Intégristes Violents (Violent Fundamentalist Environment). This unit uses a classified set of indicators to assess the degree

Sidebar 4-4: Recommendations from the British Think Tank Policy Exchange

Stop emphasizing difference and engage with Muslims as citizens, not through their religious identity. We should recognize that the Muslim 'community' is not homogeneous, and attempts to give group rights or representation will only alienate sections of the population further.

- Stop treating Muslims as a vulnerable group. The exaggeration of Islamophobia does not make Muslims feel protected but instead reinforces feelings of victimization and alienation.

- Encourage a broader intellectual debate in order to challenge the crude anti-Western, anti-British ideas that dominate cultural and intellectual life. This means allowing free speech and debate, even when it causes offence to some minority groups.

- Keep a sense of perspective. The obsession of politicians and the media with scrutinising the wider Muslim population, either as victims or potential terrorists, means that Muslims are regarded as outsiders, rather than as members of society like everyone else. (Mirza, Senthilkumaran, and Ja'far, 2007, pp. 18–19) of radicalization of a particular imam or preacher (including such variables as calls for violent Jihad, anti-Semitism, and anti-Western rhetoric) (Siegel, 2007).

CONCLUSION

The process of Islamic radicalization in Europe is ongoing and the number of radicalized Muslims seems to be increasing. In Germany, for example, the number of members and supporters of radical Islamic groups increased by 5 percent in 2009, and Germany's 29 identified radical Islamic organizations now have an estimated 36,000 members (Spiegel Online, 2010). As noted in this chapter, most European countries are still grappling with their comparatively (in historical terms) new role as countries of immigration and follow differing models with respect to their sense of national identity and the manner in which newcomers are integrated (or not) into the larger society. In addition, Muslim identity in Europe is constantly evolving—both in response to changes in attitudes on the part of the majority populations in Europe as well as to generational and other factors within Muslim minority communities. In terms of parameters of integration such as education, employment, and geographic segregation, these seem to suggest that the socioeconomic issues plaguing many Muslim communities in Europe are not likely to be solved in the foreseeable future. European countries are also grappling with issues related to Islamic education, the regulation of imams and mosques and the general role of the religion of Islam in societies that are, for the most part, profoundly secular. As noted above with respect to a handful of examples, a number of countries have been developing strategies to counter radicalization through interventions at various stages of the radicalization process. Thus far, the jury is still out with respect to the ultimate effectiveness of such efforts and whether or not Islamic radicalization in Europe has crested.

ISSUES TO CONSIDER

- What are the conditions under which Muslim immigrants to Europe arrived on that continent, and how has this contributed to radicalization among a minority within that community?
- What are the different models of national identity, and why have some European countries subscribed to one or another of them?
- How has the generation gap between first-, second-, and third-generation Muslim immigrants in Europe been expressed?
- How have European governments responded to the threat of radicalization?

CHAPTER 5

THE ROLE OF THE MILITARY IN SECURITY AND SUPPORT FOR CIVIL AUTHORITIES

In this chapter we look at the role that the military plays in the provision of domestic security and assistace for emergency response, disaster relief, and other forms of support to civil authorities in a handful of countries. Unlike the United States (with the restrictions on the domestic use of the military through *posse comitatus* and other legislation), most democratic countries have far less reluctance regarding employing the military domestically. All the countries surveyed in this chapter allow for some military role in homeland security missions. These can range from a permanent military law enforcement policing role (as with the French *gendarmerie* or Italian *carabinieri*) to military control over response efforts in crisis situations (as with the Israel Defense Force's Homefront Command) to the military playing an active role in domestic counterterrorism operations (as in Northern Ireland, the West Bank and Gaza Strip, or, very briefly, Quebec). Naturally, as democracies, all of these countries recognize the primacy of elected civilian leadership over the military, but as will be shown, none of the countries surveyed are as restrictive as the United States in terms of employing the military for domestic operations. Many countries also have the military perform specialized tasks, such as Chemical, Biological, Radiological and Nuclear (CBRN) detection and response, or take advantage of the military's ability to deploy large numbers of personnel quickly to provide disaster relief. In this chapter, we look at approaches toward the issue and practices followed by a number of countries by first looking at the legal framework in which domestic military operations occur and then focusing on the domestic role of the military, types of forces and missions, and the role of the respective national military establishments

Comparative Homeland Security: Global Lessons, First Edition. Nadav Morag.
© 2011 John Wiley & Sons, Inc. Published 2011 by John Wiley & Sons, Inc.

in disaster response and infrastructure protection. In this chapter we look at these various issue areas on a country-by-country basis with a slightly different mix of countries per issue area.

LEGAL FRAMEWORKS

Germany

In Article 87a of the German constitution, the Grundgesetz (Basic Law), the military is given the power, in a "state of defense or a state of tension" and when properly authorized, to "support police measures for the protection of civilian property..." [Germany, Basic Law, Article 87a(3), n.d.]. The *Basic Law* also authorizes the military, in situations where the police forces and federal border police are overwhelmed, to intervene against "organized armed insurgents" [Germany, Basic Law, Article 87a(4), n.d.]. The German constitution does not, however, clearly spell out what a "state of tension" consists of, although it has been pointed out that most commentators agree that this constitutes a stage where a military attack is imminent (Klose, 2006, p. 42). As Klose (2006) notes: "All the provisions in the Basic Law concerning the armed forces make very clear that they were made exclusively to enable the German armed forces to conduct the defense of German territory, together with the Allies, and for no purpose beyond these. Security against abuse was always the first priority" (p. 43). Article 87a of the Basic Law does allow

FIGURE 5.1 German soldiers on parade. This Wikipedia and Wikimedia Commons image is from the user Magnus Manske and is freely available at http://commons.wikimedia.org/wiki/File: Bundeswehr_-_10th_Anniversary_of_Multinational_Corps_Northeast.jpg under the Creative Commons Attribution-Share Alike 2.0 Generic license.

employment of the military to support police forces in the event of a clear and present danger to the democratic order of the federal republic or one of its states (Germany, Ministry of Defense, 2006, p. 56). As of this writing, there has never been an incident in Germany that required the domestic deployment of the German military. The German military is allowed to provide emergency aid and equipment but only small numbers of personnel and minimal quantities of equipment. Even these can be provided for only limited time periods and must be withdrawn once civilian resources are able to handle the situation. Also, when limited numbers of military personnel are deployed in an emergency, they cannot fill any law enforcement functions (Klose, 2006, p. 45).

At the same time, Article 35 of the Basic Law requires that federal and state (Land) authorities provide mutual assistance during disasters. Under this provision, state authorities may, in situations where their resources are overwhelmed, enlist the support of other state police forces, the federal police (Bundespolizei), and the armed forces. Moreover, if a serious incident (including a terrorist attack) occurs in the territory of more than one state, the federal government has the power to deploy the federal police or the military—although in both of the cases above, military support must be rescinded as soon as feasibly possible and the upper house of the German parliament, the Bundesrat, has the authority to order the troops to return to barracks at any time [Germany, Basic Law, Article 35(1–3), n.d.]. Even when the military is deployed under Article 35, it must remain under the direction of civil authorities with respect to disaster relief and under the authority of the regional police force, and its rules of engagement would consist of the laws of that state (Klose, 2006, p. 47). The military, in this context, is also restricted to using only weapons that the police are permitted to use (Germany, Ministry of Defense, 2006, p. 57). Given the foregoing restrictions, it is not surprising that the federal government has been working to try to expand the constitutional framework governing the deployment of the military, due to increasing fears of terrorism and other violent acts by nonstate actors.

Italy

Italy, on the other hand, has few legal apprehensions about using the military for domestic missions. For example, Law 78/2000 gives Italy's fourth military branch, a gendarmerie corps known as the Arma de Carabinieri, policing responsibilities in the areas of criminal investigation and law enforcement, maintenance of public order, and disaster relief. Under Italian law dating back to the beginning of the twentieth century, internal security, at the local level, is the responsibility of the district representative of the national government, the prefect. The prefect acts, in effect, as the representative of the Ministry of the Interior, which has the national responsibility for home-land security. The prefect thus has the legal authority to request military support (Italy, Law 690, Article 39, 1907).

France

In France, military forces outside the gendarmerie (discussed below) do not operate under specific legislation designed to allow regular military forces to carry out non-war-fighting domestic duties. This means that, at least theoret-ically, they can use their weapons only for self-defense and when witnessing criminal activities during the course of discharging their duties, they are obligated to act in the same manner as the average citizen who witnesses a criminal act (Vaultier, 2006, p. 207). As noted in Chapter 1, the constitution of the French Fifth Republic (promulgated on October 4, 1958) makes the president of the republic commander-in-chief of the armed forces and gives the president the authority to determine the broader policy directives which his/her prime minister then implements. Circular 500 of May 1959 deals with the role of the armed forces in domestic security and empowers civilian authorities to request military support under two different models: a requisition and a request for cooperation (Vaultier, 2006, p. 208). A requisition involves the obligatory mobilization of the military for domestic security roles. This can involve either a general requisition, in which a certain number of military personnel are put at the disposal of civilian authorities; a specific requisition, in which troops are mobilized for a precise mission; and a special complementary requisition, in which troops are mobilized and given the express authority to use weapons (Vaultier, 2006, p. 209). Requests for cooperation involve a voluntary military deployment that can run the range from policing duties to disaster relief.

Australia

In Australia, the Defence Act of 1903 gives the relevant cabinet ministers the power to call out the Australian Defense Force (ADF) if they feel that domestic violence is likely to occur (or has occurred) and that it will affect common-wealth or state/territory interests (Australia, Defence Act, Part IIIAAA, Division 1, Section 51A, 1903). This law does, however, require that when the military is called out to assist state governments (on the basis of a written request from those governments), it must cooperate with the police force of that state, although there is no requirement that the military force operating in response to that request put itself under state command (Australia, Defence Act, Part IIIAAA, Division 1, Section 51E, 1903). The military is, however, specifically proscribed from being called out to restrict protests, meetings, union action, and so on, unless those activities are likely to result in death, serious injury, or significant property damage.

Canada

In Canada, the military had been called out to deploy in the province of Quebec in October 1970 to shore up the exhausted police force in its attempt to combat attacks by the Front de Liberation du Quebec. Then prime minister Pierre Trudeau authorized the use of the War Measures Act to deploy troops in Quebec and establish martial law. After the Quebec crisis and the creation of specialized Royal Canadian Mounted Police (RCMP) emergency response teams, and in light of the problems associated with the British army's deployment in Northern Ireland (which the Canadians, naturally, wanted to avoid), the military's role in law enforcement and the preservation of order were downgraded (Lerhe, 2004, p. 11). In 1988, Parliament decided to limit some of the broad powers afforded the military in responding to domestic incidents by the War Measures Act. The resulting Emergencies Act creates four categories of emergencies that would warrant the domestic use of military force. The first, a *public welfare emergency*, involves military support to civil authorities coping with a natural disaster. The second, a *public order emergency*, involves military support to civil authorities dealing with an internal security program. In both of these types of emergencies, the military no longer enjoys an unlimited right to search and seizure. The third type

FIGURE 5.2 Canadian soldiers during an amphibious exercise. U.S. Navy photo, ID 090425-N-2821G-192 @ Wikimedia Commons.

of emergency is an *international emergency*, which involves a conflict with a foreign power that does not involve a full-blown war. The fourth type of emergency is a *war emergency*, which, as the name suggests, involves open hostilities with another country. In the latter two types of emergencies, the military does enjoy unlimited search and seizure powers as well as the right to control the economy and requisition private property. The Emergencies Act also gave the provinces the right to decline military support unless an emergency affects two or more provinces (Lerhe, 2004, p. 11). The act defines a national emergency as "an urgent and critical situation of a temporary nature that (a) seriously endangers the lives, health or safety of Canadians and is of such proportions or nature as to exceed the capacity or authority of a province to deal with it, or (b) seriously threatens the ability of the Government of Canada to preserve the sovereignty, security and territorial integrity of Canada and that cannot effectively be dealt with under any other law of Canada" (Canada, Emergencies Act, Section 3, 1988).

OVERVIEW OF THE MILITARY ROLE IN HOMELAND SECURITY

Israel

In Israel, the military plays a very wide-ranging role in homeland security. This varies from counterterrorism operations in the West Bank and Gaza Strip (see Chapter 2) to organizing and overseeing national preparedness and response efforts (see Chapter 8). The Israel Defense Force (IDF) was seen by Israel's first prime minister, David Ben-Gurion, as fulfilling a role far beyond the function of providing security from external enemies. The military was seen as a public service and an integrative force for an immigrant nation, not just an armed force. During different periods of its history, the IDF was involved in agricultural enterprises, immigrant absorption, providing teaching services to disadvantaged communities, and other such public services (Tirosh, 2003, p. 342).

Apart from pitched battles within the country during Israel's War of Independence (May 1948–January 1949), the civilian sector was largely exempted from military attack (although not terrorism) during Israel's first decades. Israel's wars with its Arab neighbors in 1956, 1967, 1969, and 1973 were confined almost exclusively to battlefield and border regions, and consequently, the military was focused on defending the borders and fighting wars on enemy soil—in keeping with Israel's forward battle space doctrine, which called for fighting and maneuvers outside Israel's borders due to the country's small size. Civil defense efforts on the part of the military were therefore limited. When Israel signed a peace treaty with its most formidable enemy, Egypt, in 1979, the Arab world effectively lost even the theoretical possibility of defeating Israel through military force because Egypt possessed, far and away, the largest and most effective Arab military force. Defeating Israel on the battlefield without Egyptian participation was therefore inconceivable (although for a time, Syria toyed with the idea of achieving military parity with Israel but never really came close). At the same time, deployment to the Middle East of long-range/high-payload surface-to-surface missiles gave those neighbors of Israel still in a formal state of hostilities with it (Syria, Iraq under Saddam Hussein, and Hizballah in Lebanon) the option of firing missiles at the civilian population as a way of deterring Israel or undermining its war effort in the event of another Middle Eastern war. Somewhat ironically, then, Israel's military supremacy on the battlefield brought less security for the civilian population because it made that population the primary target since it was no longer possible to realistically hope for an Israeli battlefield defeat. Israel's civilian population received its first taste of this new reality during the Persian Gulf War of 1991 when Saddam Hussein's forces fired 39 SCUD missiles at Israeli cities. More recently, Israel's adversaries (i.e., Hizballah and Hamas) have followed the same strategy, using primarily short-range/low-payload rockets. In the wake of the 1991 Gulf War, the IDF recognized that the civilian sector had come to be part of the battle space (if not, indeed, the primary battle space) and created a fourth regional command (in addition to the Northern, Central, and Southern Commands), the Homefront Command (HFC—Pikud Ha'Oref, also known by its Hebrew acronym, Pakar; see Sidebar 5-1). The HFC was created to improve interagency cooperation between the military, first responders, and government ministries; to free the three IDF regional commands to focus exclusively on the front lines; to provide military resources to the civilian sector (i.e., capabilities such as search and rescue, CBRN detection and response, etc.); and to enable the centralization of response efforts.

The HFC has five operational areas broken down into subareas that are geographically aligned to police districts and subdistricts. In the northern and southern districts, which border Lebanon and the Gaza Strip, respectively, the HFC does not have operational control of its units; rather, they are under the operational control of the commanding officers of the IDF northern and southern commands respectively, whereas the HFC has operational control of its units in the other districts. This is because the threats in the north and south of the country are more acute and require all military resources in those areas to be put at the disposal of the IDF regional commanders. The HFC also has direct control over specialist units that deal with nonconventional weapons attacks, urban search and rescue, and civilian casualty identification and management.

In normal times, the HFC is responsible for establishing emergency procedures, supervising preparedness exercises, and monitoring the

---

Sidebar 5-1: Israel Defense Force Homefront Command

The IDF Homefront Command (established in 1992 in the wake of response problems during the Iraqi SCUD attacks of the 1991 Gulf War) has the following objectives:

- Define the civilian defense concept,
- Steer, direct and prepare the civilian population for a state of emergency,
- Direct and guide all civilian systems, auxiliary organizations, the Israeli police and the military systems,
- Prepare the homefront for a state of emergency, according to the Civil Defense Law,
- Serve as the primary professional authority in the IDF for civil defense,
- Serve as a territorial command.

The responsibilities of the Homefront Command include:

- Command and coordinate all first responders involved in the incident,
- Develop a unified doctrine for all first responders,
- Carry out a unified training program for all levels of command,
- Inform and instruct the civilian population in personal and collective protection issues,
- Plan and deploy warning systems,
- Be in continuous readiness to assist police forces during terrorist incidents. (Nuriel, n.d.)

---

preparedness of the health system, municipalities, the transportation system and critical infrastructures. The HFC also operates national warning systems based on air raid sirens and, as of 2011, on a system that pushes out text messages to all cell phone subscribers. In the past, the HFC also distributed gas masks and atropine injectors to the population, although the Israeli postal service is now responsible for distribution. During normal times, the HFC is responsible for the preparedness of first responders within each district and subdistrict, although it does not have operational command over these. During periods in which Israel is facing an active wartime scenario (or, potentially, a WMD terrorist attack or other mass casualty event), the cabinet can (as noted in Chapter 2) declare a limited state of emergency, whereupon the HFC is given command and control over the other response agencies.

As noted in Chapter 3, Israel also has a quasi-gendarmerie force known as Mishmar Ha'Gvul or by the acronym Magav (border guard). This service is actually part of the Israeli police and not a military policing agency, but it does induct conscripts into its ranks (these conscripts can request to serve in the border guard or the regular police rather than the military) and as noted earlier, when its units operate in an area under military control (such as the West Bank), its forces are put under the command of the Israel Defense Force.

United Kingdom

In the UK, part of the mission of the military is to "...deliver security for the people of the United Kingdom and the Overseas Territories by defending them, including against terrorism..." (UK Ministry of Defense, 2007, p. 1.4). Nevertheless, despite this view, it is generally considered acceptable to use the military domestically only in support of civilian authorities (even when the military's role is to help maintain law and order, as in Northern Ireland, where the military's mission includes support for the police service of Northern Ireland). This had not always been the case. From 1969 to 1976, the British army played a dominant role (particularly given the collapse of the Northern Ireland police force, then known as the Royal Ulster Constabulary—RUC) in attempts to squash what was effectively an insurgency in many parts of the province. Only after 1976 did British authorities reestablish the principle that terrorism in Northern Ireland should be dealt with via law enforcement, and the RUC was once again given the lead role

FIGURE 5.3   British territorial army soldier practices target shooting. U.S. Army photo @ Wikimedia Commons.

in combating terrorism, although with the British army (which maintained between 20,000 and 30,000 troops in Northern Ireland) continuing, as noted above, to provide support including patrols and covert operations (Stevenson, 2006, p. 23).

In principle, the British military can be used for five types of supporting missions in the homeland security context and as part of the CONTEST strategy (although the military also has other missions, including fisheries protection, marine surveying, diving, etc.). These missions are hostage rescue, marine counterterrorism operations, contraband interdiction operations, explosive ordinance disposal, and search and rescue (UK Ministry of Defense, 2007, p. 2.23). The Royal Air Force also maintains the capability to shoot down rogue or hijacked aircraft.

The UK military's doctrine for provision of military aid to civil authorities (MACA) breaks such aid down into three categories of aid: military aid to government departments (MAGD), military aid to the civil power (MACP), and military aid to the civil community (MACC). The provision of MACA is based on three criteria: (1) military aid should always be used as a last resort; (2) the civil authority lacks the capacity to cope with a crisis, and it is prohibitively expensive and unreasonable to expect it to develop such a capacity; and (3) the civil authority is capable of responding adequately to a crisis, but the need to respond is urgent and the civil authority lacks the ability to deploy quickly enough (UK Ministry of Defense, 2007). MAGD aid involves ensuring continuity of operations and precludes the military from carrying out law enforcement activities or even of being armed. MAGD operations can include support for civil authorities during industrial disputes, natural disasters, and animal disease outbreaks. MACP assistance is provided for the purpose of law enforcement and internal security and consequently includes the authority to use force. MACP activities include counterterrorism operations (as in Northern Ireland), drug interdiction, and other such operations and continue only until the civil authorities are able to take charge. Military support consists of three types of aid: category A assistance for a natural disaster or major accident, category B assistance for special projects or events, and category C assistance, involving the support of individual volunteer soldiers for social services (Rollins, 2005, pp. 38–40). In category A situations under the Civil Contingencies Act, civilian authorities may request military aid if they deem local civil resources to be inadequate or that the military can respond more rapidly with particular resources needed to safeguard lives (indeed, the military can be called upon under these conditions only when lives are in acute danger). These incidents must be truly urgent in order to qualify for category A status since category A assistance is neither military aid to the civil power (MACP) or military aid to other government departments (MAGD). Local commanders (usually the commander of a local unit or a local military installation) have the authority to provide military assets to the civil authorities prior to waiting for approval via the military chain of command (although the senior leadership of the military and the cabinet must be notified whenever category A assistance is provided). Category B assistance is considerably more routine and is undertaken if that activity provides social value, provides military training value, and is sponsored by a particular civilian authority (which reimburses the military, provides appropriate levels of liability coverage, and ensures the health and safety of military personnel). Moreover, such support can be extended provided that it does not hamper military preparedness, harm the private sector (through use of the military instead of private service providers), lead to undesirable press coverage for the military, or is associated with a political organization (UK Ministry of Defense, 2007, p. 5.5). Category C aid is provided on a case-by-case basis, depending on the suitability of the volunteer and military needs.

Much of the interface between the military and civil authorities is handled at the regional level. The British military appoints joint regional liaison officers (JRLOs) to interface and participate in planning with civilian authorities in a given geographic region. The JRLOs are supported by military liaison officers (MLOs), who are usually the commanding officers of military installations or units based in the particular region in question. These officers are supported, when needed, by single-service liaison staff, who can provide specialist advice relating to things such as search and rescue, nonconventional weapons attacks, and explosives (UK Ministry of Defense, 2007, p. 2.12).

The role of the JRLO is to

1. Coordinate the operations of the three military services in support of the civil authorities.
2. Fully brief personnel in civilian response and emergency management agencies with respect to the provisions of MACA and the capabilities, procedures, structures, and limitations of the military.
3. Ensure that the military role is included in planning and exercising (UK Ministry of Defense, 2007, p. 2.13).

The Association of Chief Police Officers, Terrorism and Allied Matters Committee (ACPO TAM) provides funding for a police chief superintendent to serve as a police military liaison officer (PMLO) and be the primary point of contact for the police and the military at the national level. A similar function is performed by two senior police officers in Scotland and one who is part of the police service of Northern Ireland. In addition to liaison duties with the military at the national and regional levels, the PMLOs are responsible for police–military integration during a response and developing joint police–military exercise training programs (UK Ministry of Defense, 2007, p. 2.16).

In addition to the role of the military on land, officers of the Royal Navy actually enjoy legal powers as customs and excise officers and have jurisdiction on UK territorial seas as well as British-flagged ships in international waters. However, these powers are rarely used and Royal Navy ships will usually have police officers from the Serious and Organized Crime Agency (SOCA) on board when carrying out customs and interdiction missions (UK Ministry of Defense, 2007, p. 2.19).

Italy

In Italy, national security is based both on military and civilian defense and on a government entity known as the Agency for Civil–Military Cooperation (COCIM), which acts as the structure for crisis coordination between the military and civilian agencies. Nationally, coordination of civil defense matters is under the authority of the minister of the interior, who chairs an interministerial technical committee for civil defense. However, the actual implementation of civil defense measures is the responsibility of the prefects of Italian provinces. The military responds to major domestic incidents within the structure of the National Service of Civil Protection (PROCIV). PROCIV is an operational department under the authority of the prime minister. PROCIV can activate the military only if civilian resources are overwhelmed. PROCIV has the authority to issue emergency ordinances that allow it to carry out requisitions and expropriations, limit freedom of movement in affected areas, mobilize medical and support personnel, and so on (Cabigiosu, 2006, p. 90).

The Italian military can operate to provide disaster and other types of domestic assistance either via PROCIV or, in exceptional circumstances (where there is immediate danger to human life), on the authorization of a local military commander (Cabigiosu, 2006, p. 90). In the realm of terrorist attacks, the military will respond in rescue operations only if civilian institutions cannot cope with the situation. The only area in which the military has the lead in responding to terrorist attacks is in the case of airborne attacks. The military also has a long history of providing fairly frequent temporary support to civilian law enforcement agencies. When authorized by the Ministry of the Interior and the national police, the military can be deployed in two ways (each with a different legal and jurisdictional status). In the most basic type of deployment in support of law enforcement, soldiers must be accompanied by police officers (who exercise most of the police powers),

FIGURE 5.4   Italian army regiment on parade. This Wikipedia and Wikimedia Commons image is from the user Jollyroger and is freely available at http://commons.wikimedia.org/wiki/File:2june_2007_316.jpg under the Creative Commons Attribution-Share Alike 2.5 Generic license.

whereas in the second type of deployment, troops are given full policing powers. These deployments can last for a few weeks to several years (Cabigiosu, 2006, p. 92). The military also plays an important role in providing security for special events.

Overall, there are three different categories in which the Italian military can operate with respect to homeland security operations: as an auxiliary police force, providing direct support to the police, or providing indirect support to the police (see Sidebar 5-2). In March 2001, an internal security law (Law 128-26) was promulgated which authorized the armed forces, in specific and exceptional situations, to deploy soldiers (at the request of provincial prefects and for periods of up to six months at a time) to replace police officers in surveillance and security tasks so that the police could focus on fighting crime (Cabigiosu, 2006, pp. 102–103). The government thus has the legal authority, with a specific decree for a stipulated period of time and geographic area, to grant the military law enforcement powers and allow it to operate independently as an auxiliary police force (although the prefect in the area must validate the military's rules of engagement) (Serino, 2003, p. 5).

In addition, the military can be deployed under this legislation and Ministry of Defense regulations to carry out joint operations with the police whereby the military provide control and security over the broader area in

Sidebar 5-2: Operation "Forza Paris" in Sardinia (July–September 1992)

After the so-called "anni di piombo," i.e. the years marked by the terrorist threat when Army Units had been employed to co-operate in controlling lines of communication and after the Gulf War when they had been tasked with garrisoning sensitive targets, Army Units underwent a cycle of training activities, mainly in Aspromonte, aimed not only at improving the units' preparation but also at "bringing life" to thinly populated areas where organised crime had always been deeply rooted.

These training cycles were aimed at co-operating with Police Forces—though not directly—in that the very presence of the military helped to stress that the State was actively present in that area both physically and symbolically.

The most important exercise was held in Sardinia in the summer of 1992 and represented a real step forward in the employment of the Army for internal security tasks.

Immediately after a little boy, Faruok Kassam, had been kidnapped and mutilated, the decision was taken to hold a wide exercise to control Sardinian territory. This kidnapping roused a deeply felt indignation in the public at that time.

This exercise and the following re-deployment of powerful military units in Sardinia were not only meant at finding child Farouk. The aim was to carry out training activities to thoroughly control the territory in close co-ordination with police forces.

The goal of the exercise "Forza Paris" (from July 15 to September 22, 1992) was to train soldiers to control a wide and inaccessible area, indirectly co-operating with Police Forces and to highlight the presence and solidarity shown by the State to Sardinian people also through social events.

The deployment of Large Units in Sardinia proved to be an important and meaningful way of testing the units' combat readiness; in fact it took place after extremely short preparation cycles (5–6 days) for each Brigade.

The first unit to enter the Mamoiada–Oliena zone of employment was the airborne tactical group "Susa," an AMF(L) assigned unit which was therefore at the highest national combat readiness level. All other Battalion level units were successively deployed.

In all, 24 tactical groups took part in the exercise employing nearly 8,000 men, 2,000 vehicles and 3,000 tons of materiel airlifted to the island with 65 planes and 16 ships. To reach the major goal set for "Forza Paris," that is the control of the territory, the area of the exercise had been divided into three sectors: two at Brigade level and one at Regiment level.

Different areas of responsibility have then been identified within each sector where each Battalion carried out its training activities based on realistic criteria following counter area interdiction procedures.

It was mainly a matter of carrying out mopping-up actions in those areas assigned to each platoon level unit based on the "chessboard" subdivision of the area of responsibility of all Companies and Battalion. This way, each platoon was able to check a 3 km^2 area per day on the average.

Mopping up operations have been completed by patrolling actions (reconnaissance, scout, liaison and security) along established routes. As a rule,a representative of the Carabinieri Corps served in each platoon level to carry out police tasks which cannot be performed by military forces.

Each activity implied creating observation and warning posts at every key point. Concurrently, technico-tactical combat training activities and firing exercises were held at Macomer and Sassari ranges. To effectively monitor all ongoing operations, forces engaged in the exercise were put under OPCOM of the Sardinia Miltary Region Commander who availed himself of an ad-hoc co-ordination centre at Nuoro under the Military Region Deputy Commander supervising all Brigade Commanders.

Together with these training activities, a great number of social events involving Sardinian people were held, aimed both at providing support and promoting the image of the Armed Forces.

As for the results obtained, the exercise allowed for a remarkable reduction of petty crime in the Barbagia region. In 1995, in particular, from July to September, the following events took place:

cattle-stealing: 43 cases, with an 88% reduction compared to the same period of the previous year;

arson: 101 cases with a 53% reduction;

bomb attacks: 37cases, with a 76% reduction.

Moreover, in few months, the soldiers were not only able to complete road network and water supply works, recovery of fire damaged areas, disinfestation and disinfection operations; they were also able to overcome all formal obstacles, gathering consensus and winning the friendship of the traditionally marginalized local population. And the most important show of this unity of feeling was the mobilization of the majority of soldiers to collect blood units: resulting in 2,235 voluntary blood donors during the seventy days of the exercise. (Italy, Ministry of Defense, 2009)

which an operation is being conducted while the police carry out arrests, investigations, and other targeted activities (Cabigiosu, 2006, p. 103). In this case, military personnel are not given police powers, and thus every military unit deployed must include a police officer and the military operate under police command (Serino, 2003, p. 6). The provision of indirect support to the police essentially involves the military carrying out training exercises in areas used as sanctuaries by organized crime. This deployment does not involve the military exercising police powers but is designed to help deprive organized crime of the freedom of movement in carrying out criminal acts (mirroring in some ways aspects of a counterinsurgency operation) (Serino, 2003, p. 6). In all cases where the regular military is deployed, there can also be detachments of the carabinieri, in which case those personnel naturally have full police powers.

The Italian military has considerable experience in domestic deployment and law enforcement. In 1992, in response to mafia attacks that killed two investigate judges, Italy launched its largest homeland security operation since the end of World War II, deploying a total of over 150,000 troops over a period of six years on the island of Sicily to combat organized crime. To provide legal sanction for military law enforcement vis-à-vis civilians, the government approved Law 349, providing soldiers with police powers (Serino, 2003, pp. 3–4). The military has thus been employed for law enforcement–related missions that range from conducting surveillance, protecting sensitive facilities and targets, manning checkpoints, maintaining cordons around operational areas, conducting sweeps in rural areas, to patrolling along the railways and highways.

France

In France, the objective of increasing domestic security involves an all-hazards approach that includes an intensive and regular use of military resources. The French authorities believe that distinctions between internal and external security are disappearing, particularly in the context of dealing with terrorism, and that the military must play a central role in the protection of populations, institutions, territory, major economic activities, and "essential cultural values" (France, Ministry of Defense, n.d., p. 8). A recent French government white paper on defense noted that protecting the French population and French territory requires, among other things, the permanent surveillance of the country and its approaches via sea, land, air, and space assets (some of which are military) as well as the preparation of a reserve of some 10,000 troops to support civil authorities in disaster management (French President, 2008, pp. 11–12).

In France, the military plays a role in all three homeland security missions: domestic security (sécurité intérieure, which involves protection of persons and property), civil security (sécurité civile, which involves disaster management), and defense of the national territory (défense sur le territoire) (Vaultier, 2006, p. 203). All of these missions fall under the purview of the Ministry of the Interior. The ministry is responsible for public order and the protection of persons, property, and critical infrastructures and resources.

In accomplishing this mission, the minister of the interior is supported by the Ministry of Defense and the armed forces. Consequently, the minister of the interior is in charge not only of the national police (Police Nationale) and civil defense departments but also the military policing force, the Gendarmerie Nationale (Vaultier, 2006, pp. 204–205). The prime minister of the republic has the authority to involve the regular military (outside the gendarmerie) in any homeland security mission (usually under the command of the police or gendarmerie). In most cases, this involves supporting law enforcement bodies (civilian and gendarmerie) with joint patrols and similar activities (Vaultier, 2006, p. 207). In all cases of military intervention, the military forces follow their own command structure, but their missions are defined by the competent civil authority (whether at the national or prefect level), including the authority, or lack thereof, to use weapons (Vaultier, 2006, p. 209).

Australia

In Australia, the Australian Defense Force (ADF) can be used domestically only when civil authorities are unable to cope with an incident and use of the ADF is a last resort. Nevertheless, the ADF is often deployed to cope with natural disasters, and each year there are roughly 30 instances in which the military provides assistance to civil authorities (Yates and Bergin, 2010, p. 2). There are two overarching frameworks in which the ADF can be used domestically. The first, the Defense Force Aid to Civil Authorities (DFACA), involves assistance to law enforcement authorities in the performance of law enforcement tasks (usually, counterterrorism missions), and in certain cases ADF personnel employed on DFACA missions may be armed and authorized to use deadly force (Government of Tasmania, 2006, Section 5). The second, Defense Assistance to the Civil Community (DACC), is provided via three categories of support. Category 1 assistance occurs when immediate action is needed to safeguard life or prevent significant physical damage and the events precipitating this response are not expected to last more than 24 hours. These types of events include flooding, brushfires, and urban fires. Under these conditions, the local ADF commander has the authority to activate his/her forces and provide category 1 assistance. Category 2 assistance is brought to bear in life- or major property-threatening events that are expected to last for longer than 24 hours. This type of longer-term assistance request comes from the relevant state or territorial emergency authorities to the Ministry of Defense. Finally, category 3 assistance involves recovery efforts such as provision of supplies to an affected area, environmental cleanup, repair efforts, and road and bridge restoration. Such assistance also requires a request from the state or territory government to the Ministry of Defense (Yates and Bergin, 2010, p. 6).

FIGURE 5.5   Australian Defense Force medics during a training exercise. U.S. Navy photo @ Wikimedia Commons.

If the ADF is deployed, it retains its military chain of command but is subject to the overall authority of the relevant civil authority. As noted in the 2009 Department of Defense white paper, "The ADF will also need to be able to respond to an increasingly complex domestic security environment, in which the lines between traditional concepts of external and domestic security are increasingly blurred. In this context, the ADF has to be able to contribute to the deterrence and defeat of attacks by nonstate actors with strategic capabilities, especially should such groups ever acquire WMD, and to support civil authorities in relation to domestic security and emergency response tasks" (Australia, Department of Defense, 2009, p. 54).

Canada

In Canada, the military is seen as having a clear homeland security role. The 1994 White Paper on Defense notes that the country's armed forces are expected to:

- Demonstrate, on a regular basis, the capability to monitor and control activity within Canada's territory, airspace, and maritime areas of jurisdiction;
- Assist, on a routine basis, other government departments in achieving various other national goals in such areas as fisheries protection, drug interdiction, and environmental protection;

- Be prepared to contribute humanitarian assistance and disaster relief within 24 hours, and to sustain this effort for as long as necessary;
- Maintain a national search and rescue capability;
- Maintain a capability to assist in mounting, at all times, an immediate and effective response to terrorist incidents; and,
- Respond to requests for Aid of the Civil Power and sustain this response for as long as necessary. (Canada, Ministry of Defense, 1994, pp. 3–4)

Somewhat uniquely in the Canadian case, given the vast distances and remote regions of that country, the military is tasked with asserting Canadian sovereignty and enforcing Canadian law in remote areas—something that units like the Canadian Rangers are tasked to do. The military in Canada are also tasked with assistance to civil authorities in the area of air search and rescue and provide assistance to the Canadian coast guard with respect to marine search and rescue. They also assist in land search and rescue and have three coordination centers that respond to distress signals. The Canadian armed forces are also involved in homeland security missions as disparate as drug smuggling interdiction, fisheries protection, and environmental protection. In short, the armed forces in Canada are heavily involved in provision of support to civil authorities on a regular and ongoing basis in a broad range of areas.

TYPE OF FORCES AND MISSIONS

United Kingdom

In the UK, the reserve army is responsible for civil response. There are currently 14 specially created civil contingency reaction forces (CCRFs), which correspond to each one of the country's territorial army brigade regions. Each CCRF consists of 500 army reservist volunteers who can be mobilized on short notice and serve in the CCRF in addition to their normal reservist role (UK Directorate of Reserve Forces and Cadets, 2005, p. 3). Unlike regular reservists who have graduated from full-time military service, volunteer reservists in the UK enter the military directly from civilian walks of life. British military doctrine does allow for the use of any regular and reservists forces domestically in the event of a severe crisis (Stevenson, 2006, p. 26). Overall, however, the preference is to utilize the capabilities of the military for niche activities (i.e., CBRNE, maritime and air missions, etc.) because the use of such capabilities is easily understood and accepted by the media, whereas employing a massive and across-the-board military response might be seen by the media and the public as "calling in the cavalry," thus creating alarm and suggesting that the civilian authorities are unable to cope (House of Commons, Defense Committee, 2009, p. 9). Military participation in emergency events requires the request of the chief constable (police chief) in whose jurisdiction the event has occurred through the Home Office and also requires the approval of the defense secretary (UK House of Commons, n.d.).

Germany

In Germany, the territorial army (Territorialheer), which consists almost entirely of reservists, is tasked with providing military support for disaster relief operations (although its primary mission is still the Cold War mission of resisting a foreign invasion of German territory). The German authorities consider anyone who has ever served in the military as a reservist (and was until recently in one of the few NATO countries that maintained a system of mandatory conscription for males—albeit with the alternative of civilian public service for those who did not wish to wear the uniform) (Weitz, 2007, pp. 49–51). The Bundeswehr reserve thus sees its role as contributing to deployments to cope with threats to the citizenry and critical infrastructure, the restoration of social order, the ongoing surveillance of airspace and maritime waters, and rescue and evacuation operations (Germany, Ministry of Defense, 2003, pp. 9–10).

FIGURE 5.6 Deployment of carabinieri corps troops. This Wikipedia and Wikimedia Commons image is from the user JoJan and is freely available at http://commons.wikimedia.org/wiki/File:Firenze.Carabinieri01.JPG under the GNU Free Documentation license.

Italy

In Italy, in addition to the air force, navy, and army, there is a fourth military branch, the gendarmerie corps (known as the Arma de Carabinieri). The majority of Carabinieri units are responsible for law enforcement missions and the maintenance of public order and also focus on specialized law enforcement activities such as fighting organized crime and the drug trade. The Carabinieri also conduct military duties such as military police and security tasks as well as oversees policing deployments. They have a hybrid command in that they report to the minister of the interior with respect to their law enforcement and public security tasks and to the military chain of command in the context of their military duties. There are some 5000 Carabinieri stations across the country as well as mobile territorial battalions. Maritime, border policing duties, and customs enforcement are carried out by another force, the financial police (Guardia di Finanza), which, although under the authority of the Ministry of the Economy and Finance, is a military force. Even Italy's forestry police is a military policing organization, and in fact, the country's national police, which is the only civilian policing entity, was, until the early 1980s, also a military policing entity.

France

In France, the primary law enforcement and public security body in 95 percent of French territory is a military policing force, the Gendarmerie Nationale. Separate military units are in charge of the firefighting units in France's two largest cities, Paris and Marseille, with engineering units responsible for fighting fires in Paris and naval units responsible for fighting fires in Marseille. The Gendarmerie includes naval, helicopter, counterterrorist, riot police, and other specialized units, and its mission includes the maintenance of public order, criminal investigations, intelligence gathering, policing the roads, protecting critical infrastructure targets, responding to WMD attacks, and protecting the country's nuclear weapons (Vaultier, 2006, pp. 207–208). The Gendarmerie has two primary components, the Gendarmerie Départmentale and the Gendarmerie Mobile. The former is the primary policing force outside the major metropolitan areas (where the Police Nationale has primary jurisdiction) and the latter plays the dominant policing role in the counterterrorism context. The Gendarmerie is also the prime law enforcement agency tasked with flight and airport security and posts sky marshals on sensitive flights. Approximately half of the French army's reserve forces are reserve troops of the Gendarmerie (although these generally do not enjoy full police powers). Those military reservists who do not serve in the Gendarmerie reserves can still be called up for homeland security

FIGURE 5.7 French gendarmes. This Wikipedia and Wikimedia Commons image is from the user David Monniaux and is freely available at http://commons.wikimedia.org/wiki/File: Gendarmes_501585_fh000019.jpg under the Creative Commons Attribution-Share Alike 3.0 Unported license.

duties. Military personnel discharged from regular army service are required to serve an additional five years in the Operational Reserve (Réserve Opérationnelle). During crises, the Operational Reserve can be called upon to assist the population and assist in maintaining continuity of governmental operations. Moreover, in times of extreme crisis, these reservists can be used to protect public facilities, secure the borders, and assist in provision of domestic security (Weitz, 2007, p. 39).

Canada

In Canada, the military is broken down into the regular force and the reserve force. The reserve force consists of four components, of which three, the primary reserve, the supplementary reserve, and the Canadian rangers play prominent homeland security roles. The army component of the primary reserve (known as the "militia") supports regular army forces in crises, and the naval reserve performs costal security operations that are not part of the mission of the active-duty navy (Canada does not have a coast guard), including port security and control of shipping (Weitz, 2007, pp. 59–60). The naval reserve operates Canada's 12 coastal defense ships and possesses port security/harbor defense units that can be deployed anywhere in the country. The supplementary reserve forces can be called up for periods of limited duration to supplement either the regular force or the primary reserve and thus can serve in a homeland security capacity. Finally, the Canadian rangers play what is perhaps the most regular and intensive homeland security role of all the reserve components, as their mission is to provide a military presence in the isolated and sparsely settled far north of the country in areas that are so isolated that the regular force does not operate there (Weitz, 2007, p. 61) (see Sidebar 5-3). Consequently, the rangers are often the first responders when these isolated areas experience natural disasters. The rangers, the majority of whom are volunteers, are formally tasked with reporting unusual activities, collecting data in support of military operations, conducting surveillance and sovereignty patrols, providing local expertise to reserve and regular army units, and providing local assistance to search and rescue activities (*Canadian Rangers*, n.d.). After 9/11, the role of reservists in coping with domestic emergencies was enhanced and the government has repeatedly mobilized reservists to deal with weather crises, plane crashes, and other disasters.

Sidebar 5-3: The Canadian Rangers

Canada's vast northern expanse and extensive coastlines have represented a significant security and sovereignty dilemma since the Second World War. With one of the lowest population densities in the world, and one of the most difficult climatic and physical environments to conduct operations, a traditional military presence is prohibitively costly. As a result, the Canadian Rangers, a little-known component of the Reserves, have played an important but unorthodox role in domestic defence over the last sixty years. This component of the CF Reserves, managed on a community level, draws on the indigenous knowledge of its members, rather than "militarizing" and conditioning them through typical military training regimes and structures. Embodied in its communities and peoples in isolated areas, the Canadian Forces continue to benefit from the quiet existence of the Rangers. While commentators typically cast the Canadian Rangers as an arctic force—a stereotype perpetuated in this article—they are more accurately situated around the fringes of the country. Their official role since 1947 has been "to provide a military presence in those sparsely settled northern, coastal and isolated areas of Canada which cannot conveniently or economically be provided by other components of the Canadian Forces." They are often described as the military's "eyes and ears" in remote regions. The Rangers also represent an important success story for the Canadian Forces as a flexible, inexpensive, and culturally inclusive means of "showing the flag" and asserting Canadian sovereignty while fulfilling vital operational requirements. They often represent the only CF presence in some of the least populated parts of the country, and serve as a bridge between cultures and between the civilian and military realms. The Rangers represent an example of the military successfully integrating national security and sovereignty agendas with community-based activities and local management. This force represents a practical partnership, rooted in community based monitoring using traditional knowledge and skills, which promotes cooperation, communal and individual empowerment, and improved cross-cultural understanding.

The Canadian Rangers: An Overview

The Canadian Rangers were first conceived amidst the modern realities of the Second World War and the Cold War. The force was originally modeled after the Pacific Coast Militia Rangers (PCMR), a home guard established along the West Coast in 1942 to meet potential Japanese incursions. The PCMR was predicated on the idea that unpaid volunteers, often too old or too young to serve overseas, could perform useful military functions while carrying out their everyday civilian lives on the land and sea. Given their intimate knowledge of local areas, they could provide intelligence, act as guides, and delay an enemy advance using guerrilla tactics. All told, more than 15,000 British Columbians served in the PCMR before it was stood down in late 1945. By 1947, chilly superpower relations and a new focus on northern security, coupled with renewed sovereignty concerns related to a US military presence in the North, led the government to establish the Canadian Rangers as a Corps of the Reserve Militia. This force would be unpaid, provided with armbands, a .303 rifle, and 200 rounds of ammunition a year. In war, they would serve as coast watchers and guides to regular troops, assist authorities in reporting and apprehending enemy agents and saboteurs, provide local defense against small enemy detachments, and undertake ground search and rescue (GSAR) operations. Their peacetime roles were similar, focusing on guiding troops on exercises, collecting detailed information about their local areas and reporting any unusual activities, and providing GSAR parties when tasked. They were recruited from local areas, commanded by civilian leaders from their communities, and carried on their daily lives. The Rangers survived the oscillating cycles of military concern about the North through the second half of the 20th Century. Military and political interest in the Rangers diminished by the late 1950s, when technological solutions like the Distant Early Warning (DEW) Line were conceived to secure the continent. Although the Rangers were left to "wither on the vine," they did survive—largely because of the extremely small price tag attached to them. During the 1970s the "Northern" Rangers enjoyed some growth as a sovereignty-bolstering measure but it was not until the mid-1980s, when the voyage of the U.S. Coast Guard vessel *Polar Sea* renewed sovereignty concerns related to the Northwest Passage, that the Rangers underwent dramatic growth. By 1992 the national strength of the force rose to 3200 (and doubled in the territorial north). The Rangers grew "North of 60" after 1970 because the basic structure already existed and was very inexpensive, but also because a "new security discourse" emerged. Military activities in the arctic could not longer be divorced from domestic socio-economic, cultural, and environmental health issues. Aboriginal leaders repeatedly called for the demilitarization of the arctic on social and environmental grounds, and construed the military presence as a threat to their peoples' security. These pressures encouraged program assessment using both state-centred security and broad social criteria. Military officers noted that the public and Native leaders took great interest in the Rangers, and that "while their motivation and enthusiasm may not be entirely military oriented, it is genuine and perhaps it is an excellent opportunity to seriously consider realistic and practical improvements in the Ranger force." Beginning in the late 1980s, explicit government statements increasingly stressed the socio-political benefits of the Rangers in Aboriginal communities and the force underwent remarkable growth during a general era of fiscal and personnel downsizing in the Canadian Forces. The Rangers were politically and publicly marketable as a military success story. There are currently 4000 Rangers in 165 patrols across Canada. Overall command is centralized at National Defence Headquarters, administered by the DCDS, while operational and administrative control of Canadian Rangers in the field is delegated to the Commander of Canadian Forces Northern Area (CFNA) and to the Commander of Land Force Command (LFC).

In 1998, five Canadian Ranger Patrol Groups (CRPGs) were formed to coordinate the activities of Ranger patrols in their respective areas of responsibility. Until 1998, the Rangers existed as a subcomponent of the reserves. Reorganization into CRPGs made them a total force unit, with each Patrol Group commanded by a major (CO) and a captain (DCO). Military sovereignty patrols travel in very austere areas in extreme conditions to assert Canadian sovereignty. The core concept is that citizens in isolated and coastal communities, far from the main southern belt of population, can serve as the military's "eyes and ears" during the course of their everyday lives. Rather than asking these individuals to leave their communities to join the Regular Forces or Primary Reserves, they can make meaningful contributions to their country at home. The perceived value of individual Rangers is directly linked to their civilian experiences and practices. First and foremost, a Ranger has usually lived in an area for a long time and is intimately familiar with the local people, terrain, and weather conditions. Second, he or she is, ideally, at least, working on or near the land or sea, and thus in a position to observe unusual incidents. Third, a Ranger possesses certain skills and expert local knowledge that supports the force's role in the CF. Correspondingly, membership in the Canadian Rangers is distinct from the regular force and other reserve force units. The only formal entry criteria is that men and women who join are over eighteen years of age, are Canadian citizens or landed immigrants, in good health, and willing to be members of the Canadian Forces. There is no upper age limit. So long as an individual can still perform his or her duties, he or she can remain a Ranger. Some anecdotes are truly amazing—74-year-old Ranger Peter Kuniliusie of Clyde River, Nunavut, retired in November 2004 after *52 years* of continuous service with the force. Indeed, it is accommodation and acceptance of social diversity and experience that makes the Ranger concept unique and feasible. Apart from annual Ranger training exercises conducted by Regular or Reserve Force instructors, ongoing Ranger activities are often indistinguishable from civilian practices. An excellent example is ground search and rescue (GSAR). Rangers often participate in ground searches for lost individuals or groups without the prior knowledge of their group headquarters. As the only orga-nized group in many isolated communities, the Rangers are singularly equipped to assist SAR specialists, and their contributions generate significant media attention. In 1999–2000, for example, Rangers and personnel from 1 CRPG took part in 164 volunteer GSAR missions, one medevac, and one emergency rescue. Without official direction, however, the Rangers, even if they are wearing their uniforms, are not performing the task as Canadian Rangers *per se*; they are acting as private citizens and are not paid. Although this blurred line between their "civilian" and "military" identities remains vague, in emergencies individual Rangers act first and foremost as community members. The Rangers also represent an important means of sharing knowl-edge within Northern communities. The potential loss of traditional Canadian Ranger patrols, by virtue of their locations and largely Aboriginal composition, are representative elements of the CF in this respect. All members of the Canadian Rangers are Canadian citizens. Nonetheless, their diversity embodies the country's multicultural identity. Although there are no official statistics generated, the CRPG patrols are representative of the diverse ethnic composi-tion of the North. The majority of Rangers in the Yukon are "White" (as is the population itself). In the Northwest Territories the patrols reflect the geo-graphic and linguistic dispersion of Northern peoples. Most of the Rangers patrols south of the tree line are comprised of members of Gwich'in, Dene, MÕtis, and "White" communities. North of the tree line, most of the patrols are Inuit. In Nunavut, the Rangers are almost entirely Inuit and most operations are conducted in Inuktitut.

A Ranger patrol is rooted in its community, and operates on a group (rather than individual) basis. Each Ranger patrol is led by a sergeant, who is seconded by a master corporal, both of whom are *elected* by the other members of the patrol. So too are Ranger corporals, who command sections of a patrol at a 1:10 ratio. Elections are held in patrol communities on an annual basis and exemplify the self-administering characteristics of the Rangers. Patrol leaders are the only members of the CF who are elected to their positions, and therefore are directly accountable to their "subordinates" in a unique way. Furthermore, while "hierarchical" on paper, Ranger "command" can be less rigid in practice. Decision-making in arctic communities is based upon consen-sus, and this is reflected in the patrols themselves. For example, instructors explained that when they ask a Ranger sergeant a question in some Nunavut communities, he or she will turn to the elders in the patrol for guidance prior to responding. In this sense, while the sergeant is theoretically in charge of a patrol, the practical "power base" may lay elsewhere. As a result, instructors must be prepared to present their plans to the entire patrol: the best way to approach any challenge is to sit down and discuss it with a patrol, offering more explanation than would be typical in the south. Warrant Officer Kevin Mulhern suggested that the "mission-focus" mentality should be reversed when dealing with the Rangers: it was often better to explain what the military wanted to accomplish with the Rangers, and then figure out with them what should be done in terms of a mission. In practice, patrols are not tasked out of an expectation that each individual can do everything, or that a leader possesses the strongest skill set, but that someone in the patrol has the skill set to conduct the patrol while it completes a given activity. As a result, individual testing is limited as an indicator of a patrol's competencies. These units tend to respond better to communal efforts.

The Rangers are seen as an integral component of the government's strategic vision. Their official task list includes the following:

1 Conduct and provide support to sovereignty operations.

    a Conduct surveillance and sovereignty patrols (SOVPATs) as tasked 1.1.1.1. In 2003–04, for example, the Rangers conducted over 162 patrols of various types in the arctic, which contributes to CFNA's mandate to provide surface surveillance in its area of operation. SOVPATs also confirm that Ranger patrols can successfully plan and complete relatively complex tasks without direct supervision by a Ranger instructor. Therefore, they help to build confidence for patrols.

    b Participate in CF operations, exercises and training. Rangers help other CF elements prepare for arctic exercises or operations, provide local guidance, and teach traditional survival skills. Ranger participation in sovereignty operations contributes directly to re-establishing the diminishing Land Force operational capabilities in the North.

    c Report suspicious and unusual activities that are out of character with the routine of an area. For example, Rangers have reported several submarine sightings since 1997 that have drawn significant media interest.

2 Conduct and provide assistance to CF domestic operations.

    a Conduct territorial, coastal and inland water surveillance as required/tasked.

    b Provide local knowledge and expertise.

Rangers have recently acted as observers and guides during West Coast operations to counter illegal immigration, and served as advisers during *Exercise Narwhal* around Pangnirtung and Cumberland Peninsulain August 2004.

   c Provide assistance to other government departments.

   d Provide local assistance and advice to Ground Search and Rescue operations.

   e Provide support in response to natural disasters and humanitarian operations.

Although not intended as a "force of first resort," such as police, fire and medical specialists, Rangers continue to support their communities in cases of domestic emergency.

In 1999, members from eleven of the fourteen Canadian Ranger patrols in Nunavik (northern Quebec) arrived in Kangiqsualujjuaq in response to the massive avalanche. The extraordinary display of Ranger co-operation resulted in Chief of the Defence Staff awarding a Canadian Forces Unit Commendation to 2 CRPG. Potential emergencies that Rangers prepare to encounter include a major air disaster or a cruise liner running aground.

Several omissions are worth noting. Although the original 1947 list of Ranger tasks included tactical actions to delay an enemy advance, this expectation has been officially dropped. The CF no longer expects the Rangers to engage with an enemy force: indeed, they are explicitly told not to assist "in immediate local defence by containing or observing small enemy detachments pending arrival of other forces" nor to assist police with the discovery or apprehension of enemy agents or saboteurs. Presumably, such tasks would put the Rangers at excessive risk given their limited training. Furthermore, the Rangers cannot be called out in an Aid to the Civil Power capacity, given training limitations and the civil-military identities embodied in the force. Given the positive working relationship that the Rangers embody between the CF and Aboriginal communities, for example, a situation resembling the Oka Crisis could place the Rangers in a confrontation with Native militants and would have a severe, deleterious impact on their credibility. The final Ranger task is the most general and basic—to maintain a CF presence in the local community. This is fundamental, given the reductions in Northern military operations over the last several decades and the DND's commitment to having a "footprint" in communities across the country. The Rangers represent more than 90 percent of CF representation north of the 55th parallel, and provide a special bond with their host populations. They are far more than the military's "eyes and ears"; they are an organized group that communities can turn to for numerous activities. Unorthodox roles, such as breaking the Yukon Trail for dog mushers, ensuring that polar bears do not attack unsuspecting trick-or-treaters in Churchill, and welcoming dignitaries, bring favorable media attention. Their participation in Remembrance Day parades reinforces the intimate and continuing, positive military presence in Canadian life. They are simultaneously citizen soldiers and citizen servers, intimately integrated into local community activities, ensuring that the CF is not socially isolated or structurally separated from Northern societies. (Lackenbauer, 2005–2006)

Australia

The Australian Defense Force (ADF) maintains forces for domestic exigencies. The ADF's Incident Response Regiment, which was created in 2002, has the capacity to respond to WMD incidents domestically and overseas. The regiment includes intelligence, signals, medical, ordnance, transport, electrical and mechanical engineers, and scientists. The ADF ground forces also operate an emergency response squadron, which provides the army with an emergency response capability in the areas of fire fighting and search and rescue (Commonwealth of Australia, Incident Response Regiment, 2005, p. 18). The tactical assault group can respond on short notice to state requests for commando operations to free hostages, capture facilities, and conduct high-risk searches (Australian Government, 2006, p. 71). Finally, the Reserve Response Force (RRF) serves as the Australian army reserves counterterrorism force. The RRF is tasked with supporting the regular army in response to terrorist attacks and civil emergencies and will be able to conduct security tasks (i.e., cordoning, searches, traffic control, cite protection, etc.) in support of the active-duty military (Burton, n.d.).

DISASTER RESPONSE AND INFRASTRUCTURE PROTECTION

United Kingdom

In the UK, the military can be called upon by a chief constable (via the Home Office) with the consent of the minister of defense to provide aid in bomb disposal, dealing with CBRN materials, and other tasks related to large-scale disasters, such as establishing cordons, evacuation, provision of temporary housing, and provision of food and medical care to disaster victims (Stevenson, 2006, pp. 29–30). British military doctrine provides for three types of aid to civil authorities: the provision of military personnel and equipment for emergencies and routine situations, the use of military forces for nonmilitary governmental duties, and the provision of military aid to restore law and order in the wake of a civilian request for assistance. The British army plays an important role in protecting such crucial sites as nuclear facilities, conventional power plants, communications and computer networks, and ports and airports. In February 2003, over 1000 troops were deployed in and around Heathrow airport outside London, due to fears that Islamist radicals were planning to fire a surface-to-air missile at an airliner (Stevenson, 2006, pp. 26–27).

Italy

In Italy, in principle, the military can intervene in all areas of civil support. This includes contingencies such as disaster relief and military support of law enforcement. The Italian army plays a role in supporting police forces (including the carabinieri) in protecting critical infrastructure targets in cases of a terrorist threat. In most cases, soldiers deployed on these missions are given police powers to detain persons and search them and their vehicles (Cabigiosu, 2006, p. 112).

France

In France, the Civilian Protection Agency (Protection Civile) is the primary agency for organizing disaster response and includes several military units. In France, the Gendarmerie Nationale patrols 90 percent of the road network and the railway network outside the greater Paris region and are responsible for protecting critical infrastructure targets, including government buildings, defense infrastructure, energy and communication facilities, and France's nuclear weapons (Vaultier, 2006, pp. 219–220).

CONCLUSION

The countries discussed briefly in this chapter all share a willingness to employ the military in a domestic context but differ in terms of the extent and role that the military plays in these types of missions. Arraying these countries along a continuum in terms of the degree to which the military plays a homeland security role is difficult because there are so many different ways that the military can be and is employed in the various countries, and the missions are often so disparate as to make comparisons difficult. In general, the Germans, British, Australians, and Canadians restrict the use of the military to clearly defined crisis situations, with several layers of military response coming into play that are dependent on the degree of the crisis and other factors. However, given its huge geographic expanse and the fact that much of its territory is extremely sparsely inhabited, Canada has found that it needs to use the military (in the form of the Canadian Rangers) for provision of day-to-day community services and the assertion of Canadian sovereignty in these isolated areas. France and Italy use their respective military establishments much more extensively in the domestic sphere since these (albeit under the authority of their respective ministers of the interior and not ministers of defense) operate as the prime law enforcement agencies in rural and semirural areas as well as with respect to specific types of law enforcement in urban areas. Moreover, both countries have used military forces outside the Gendarmerie and Carabinieri for domestic policing and disaster response missions. Finally, Israel does not normally employ the military domestically, but the military does have oversight and authority in matters of training, doctrine, and preparedness and, in certain circumstances, can be given authority over all the first-responder agencies in order to coordinate efforts in the case of war, large-scale disasters, or mass-casualty WMD terrorist attacks. In Israel, the military also plays a prominent role in border security patrolling most of the countries borders as well as the Separation Barrier with the West Bank. Border security issues will be addressed in the next chapter.

ISSUES TO CONSIDER

- How does the approach of common law countries (UK, Canada, Australia) in the domestic use of the military compare with that of countries such as France and Italy?
- What are the similarities and differences in the use of the military for civil support in the UK, Canada, and Australia?
- What role does the military play in the vast and largely empty northern reaches of Canada?

CHAPTER *6*

BORDER SECURITY AND IMMIGRATION POLICIES

In this chapter we focus on border security and immigration policies and survey approaches, strategies, and institutions in a number of countries as well as, and primarily with respect to, the supranational European Union (EU) and the affiliated Schengen system. The EU is of particular interest because in the context of border security it acts, to some degree, as a federative body and thus is of more relevance to the United States, which also has to deal with overlapping sovereignties and corresponding entities, particularly in the areas of maritime and border security but also with respect to some aspects of immigration enforcement (a federal duty, but one with which local law enforcement entities frequently have to interface). Like the United States, the EU as a whole faces challenges in terms of illegal immigration over land and maritime borders.

THE EU: OVERVIEW

The European Union encompasses nearly 4.5 million square kilometers and has the seventh largest territory in the world: extending from Finland and Ireland in the northeast and northwest to Portugal, Spain, and Cyprus in the southwest and southeast, respectively. It shares land borders with Norway, Switzerland, Russia, Belarus, the Ukraine, and Turkey (as well as a number of states in the former Yugoslavia and Albania that are surrounded by EU states). EU member Spain also shares land borders with Morocco, due to its possession of two enclaves on the Moroccan coast. The EU also possesses an extensive southern maritime littoral with North Africa as well as islands (in particular the Canary Islands and Malta) off the coast of Africa.

Comparative Homeland Security: Global Lessons, First Edition. Nadav Morag.
© 2011 John Wiley & Sons, Inc. Published 2011 by John Wiley & Sons, Inc.

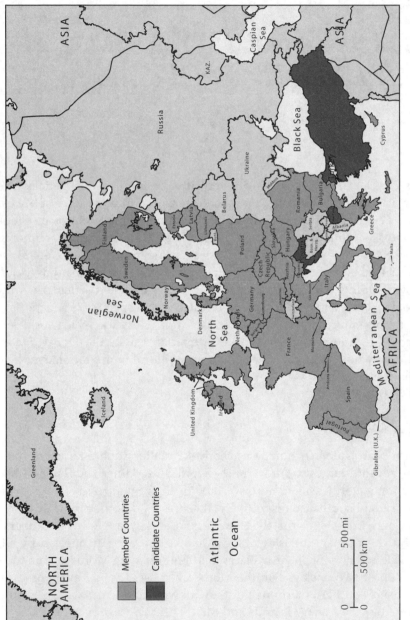

FIGURE 6.1 European Union.

The European Union has its antecedents in an organization established in 1950, the European Coal and Steel Community (ECSC), which was designed to bind France and Germany together (although also including the Benelux countries—the Netherlands, Belgium, and Luxembourg—as well as Italy) and establish multinational control over what were then two critically important precursors to military armament and war: coal and steel. The ECSC served as a useful way to control the use of coal and the production of steel in what was then West Germany (thus reducing the postwar likelihood of German rearmament—something that was of concern to Europeans as it was only five years after the end of World War II) but also established the principle that Europe could break out of the cycle of self-destructive wars only through economic and political unification. The arrangement was broadened and expanded upon with the signing of the Treaty of Rome in 1957, which established the European Economic Community (EEC, also known as the European Common Market) and the European Agency for Atomic Energy (Euratom). The three institutions were collectively known as the European Community (EC). Throughout the 1960s and 1970s, the EC expanded both in terms of its authority to determine matters of policy and in terms of the number of countries joining the arrangement. In 1992, the Treaty on European Union was signed, formally creating the European Union and setting rules for a single currency, closer cooperation in law enforcement and domestic affairs, and further strengthening the already existing provisions for unregulated trade between member states. At present, there are 27 EU member states.

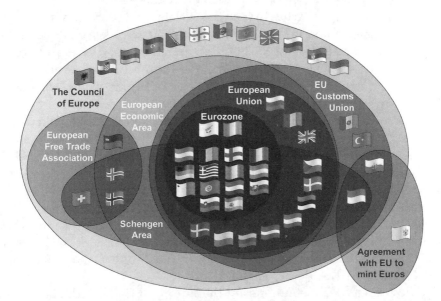

FIGURE 6.2 Supranational European bodies.

The EU is not a federation but neither is it a confederation of completely independent national governments; rather, it is a hybrid structure that has elements of both a single federal government and independent national governments in a confederative relationship. It is true that the member states of the EU remain technically independent and participate in the EU at will, but the economic costs of pulling out of the EU are such that no member state has thus far seriously contemplated this option and, over the years, EU institutions have become stronger and a series of treaties have expanded the power of the EU at, in effect, the expense of the sovereignty of the member states. EU institutions thus enjoy authority and even precedence over national governments in those areas of law and policy delegated to the EU by its member states. Entire areas of policy (including border security and significant parts of immigration policy, which are the subject of this chapter) are under the authority of Brussels (the EU's de facto capital by virtue of the fact that the offices of most of the EU bureaucracy are located in the Belgian capital). As noted above, the EU governs on the basis of a series of treaties (including those mentioned above) that act as *primary legislation*, outlining the powers and responsibilities of EU institutions and thus determining which areas of law and policy member states have decided to turn over to EU control and responsibility. On the basis of these treaties, a large body of *secondary legislation* has developed over time that consists primarily of regulations, directives, and recommendations adopted by EU institutions (European Union, 2010).

According to the 2008 Treaty on the Functioning of the European Union (TFEU), the EU has the exclusive right to legislate in those areas over which it has been granted authority, and member states may legislate on these issues only if empowered to do so by the EU. In areas in which governance authority is shared between the EU and member states, member states are allowed to legislate and implement policy in the absence of EU legislation and policy (TFEU, Title 1, Article 2, 2008). The EU enjoys exclusive authority over customs policy, monetary policy, and common commercial policy and exercises extensive authority (although shared with the member states) over regulation of the European market, agriculture and fisheries policy, environmental policy, transportation policy, economic policy, energy policy, safety regulations, research and development, and the regulation of borders, immigration, and social policy (TFEU, Title 1, Articles 4 and 5, 2008). Finally, the TFEU gives the EU a support and coordination role, supplementing the policies and actions of member states in the areas of health policy, industrial policy, tourism, education, culture, and civil protection (TFEU, Title 1, Article 6, 2008).

The primary EU institutions are the European Council, the European Commission, and the European Parliament (EP). The European Council

FIGURE 6.3   European parliament. This Wikipedia and Wikimedia Commons image is from the user Cedric Puisney and is freely available at http://commons.wikimedia.org/wiki/File:European-parliament-strasbourg-inside.jpg under the GNU Free Documentation license.

(also known as the Council of Ministers) is the chief decision-making body of the EU and consists of one minister from each of the member states. The decision as to which minister from each member state will attend a given meeting of the Council depends on the issues on the agenda (e.g., agriculture ministers comprise the Council when that body meets to discuss the EU's agricultural policies). The Council's role is to represent the interests of the member states, and it shares legislative power with the European Parliament (EP) (in a system known as the *co-decision procedure*) so that either the Council or the EP can legislate but each must ratify the legislation passed by the other. The Council makes decisions based on either majority vote, qualified majority vote (in which larger countries have more votes than smaller ones), or unanimous approval, depending on the issue at hand, as different voting rules apply for different issues (European Union, 2010).

The European Parliament, based in Strasbourg, France, is the EU's direct representation body, and its membership is, accordingly, elected directly by the citizens of EU member states (via universal suffrage, for five-year terms). The EP used to be a largely consultative body, but in the wake of the 1992 Treaty of Maastricht, it was empowered through the introduction of the aforementioned co-decision procedure, whereby it is now on par with the EU

Council with respect to legislative powers. The EP also shares equal responsibility with the Council over adoption of the annual EU budget, which is proposed by the EU Commission and then debated and voted on by both the Council and the European Parliament. Members of the European Parliament (MEPs) also have the power to put oral and written questions to the Commission and Council and to supervise the day-to-day management of EU policies (European Union, 2010).

The last of the primary EU institutions is the European Commission, which, like the Council, is primarily based in Brussels. The Commission serves as the executive and bureaucratic arm of the EU responsible for managing and running the EU. The Commission consists of one commissioner from each member state, who is appointed for a five-year term by agreement between the member states and subject to the approval of the EP. The Commission proposes EU legislation, administers EU directives and regulations, and generally ensures the smooth functioning of this supranational entity. The Commission's job is to uphold the "common interest" and not favor any particular national government and, as "guardian of the treaties" the Commission must ensure that all EU regulations and directives are being implemented by the member states (European Union, 2010).

The EU also possesses a High Court of Justice, based in Luxembourg, which interprets EU legislation and ensures that member states comply with EU legislation. It can also override national legislation and, in certain situations, can hear cases from EU citizens against their own governments for failing to comply with EU law. The court is made up of one judge from each member state appointed by joint agreement of the member states for a renewable term of six years. There are also a number of additional EU institutions, including a Court of Auditors (audits spending in the EU), a European Economic and Social Committee (made up of civil society nongovernmental organizations that represent various economic and social interest groups), a Committee of the Regions (which represents regional and local governments in the EU), a European Investment Bank (which provides loans for less-developed regions of the EU), and a European Central Bank (responsible for monetary policy and the Euro) (European Union, 2010).

The EU framework is highly useful, from a homeland security standpoint, in that it provides advantages in six spheres. First, the EU framework allows for the informal and formal exchange of knowledge, data, and best practices between member states via meetings, conferences, and exercises that take place under EU auspices. Second, the EU provides significant amounts of funding for research in technological and policy realms designed to solve existing problems. Third, alert systems within the EU help transmit information on unfolding crises between member states and help them to transmit information on threats. Fourth, the EU creates strong conditions for cooperation between member states on the institution of common standards and transnational coordination in crisis management (helping member states cope with issues such as pandemics and instituting common standards of airport and port security). Fifth, member states can turn to the EU to help facilitate support and resources during a crisis from other member states (e.g., aircraft to fight wildfires) (Larsson, Frisell, and Olsson, 2009, pp. 7–8).

THE EU: BORDERS AND IMMIGRATION

One of the areas in which EU institutions enjoy wide authority is with respect to border control policy and immigration policy. The adoption of the Single European Act in 1986 made the freedom of movement of community citizens across the borders of member states a requirement, thus helping to create the concept of a common European citizenship. This was seen by many as the logical outgrowth of the increasing power and authority of European institutions and of the concept of a shared European destiny. In practical terms, however, implementing the approach called for in the Single European Act meant that the member states would have to design a policy that abolished borders between member states (*internal borders*), created common standards for control over borders between member states and nonmember states (*external borders*), and initiated a common immigration policy, at least insofar as that policy dealt with immigrants from other member states, as immigration policies with respect to citizens of nonmember states is still under the authority of the respective national governments that make up the EU. The process of implementing these changes was a difficult one because of the different degrees of sensitivity among member states with respect to the internal security challenges presented by the abolishment of border controls. The member states could not agree on a basic definition of what constituted an acceptable level of domestic security, and given that the modern state's legitimacy is based, to a significant degree, on provision of security, this issue was clearly of critical interest to the respective governments involved. The absence of agreement on this matter meant that the member states chose to follow two separate tracks: one within the EU (then the EC) and the other based on intergovernmental agreements (Gogou, 2006). These two approaches were later amalgamated in the EU's Amsterdam Treaty of 1997, which incorporated both tracks into EU law. Until that point, the more substantial of these tracks was the intergovernmental track, which produced the 1985 Schengen Agreement.

FIGURE 6.4 Border sign on the Austrian–German border. This Wikipedia and Wikimedia Commons image is from the user BlueMars and is freely available at http://commons.wikimedia.org/wiki/File:SchengenGrenzeBayern-Tirol.jpg and has been released into the public domain.

The Schengen System

The Schengen Agreement and subsequent Convention Implementing the Schengen Agreement came fully into force in 1995. They abolished checks at internal borders of the signatory states and created a single external border for the purpose of immigration and customs checks, which were to be carried out by the national authorities according to identical procedures. In addition, a series of compensatory measures were instituted to balance the new freedom of movement within the Schengen area with better coordinated law enforcement, internal security, and judicial efforts. For example, a Schengen Information System (SIS) database was set up that allows information sharing across signatory states with respect to certain types of data on persons and goods (European Union, 2009). In addition, it was determined that border crossings into the Schengen area would occur only during fixed opening hours, would involve the same security procedures, and foreigners would be limited to stays of up to three months in the common area (i.e., outside the country granting the visa). At airports and ports, arrivals are separated into those traveling within the Schengen area and those arriving from countries outside the Schengen area.

The Schengen area was gradually extended to include almost all of the member states, although the United Kingdom and Ireland have opted to maintain border controls with other EU countries and thus participate in only some aspects of the Schengen Accords (particularly with respect to police and judicial cooperation and the SIS), and Bulgaria, Cyprus, and Romania have not yet been granted full-fledged membership in the Schengen area. While Schengen is associated with the EU through the Treaty of Amsterdam, it is still an intergovernmental agreement affiliated with the EU rather than an EU institution.

In addition to the abolition of border checks at internal borders and the creation of a common EU border on the edge of the Schengen area, the agreements also provide for harmonization of training of border security staff and the embedding of liaison officers from other signatory states in border policing agencies at the external borders of the Schengen area. Moreover, the accords also drew up common rules for dealing with asylum seekers (based on the 1990 Dublin Convention and the 2003 Dublin II Regulation), the institution of cross-border surveillance and hot pursuit rights for police forces in signatory states and the creation of a faster extradition system between Schengen countries (European Union, 2009).

A number of non-EU countries (such as Switzerland, Norway, and Iceland), known as *Schengen-associated countries*, have agreements with the EU whereby they essentially follow the same measures with respect to their borders external to the Schengen area as do Schengen signatories, and consequently there are no border checks between those countries and Schengen area countries, and those countries are considered part of the Schengen area. To join the Schengen system, new EU member states (or other European countries with strong links to the EU) must undergo a Schengen evaluation by experts appointed by the European Commission to verify that the candidate country is able to adequately control its land and sea borders, as well as airports, and that it has the right procedures in place to issue visas, ensure police cooperation with other member states and Schengen signatories, and is ready and able to connect to the SIS and assure data protection (European Commission, 2010).

The abolition of internal border checks between Schengen signatory countries does not, of course, mean that these national borders have ceased to exist. In the words of Article 2 of the agreement: "Where public policy or national security so require a Contracting Party may, after consulting the other

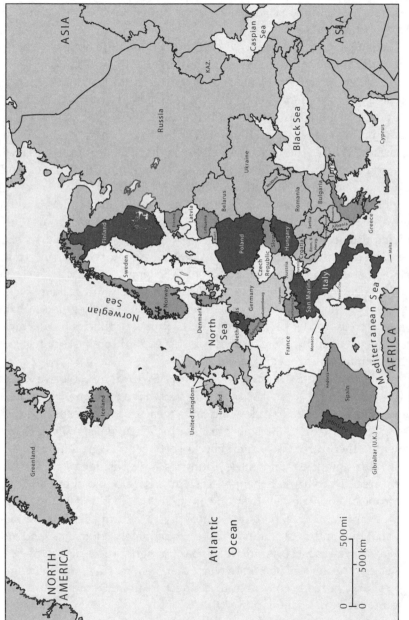

FIGURE 6.5 Schengen Agreement.

Contracting Parties, decide that for a limited period national border checks appropriate to the situation shall be carried out at internal borders. If public policy or national security require immediate action, the Contracting Party concerned shall take the necessary measures and at the earliest opportunity shall inform the other Contracting Parties thereof" (Convention Implementing the Schengen Agreement, Chapter 1, Article 2, 1985). Consequently, signatory states have the ability to exercise control over their borders with other Schengen countries in the event of a threat to homeland security from terrorism, pandemics, and so on. Article 5 of the Schengen Agreement requires that signatory states admit only foreign nationals that are not "considered to be a threat to public policy, national security or the international relations of any of the Contracting Parties" (Convention Implementing the Schengen Agreement, Chapter 2, Article 5, 1985). Article 10 allows for the institution of a common visa with a maximum stay of three months. Border authorities at EU external borders operate according to a document known as the Common Manual and thus follow EU-wide regulations with respect to border controls.

In keeping with the intergovernmental approach of Schengen, individual European governments have set up bilateral and multilateral mechanisms to enforce Schengen arrangements. For example, the French and German authorities have set up joint enforcement centers that allow for joint planning and operations along the Franco-German border and which include representatives from the French national police and Gendarmerie as well as German state police from the states of Baden–Württemberg and Rhineland–Palatinate and the German Federal Police. A similar, trilateral arrangement exists under the Police Euregion Maas-Rhine, which includes law enforcement officers from Belgium, Germany, and the Netherlands (Hobbing, 2005, p. 15). Such intergovernmental arrangements, coupled with hot pursuit rights, mean that many member states law enforcement bodies enjoy unprecedented rights to operate in neighboring member states. For example, German law enforcement officers routinely patrol areas in the Netherlands that are popular with German tourists during the tourist season (Briefing by senior German Federal Police official, 2010).

EU Integrated Border Management

The European Agency for the Management of Operational Cooperation at the External Borders of the Member States of the European Union (FRONTEX) was created in October 2004. The agency's role is to coordinate operational cooperation between member states in the management of external EU borders (land and sea), to help create common training standards and train national border protection forces, to carry out risk analyses, and to organize support for member states in the event that they need additional technological or personnel resources (Amaral, 2007, p. 5). Like other EU institutions, FRONTEX does not act as a supranational entity with its own enforcement arm but, rather, acts primarily to assist national entities in enforcing national and EU laws. Member states' border services are responsible for the protection of the EU's external borders, but it is recognized that the EU's ability to regulate incoming persons and goods is only as good as the ability of its weakest external national border policing service.

One of the primary issues that FRONTEX deals with is the illegal immigration problem. The entire southern littoral of Europe as well as the Canary Islands in the Atlantic, the Spanish enclaves of Ceuta and Melilla on the North African coast, Sicily and the island of Lampedusa, and Greek islands in the Aegean act as the primary magnets for illegal immigrants. Although there are also illegal immigration routes from the east, the largest volume of illegal immigrants comes to Europe via the African continent. FRONTEX accordingly employs a surveillance system targeted at illegal immigration that consists of five (more or less) ongoing operations: Hera, Minerva, Hermes, Nautilus, and Poseidon. Each of these operations is conducted by a mix of EU member states coordinated by FRONTEX and is centered on a different geographic region from the Atlantic to the eastern Mediterranean. Ongoing maritime surveillance operations are conducted in the framework of two regional coordination centers under FRONTEX auspices: the ESBC (Eastern Sea Borders Centre, which focuses on the eastern Mediterranean) and the WSBC (Western Sea Borders Centre, which focuses on the western Mediterranean, Atlantic, and Baltic) (FRONTEX, 2006, p. 12).

In early 2008 the European Commission put forward an initiative known as the European Border Surveillance System (EUROSUR) to provide a technical platform for integrated maritime surveillance (see Sidebar 6-1). EUROSUR is envisioned as a "system of systems" which will involve, first, the upgrading of existing national border surveillance system and the interlinking of communications between national border control entities. In the second phase of the project, other surveillance tools, such as satellites, will be brought into the network, thus helping create an intelligence picture beyond the border zone. In the third and final phase of the initiative, a common information-sharing environment will be created for the maritime domain (Directorate-General for Maritime Affairs and Fisheries, 2008, p. 14). EUROSUR's role is to reduce the number of illegal immigrants who enter the EU undetected, increase the security of the EU through prevention of cross-border crime and enhance search and rescue capabilities.

In addition to the maritime domain, the primary EU land borders that act as magnets for illegal immigrants are the borders between Slovakia and Ukraine, between Slovenia and Croatia, and between Greece and Albania and

---

Sidebar 6-1: Creating a European Border Guard?

In May 2002, an EU feasibility study on the ways and means of creating a European Border Police service suggested that rather than creating a unified, EU-wide border policing agency, that national border police forces be linked by common coordinating centers focusing on different areas of border security, have a common training curriculum, share budgetary costs and share common units for special tasks (including rapid response units). The different national approaches towards the idea of a European Border Policing agency reflect a range of attitudes in Europe towards European integration. One school of thought argues that an integrated Border Policing Agency under the authority of the EU's Council of Ministers must be created and that this should gradually supplant existing national border police forces. Another school of thought argues that a network model of border policing should be created in which the EU border policing agency acts as a network of national border policing units. Accordingly, separate national forces would continue to exist but they would enjoy common training and equipment standards and follow instructions issues by the EU's Council. (Monar, 2006)

Greece and Turkey. There are also attempts to access the EU illegally via airports, particularly those in Paris, London, Frankfurt, Amsterdam, Madrid, and Milan. FRONTEX possess teams of 500 to 600 border policing experts from the member states, known as RABITs (rapid border intervention teams), which can be made available on short notice to assist states facing a major influx of illegal aliens. EU regulations empower these teams with full enforcement powers even though they are operating on the territory of another member state. There are also plans to deploy permanent joint border policing teams (known as FRONTEX joint support teams—FJSTs) drawn from member states at principal border crossings and hotspots for illegal immigration (Amaral, 2007, p. 7). FRONTEX also maintains a record of all physical assets that it can put at the disposal of a member state when required: this includes a fleet of 20 fixed-wing and nearly 30 rotary aircraft as well as 100 patrol boats.

One of FRONTEX's main missions is to assist member states in the training of their national border policing personnel and the creation of common training standards. In December 2005, FRONTEX developed a "partnership academy system" to leverage some of the 136 existing border guards training academies in the EU so that they will train border personnel according to FRONTEX common standards. FRONTEX is also involved in creating a Common Core Curriculum (CCC) to help standardize the training of border security personnel throughout the EU (FRONTEX, 2006, p. 17).

FIGURE 6.6 Illegal immigrants off the coast of the Italian island of lampedusa. This Wikipedia and Wikimedia Commons image is from the user Micniosi and is freely available at http://commons.wikimedia.org/wiki/File:Arrivo_di_clandestini_nel_mare_di_Lampedusa.JPG and has been released into the public domain.

EU Immigration and Asylum Policy

While the management of EU external borders is handled by the member states in coordination with EU institutions such as FRONTEX and according to EU guidelines, immigration policy, as noted earlier, is still primarily the purview of the individual countries that make up the EU. Of course, immigration is an EU/Schengen-wide issue because once a person gains asylum or other status in one EU and/or Schengen country, the absence of internal borders means that the person can then easily access other such countries. Nevertheless, immigration issues are usually seen as core sovereignty issues and thus governments are loath to give up their powers in this area. Whereas EU rules provide citizens from member states the right to live and work in other member states, the question of admitting non-EU nationals, as noted earlier, is left to the member states themselves.

In 2004, the EU established an agenda, known as the Hague Program, designed, among other things, to create a set of common asylum policies across the EU based on the principle of *minimum common standards*. This built upon the 1999 Tampere Program, which produced the first set of binding EU agreements on asylum (van Selm, 2005). The Hague Program does not attempt to create a European immigration policy because, under the

Schengen Convention, asylum seekers must be processed by the country in which they request asylum. Nevertheless, the nature of open borders made it necessary to address discrepancies in asylum legislation that made certain countries easier for asylum-seekers to enter. EU directives set out the lowest protection threshold that member states must satisfy with respect to refugees and other asylum-seekers, and these directives form part of what is known as the *Common European Asylum System*. Any person requesting asylum in the EU can only apply to a single member state, and if that state decides to grant asylum, that right is not recognized by other member states, although all member states will recognize and abide by a decision by one member state to deny asylum (Guild, Carrera, and Eggenschwiler, 2009, p. 2). Generally speaking, asylum seekers will be required to submit their claim for asylum to the first EU and/or Schengen country at which they arrive and will be unable to obtain asylum in another EU country if rejected by the first. However, the Treaty on the Functioning of the European Union calls for the development of a common EU asylum policy that will create a uniform status for nationals of third countries requesting asylum, a common system of temporary protection for displaced persons in the event of a mass inflow and common procedures for granting and withdrawing asylum status (TFEU, Title IV, Articles 78 and 79, n.d.).

NATIONAL IMMIGRATION AND ASYLUM POLICIES IN EUROPE

As noted in the discussion above, while border controls between Schengen countries have been abolished (although they can be reinstated for brief periods as noted earlier under Article 2 of the agreement), immigration policy with respect to non-EU country citizens is still largely within the realm of national governments. A few examples of EU member states are discussed below, including naturalization and asylum policies as well as enforcement institutions and relevant legislation.

United Kingdom

The process of obtaining citizenship is treated differently by different EU countries and ranges from a purely bureaucratic process to one that is viewed as transformative. The UK, for example, has a bureaucratic approach to the acquisition of citizenship, but this has, increasingly, been recognized as problematic. A 2001 Home Office publication noted that the UK needs to develop a sense of civic identity and shared values, but at the same time noted that at the time of writing: "It is not altogether surprising that many applicants [for naturalization] do not appear to attach great importance to acquiring British citizenship, beyond the convenience of obtaining a British passport.

We do not actively encourage people to become British citizens or prepare them in any way. Nor, unlike some other nations, do we celebrate the acquisition of citizenship. Instead, we treat it as essentially a bureaucratic process, with a certificate being sent through the post at the end of it" (UK Home Office, 2002, p. 22).

The Immigration and Nationality Directorate, a unit of the Home Office responsible for immigration policy, has developed an approach to asylum seekers that focuses on the use of induction centers as the first stage of the asylum process. Asylum seekers arriving at the induction center are briefed on the asylum process and given information on access to legal advice. Before they leave the induction center, asylum seekers sign a document confirming their understanding of the asylum process, their commitment to comply with temporary admissions and reporting requirements, their commitment to leave the UK should their asylum request be rejected, and information on obtaining assistance to return to their respective countries of origin. Asylum seekers remain in the induction centers for 1 to 7 days, depending on whether they require financial support for accommodation and are to be provided room and board (UK Home Office, 2002, pp. 53–54). When asylum seekers leave the induction centers, they are be provided with application registration cards, which are biometric smart cards that contain personal information, including photographs, fingerprints, and employment status. Those asylum seekers requiring longer-term accommodation are accepted into accommodation centers that not only provide housing and meals, but also health care and training courses in the English language and IT (information technology) skills. Residents of these centers are free to come and go, but they must register with the center on a regular basis or risk losing their financial support (UK Home Office, 2002, p. 55). To prevent those not authorized to visit the UK from reaching British soil, the UK operates an airline liaison officer (ALO) network with ALOs in 20 airports around the world. Their task is to train airline personnel in passport and visa requirements and forgery awareness and to provide on-the-spot advice (UK Home Office, 2002, pp. 92–93).

In the area of nonnationals suspected of terrorist activity, the 2001 Anti-Terrorism, Crime and Security Act allows the Home Secretary to certify a noncitizen suspected of terrorist activities as an international terrorist and to deport that person and allows appeal of such an order to be heard exclusively by a Special Immigration Appeals Commission (SIAC), which frequently hears appeals and makes decisions in secrecy (Schoenholtz and Hojaiban, 2008, p. 9). During the appeals process, the foreign national can be detained pending the outcome of the appeal. The UK authorities maintain that deportation proceedings against suspected terrorists are not criminal proceedings but, rather, administrative ones, and consequently, foreign nationals do not enjoy the same protections given to those accused of criminal acts (Amnesty International, 2006, p. 15). As noted, the SIAC appeals process often occurs in secret. The Home Secretary is allowed to introduce classified information during closed SIAC hearings in which the detainee and his/her lawyers are excluded (with the detainee being represented by a *special advocate*, an attorney with security clearance who is not allowed to report to the detainee regarding the nature of classified evidence used against him/her) (Amnesty International, 2006, p. 16). Moreover, the 2002 Nationality, Immigration and Asylum Act empowers the Home Secretary to remove British citizenship from those who previously held a foreign citizenship or who possess dual nationality if that person has acted in a manner that seriously prejudices the vital interests of the UK or any of its overseas territories (UK Nationality, Immigration and Asylum Act, Section 4/40, 2002). In addition, asylum is automatically refused to any persons suspected of association with terrorist organizations.

A number of laws, including the recent 2007 UK Borders Act, allow UK Border Agency officers to board and search vehicles, stop, question, and detain people and make arrests (either with or without a warrant). UK Border Agency officers detaining a person cannot do so for more than three hours (by which time the detainee will need to have been turned over to the police). However, UK Border Agency officers may arrest a person if there is a warrant against that person or if the person assaults or is suspected of planning to assault an officer (UK Borders Act, Chapter 30, Section 23, 2007).

In November 2007, the British government amalgamated three agencies, the Border and Immigration Agency, UK Visas, and HM Revenue and Customs into a common entity, the UK Border Agency. The UK Border Agency acts in tandem with special branch officers from the police forces that have jurisdiction over a given port of entry and ports officers (who are plainclothes "port police" officials with full law enforcement powers). UK Border Agency staff are deployed to 135 countries where they carry out checks of travel documents and passenger manifests (see Sidebar 6-2). Agency staff also conduct en-route screening (based on a risk analysis) of persons and cargo, screen cargo and persons upon their arrival at a UK port of entry and can track foreign nationals, if need be, once they have entered the UK via identity cards, E-passports, and automatic number (license) plate recognition (Police National Legal Database and Andrew Stainforth, 2009, pp. 84–85). Officers at the UK Border Agency have primary responsibility for enforcing immigration and customs law at ports and airports, while the police have primary responsibility for providing security at ports and airports (with the special branches responsible for intelligence gathering and inter-diction at the airports and ports) (Association of Chief Police Officers of England, Wales and Northern Ireland and The UK Border Agency, 2008).

Sidebar 6-2: UK E-Borders Initiative

The e-Borders program will revolutionize our capability to process passengers entering and leaving the UK by air, rail and sea. The program will use systems and technology to ensure that we know much sooner who is entering and leaving and will afford better opportunities to manage the passenger volumes.

Project Semaphore is a pilot scheme that is being used to test elements of e-Borders systems and processes. Project Semaphore was launched in November 2001 and will be superseded by the full e-Borders system, which is being rolled out over a period from 2008 to 2014.

Project Semaphore currently captures inbound and some outbound passenger information from 18 air carriers on selected routes originating from 37 non-UK airports. To date 9 million passenger movements at an annual rate of 12 million passenger movements have been captured. The passenger information received is passed through an operations center staffed by operators from the UK Border Agency and police services and checked against various agency watchlists. This increasing data capture has already provided benefits to the British intelligence community with over 100 'alerts' having been issued to them.

Project Sempahore is therefore assisting the intelligence community by:

- Increasing the effectiveness of monitoring travelers moving into, out of or transiting the UK;
- Reducing the number of routes available to those not wishing to come to the notice of UK authorities;
- Increasing the number of effective interventions against individuals of national security interest;
- Increasing intelligence on individuals of national security, immigration or criminal interest; and
- Capturing increased quantities of data for post incident or historical analysis. (UK Government, 2006, p. 23)

UK police forces deploy approximately 1500 special branch officers at ports and airports. The Serious and Organized Crimes Agency (SOCA) also plays an important role having taken on some of the tasks in the realm of customs and immigration. Additional uniformed police are paid for by airport operators, and eight seaports have special port police (Gregory, 2009, p. 2). After the 2005 terrorist attacks in London, then-prime minister Tony Blair suggested some form of border police force, but for a variety of reasons, this option has thus far been deemed undesirable, and policing at ports and airports remains decentralized (Gregory, 2009, p. 3).

Germany

Germany's approach toward asylum issues was set by the federal republic's founders, who felt that Germany bore a duty to aid persecuted people the world over in light of previous German crimes against humanity and genocide. Indeed, this right was enshrined in Article 16 of the federal republic's constitution (Basic Law). For decades, access to what was then West Germany was fairly open and the federal republic was considered to have one of the most liberal asylum policies of any country. Naturally, this meant that many economic refugees were able to pose as asylum seekers and receive residency rights in West Germany, thus leading to an abuse of the system.

In 1993, Article 16 was amended to increase the restrictions on claiming asylum rights, and the character of asylum guarantees was changed from a constitutional right to an administrative regulation (Post and Niemann, 2007, p. 12). The primary border security agency in the country is the German Federal Police (Bundespolizei), which controls access at Germany's airports and seaports and works with the Customs Administration (Zollverwaltung) to control the importation of goods that could be a threat to public safety. All of Germany's neighbors are Schengen signatories, and consequently, the federal police do not regularly patrol the country's land borders.

Germany's Anti-Terrorism Act of 2002 included significant changes with respect to asylum and immigration to ensure that terrorists were excluded from international refugee convention protections. German authorities were given the power to deny visas or asylum to applicants who (1) posed a threat

FIGURE 6.7   German-issued Schengen residency permit. German Federal Document, Free of Copyright According to Section 5 German Copyright Act, photo in the public domain @ Wikimedia Commons. http://commons.wikimedia.org/wiki/File:SchengenResidencePermit.JPG.

to democracy or internal security, (2) used violence in furtherance of political goals, or (3) were affiliated with international terrorist groups (Schoenholtz and Hojaiban, 2008, p. 8). The Residence Law (Augenthaltsgesetz) of January 2005 allows for the denial of entry or expulsion of non-German nationals who pose a threat to security or who incite hate and violence, thus making "hate preaching" grounds for deportation. The law similarly allows the authorities to prohibit or curtail political activities conducted in Germany by foreign nationals (UK Foreign Office, 2005, p. 13). Expulsion of aliens is authorized on various grounds. Expulsion is obligatory if a foreigner is sentenced to incarceration for at least three years or convicted under the Narcotics Act or sentenced to at least two years for civil disorder, rioting, or human smuggling. Expulsion is also highly recommended in the event that aliens pose a threat to security or incite to violence or hatred (Germany, Foreign Ministry, 2009). Land (state) authorities or the Federal Ministry of the Interior can issue a deportation order, which is immediately enforceable without prior warning in the event that an alien is believed to present an immediate security threat to the country.

France

In France, asylum claims are handled by the Office for the Protection of Refugees and Stateless Persons (Office Française de la Protection des Réfugiés et des Apatrides—OFPRA). Asylum can be granted to persons who are persecuted in their country of origin as well as to those who can show that returning to their country of origin would subject them to capital punishment, torture, or other inhumane treatment. Asylum applications are handled by the prefectural authorities at the location of entry of the asylum seeker.

With the exception of citizens or other legal residents of member states of the European Union and other Schengen countries and those countries with which France has a visa waiver agreement, visitors to France must obtain visas to enter the country (although this is admittedly difficult to enforce with respect to French land borders since, during the normal course of affairs, France has no land border controls, as all the countries on its borders are Schengen signatories). French consulates and embassies can legally refuse to grant a visa to people with suspected terrorist links for either long- or short-term visas (short-term visas are handled by common Schengen protocols, which, as noted above, allow for denial of visas to terrorism suspects and the listing of such people in the Schengen Information System database, which allows denial of access to other Schengen countries). Obtaining a visa does not, however, automatically grant the person access to French territory, and French border officials can refuse entry to France on the grounds of public security risks emanating from a link to terrorism. This denial of entry can also extend to foreigners holding a residence permit (carte de séjour) and can thus serve as a method for the de facto expelling of a person suspected of terrorism ties whose trip overseas can be used as a means to deny him/her return access to France. Moreover, foreigners can be deported from France if their "conduct [is] linked to activities of a terrorist nature" or if they engage in "explicit and deliberate acts of incitement to discrimination, hatred or violence" (France, Prime Minister's Office, 2007, pp. 55–56). Expulsions of people from France are governed by the Code de l'Entrée et du Sejour des Etrangers et du Droit d'Asile (CESEDA—Code on the Entry and Stay of Foreigners and Right to Asylum). Foreign terrorism suspects can be deported following a criminal conviction, known as Interdiction du Territoire Français (ITF, or ban from French territory), or they can be issued an administrative expulsion order by the Interior Ministry, known as an Arrete Ministerial d'Expulsion (AME—ministerial expulsion order). AME expulsions are subject to review by an expulsion commission, made up of two judicial magistrates and one administrative magistrate and these are tasked with evaluating whether the expulsion is necessary and proportionate based on the nature of the threat posed by the person to be expelled (Human Rights Watch, 2007, p. 19). Generally speaking, AME expulsions are used against people that the government has difficulty prosecuting for criminal offenses, particularly with respect to crimes such as incitement, preaching hatred, and/or violence, because the current law against incitement does not allow for criminal deportation of the convicted offender (Human Rights Watch, 2007, p. 20).

CONCLUSION

The matter of border security and immigration and asylum policies is complex and multifaceted and this chapter merely served to scratch the surface of this issue in the context of the EU and a handful of European countries. The EU and its various treaties and institutions serves as an interesting and unique model of reconciling national policies with a powerful supranational entity to which member states have subordinated some of their national sovereignty. As a limited version of a "United States of Europe," it is logical that the EU should set the rules (in the context of the broader Schengen arrangement) for commerce and the movement of persons within the union, and this, in turn, necessitates a common set of policies and procedures with respect to external borders, visa issuance, immigration police, and asylum policy. The discussion of British, German, and French visa and immigration policies shows that there is still considerable variability in approaches toward these policy issues on the part of the three leading member states in the European Union (though the UK, as noted earlier, is not part of most of the Schengen arrangements), and consequently, EU authorities will have some significant challenges in creating a single set of EU policies to address immigration and asylum issues.

ISSUES TO CONSIDER

- How is the European Union structured?
- What is the European Union's authority with respect to border security, and how does this relate to the Schengen agreements?
- Which steps has the European Union taken to set up a common immigration policy?
- How do immigration and naturalization policies differ across the UK, France, and Germany?

CHAPTER 7

SECURITY POLICIES

Critical Infrastructure Protection, Public–Private Partnerships, and Aviation, Maritime, and Surface-Transport Security

In this chapter we focus on security strategies, policies, approaches, and institutions with respect to a number of countries in our survey. The first few chapters of this book have dealt with preventive policies in the areas of counterterrorism, law enforcement, and mitigating radicalization. In this chapter we address primarily the issue of target hardening by looking at critical infrastructure protection (and the related topic of public–private security partnerships) and then move on to looking at policies and institutions in the realm of transportation security (aviation, maritime, and surface).

CRITICAL INFRASTRUCTURE PROTECTION (CIP): INTRODUCTION

Infrastructure provides people with basic services in the areas of supply, disposal, transportation, and communication. Most definitions of infrastructure focus on three types: economic (communications and transport networks; energy, water, and sewage supply systems; residential and commercial structures; roadways; bridges; dams; etc.); social (schools; hospitals; shopping malls; stadiums and parks; etc.); and human capital infrastructure (intangible assets such as knowledge, skills, health, and money) (Yates, 2003, p. 7). Upon closer inspection, however, the latter two are essentially dependent on the first, and thus the first category may be deemed *critical Infrastructure*. According to the United States' *National Strategy for the Physical Protection of Critical Infrastructures and Key Assets* of February 2003,

Comparative Homeland Security: Global Lessons, First Edition. Nadav Morag.
© 2011 John Wiley & Sons, Inc. Published 2011 by John Wiley & Sons, Inc.

critical infrastructures are to be found in the agriculture and food sectors, the water sector, public health and emergency services, government, the defense industrial base, information and telecommunications systems, the energy sector, the transportation sector, banking and finance, the chemical industry, and the postal and shipping sectors (Bush, G., 2003, p. 6). One of the elements that makes critical infrastructure protection different from other areas of homeland security policy described in this book is that most critical infrastructures are in private hands, and consequently this policy area requires significant coordination between government and the private sector.

In this section of the chapter we explore a handful of approaches on the part of a number of countries to safeguarding one or more of these critical infrastructures (as defined by the *National Strategy*). We explore this issue from the perspective of overall approaches and strategy with respect to critical infrastructure protection (CIP), interface between governmental security entities and the private sector, national CIP institutions and their role, and private sector–CI responsibilities and licensing issues. The intent of the chapter is not to provide a comprehensive account of CIP across each country and each sector, but rather to provide an overall sense of a few of the approaches taken overseas as well as some examples of specific policies.

OVERALL APPROACH TO CIP AND THE PRIVATE SECTOR

Israel

Israel has an interesting approach to CIP matters. Many critical infrastructures in Israel were initially government owned and many still are, although some have subsequently become privatized. Naturally, those infrastructures in which the government still has a controlling share or complete ownership can be fairly easily regulated, but wholly private industries are also significantly regulated by law and practice in the spheres of protection and civil defense. In principle, security and civil defense preparedness is the responsibility of the business itself. Each major industrial plant or other large facility is required to:

- Appoint an employee responsible for the civil defense of the facility
- Create a security system for the facility
- Install an alarm system
- Purchase emergency equipment and train employees in their use
- Prepare bomb shelters or secured spaces (Israel Defense Force Homefront Command, n.d.a.)

FIGURE 7.1   Orot Rabin power station, Hadera, Israel. This Wikipedia and Wikimedia Commons image is from the Israel Electric Company Archive and is freely available at http://commons.wikimedia.org/wiki/File:PikiWiki_Israel_6925_Rabin_Lights_power_stations_at_Hadera.jpg under the Creative Commons Attribution 2.5 Generic license.

When the government declares a state of emergency, this activates an emergency system known in Hebrew as Melah, an acronym for Meshek le Sha'at Herum (Emergency Economy). The principle behind the Melah system is that a national emergency is likely to require a significant call up of reservists, and this may seriously hamper the operation of many economic enterprises of national significance due to the military mobilization of many of their personnel. This situation can, of course, result in a dangerous drop in production and economic activity that can weaken the war effort and hurt the economy. Accordingly, during such crisis situations the economic system must be organized differently in order to coordinate production between industries, provide resources, and determine which activities should take precedence (Israel, National Emergency Economy Headquarters, 2001). The Emergency Economy system is thus designed to provide essential goods and services and it does so through the High Committee for the Emergency Economy (Va'adat Melah Elyona), which is responsible for running the critical industries during a national emergency and oversees a series of lower-level Emergency Economy committees at the level of territorial military commands, Ministry of the Interior geographic districts and municipalities, local councils, or regional councils. At the local level, every municipality or local or regional council is required to set up a local Emergency Economy committee to guarantee the provision of critical goods and services during times of national emergency. The head of this committee at the local level is the mayor or the head of the local or regional council.

Government ministries in Israel have authority over particular sectors of the economy during national emergencies. For example, the Ministry of National Infrastructures is responsible for the water, electricity, gasoline, and natural gas sectors, while the Ministry of Transport is responsible for the public works department, the merchant marine, the country's fleet of trucks, the railway system, sea and airports, and so on. The determination as to which business enterprises are to be considered critical industries is made by an Inter-Ministerial Advisory Committee appointed by the Minister of Industry, Trade and Labor, which includes representatives from the Emergency

FIGURE 7.2   Nesher cement factory, Ramle, Israel. This Wikipedia and Wikimedia Commons image is from the user Sambach and is freely available at http://commons.wikimedia.org/wiki/File: Nesher_Factory,_Ramla.jpg and has been released into the public domain.

Economy system national headquarters and representatives of the military (Israel, National Emergency Economy Headquarters, 2001). Once a particular enterprise is deemed by the relevant ministry to be a critical industry, critical personnel needed to maintain the operation of that particular factory or other facility or service are exempt from military callup and are required, under the State of Emergency Employment Law of 1967, to report to their place of employment and engage in any work asked of them. The penalty for failure to report to work when so ordered is a three-year prison term (Israel, State of Emergency Employment Law, Chapter 6, Section 34, 1967). This legislation also empowers the Minister of Industry, Trade and Labor to require that every place of employment provide information about its employees and operations for the purpose of determining whether the business and its employees are needed for the emergency economy.

Israel's Civil Defense Law of 1951 and the Civil Defense Regulations (Equipment, Factories, Institutions and Training of Personnel) of 1973 require that all privately and publically owned factories and other places of employment train their personnel in personal protection and maintain emergency equipment and first-aid supplies (Israel Defense Force Homefront Command, 2007, p. 5). Each factory or other large-scale place of employment is required to appoint someone to be responsible for civil defense at that place of employment (see Sidebar 7-1). Those persons are required to undertake training courses under the auspices of the IDF Homefront Command (HFC) and other government agencies. The HFC also sends personnel to factories and other places of employment in order to train employees in civil defense techniques.

Sidebar 7-1: IDF Homefront Command Instructions for Preparedness in Factories and Other Large Enterprises

The owner/manager of the factor or other enterprise should designate specific employees to be responsible for the following functions:

- Civil Defense—in cooperation with the HFC district command.
- First Aid
- Firefighting
- HAZMAT (for factories that maintain hazardous materials on site)
- Employee Welfare
- Bomb Shelter Maintenance

Additional functions that can be filled (depending on the nature of the factory/business) include:

- Search and Rescue
- Crowd Control
- Security

Civil Defense

The designated employee in charge of Civil Defense must be trained by the HFC and other government agencies and is expected to be a senior employee with significant experience and a good understanding of the factory's/enterprise's operations.

The employee in charge of Civil Defense will be responsible for:

- Planning civil defense systems in the factory/place of employment.
- Planning emergency rules.
- Forming a civil defense team and training and exercising this team.
- Serving as the point of contact between the HFC and the factory/place of employment.
- Becoming familiar with the civil defense equipment at the factory/place of employment.

During routine emergency periods, the employee in charge of Civil Defense will:

- Determine which employees will be responsible for particular emergency functions.
- Publicize emergency rules.
- Ensure that civil defense equipment and bomb shelters are in working order.
- Develop an action plan to maintain production/work during emergencies.
- Strengthen preparedness to deal with emergency situations.

During an actual emergency, the employee in charge of Civil Defense will:

- Prepare situation reports for the owner/manager of the factor/enterprise.
- Maintain contact with emergency services.
- Maintain the security of the factory/place of employment.
- Organize treatment of injured employees.

HAZMAT, Firefighting, and First Aid Teams

The primary missions of HAZMAT, firefighting and first aid teams are:

- Putting out fires with local resources until firefighting and emergency services arrive at the scene.
- Prompt provision of first aid, including evacuation of critically injured persons.
- Immediate response to hazardous materials spills.

During routine emergency periods, the teams are responsible for:

- Ensuring the operation and updating of equipment.
- Limiting the stocks of toxic materials.
- Maintenance of adequate signage and lighting in bomb shelters and protected rooms.
- Marking of evacuation vehicles.
- Training employees.
- Facilitating inspection of protective equipment and bomb shelters.
- Preparedness for emergencies.

During an actual emergency, the teams are responsible for:

- Dealing with hazardous materials spills.
- Establishing evacuation triage areas for the injured.
- Provision of First Aid.
- Provision of situational awareness to first responders.

Employee Welfare Team

The Employee Welfare Team is responsible for providing psychological/mental health support to employees. Proper training and preparedness of the Employee Welfare Team will engender confidence among employees that they and their families will receive support during emergencies and this will increase the percentage of employees that come to work during emergencies thus ensuring the continued operation of the factory/place of employment.

The primary missions of the Employee Welfare Team are:

- Support for employees and their families in preparing for emergencies.
- Provision of lines of communication between management and employees.
- Preparing a program for increasing personal and organizational commitment.

- Mapping out and treating groups that have a greater tendency towards anxiety and absences.
- Provision of information on civil defense to employees.
- Training management and directing it towards dealing with employee motivational problems on the one hand and employee exhaustion on the other.

During routine emergency periods, the team is responsible for:

- Instituting a support system for employees—toddler care center, kindergarten.
- Assistance with provision of basic foodstuffs/supplies for employee families.
- Assistance in maintenance of contact between employees and family members.
- Instituting psychological support groups.
- Assistance for the families of injured employees.
- Identification of signs of psychological difficulties within the employee population.
- Preparedness for dealing with emergency events.

During an actual emergency, the team is responsible for:

- Analysis of the general condition of employees.
- Assistance for employees in bomb shelters.
- Interface with the press.
- Provision of assistance to the injured.
- Facilitation of contact between employees and the National Insurance Institute.

Bomb Shelter Maintenance Team

The Bomb Shelter Maintenance Team is responsible for the maintenance and operation of the facility's bomb shelters and access routes to them. The team must be experienced in operating the bomb shelter systems (electricity, communications, plumbing, air and filtration systems). One of the team members should work in an area close to the bomb shelter so that he/she can arrive first at the bomb shelter and prepare it for the arrival of employees and visitors. The head of the team will be provided guidance by the employee in charge of civil defense and will report on any problems during periodic inspections of the bomb shelters.

During routine emergency periods, the team is responsible for:

- Ensuring there is no blockage of access routes to the bomb shelters and that the signage is clear.
- Periodic inspections of the bomb shelters and their associated equipment.
- With respect to bomb shelters that are dual use [are used for other purposes in non-emergencies], to ensure that no more than 20% of the area of the bomb shelter is not taken up by other items so that the bomb shelter can be readied for emergencies within 4 hours.
- Closing of air openings with steel caps and screws.
- Preparedness for emergencies.

During an actual emergency, the team is responsible for:

- Arriving first at the bomb shelter and making it operational.
- Ensuring that exits not used for allowing people access into the shelter are properly sealed. (Israel Defense Force Homefront Command, 2007, pp. 8–12)

United Kingdom

The British approach treats critical infrastructure operators as first responders tasked with an immediate role in crisis situations. Critical infrastructure operators are considered category 2 responders, who play a response role behind the category 1 responders, which consist of the police, emergency services, local authorities, health organizations, and similar entities. Category 2 responders include electricity, gas, water, sewage, and telephone providers as well as operators of the rail, airport and seaport infrastructures, and the highways. Utility providers in particular have a statutory obligation to provide a response in an emergency and to work toward rapid reinstatement of services (Somerset Local Authorities' Civil Contingencies Partnership, 2006, pp. 2, 9). The primary piece of preparedness and response legislation, the Civil Contingencies Act requires that category 1 and 2 responders jointly form Local Resilience Forums (LRFs) based on police jurisdictions to help coordinate the local response to emergencies.

Each critical infrastructure sector is expected to operate under existing strategies and to interface and receive instructions from the relevant lead government department (LGD). For example, the telecommunications operators work with the Department of Business Innovation and Skills (BIS) to ensure that the telecommunications system will work during an emergency and that faults will not cascade through the entire system. Telecoms (the largest of which is British Telecom—BT) must operate according to an emergency plan known as NEAT (National Emergency Alert for Telecoms), which takes the form of a conference call between providers and the BIS (and communications regulatory bodies in other parts of the government) designed to provide information about a problem and share information across government and industry. There is also an Electronic Communication–Resilience and Response Group (EC-RRG) that is responsible for administering the National Emergency Plan for the UK Telecommunications Industry and a memorandum of understanding between industry and government for cooperation in emergency situations. In addition, UK telecommunications providers are regulated by Ofcom, an independent regulatory authority responsible to Parliament and, according to licensing stipulations, must make the telecommunications spectrum available to the military for emergency communications (UK Department of Business Innovation and Skills, n.d., pp. 5–7).

Canada

Canada's approach to CIP is twofold: to protect facilities identified as critical and to ensure redundancy of services so that these services can continue to be

FIGURE 7.3 Brilliant Dam on the Kootenay River, British Columbia, Canada. This Wikipedia and Wikimedia Commons image is from the user Brian and is freely available at http://commons.wikimedia.org/wiki/File:Brilliant_Dam.jpg under the Creative Commons Attribution-Share Alike 2.0 Generic license.

provided should part of the system fail. The fundamental principles that underpin the country's CIP strategy are:

- Raising awareness of CIP among senior managers in industry and officials in government and convincing the private sector that heightened security has business value.
- Integration of CIP (including cyber security) into governmental emergency management programs and encouragement of business continuity planning.
- Ensuring that critical infrastructure operators are accountable to Canadians via legislation, regulation, and policy.
- Employing an all-hazards approach that takes into account disparate threats, including terrorism, natural disasters, accidents, computer viruses, and so on (Public Safety and Emergency Preparedness Canada, 2004, pp. 6–7).

The Canadians also set up a National Cross-Sector Forum to promote information sharing across critical infrastructure sectors and federal, provincial, and territorial government representatives (Government of Canada, 2009, p. 7).

GOVERNMENT AGENCY SECURITY INTERFACE WITH THE PRIVATE SECTOR

Israel

In Israel, legislation provides for particular designated institutions of national importance (termed *secured institutions*) to receive special security measures. The security section of the Operations Department of the Israel police is responsible for providing the following services to secured institutions:

- Screening, training, and provision of guidance to those entities' security personnel.
- Setting procedures to be followed by administrative and security personnel in those institutions.
- Provision of intelligence data, weekly intelligence reports, and professional guidelines to the security officers of these institutions.
- Supervising and monitoring existing security arrangements and installations (Geva, 1995).

United Kingdom

In the UK, full-time dedicated counterterrorism security advisors (CTSAs) exist in every one of the country's police forces. A central National Counter Terrorism Security Office (NaCTSO) trains and supports CTSAs in providing intelligence on threats to the private sector and on providing advice regarding counterterrorism practices. CTSAs dispense advice and support relating to matters such as bomb threats, building searches and evacuations, contingency planning for terrorism incidents, dealing with suspicious packages, facility security, protection of key assets, and prevention of the acquisition of dangerous materials by questionable persons. CTSAs are also tasked with identification of vulnerable sites and industries and to help choose "trusted contacts" within these locations and industries who are allowed to see more sensitive information and who can work with the police to develop contingency plans for their firm or facility (Safe Cities Project, 2004, pp. 8–10). In London, CTSAs deployed by the Metropolitan Police hold regular briefings for the business community discussing threats and appropriate security measures and also maintain an email and pager system that can provide thousands of business personnel in London with real-time instructions in the event of an attack (Ponenti, 2007). The UK Center for the Protection of National Infrastructure (CPNI) also provides advice to elements of the critical national infrastructure and other sectors via teams of sector-based specialists, advisers, training, and online and hardcopy advisory products (Center for the Protection of National Infrastructure, 2010).

Australia

In Australia, the Australian Security Intelligence Organization (ASIO) maintains a database of all critical infrastructures with entries marked for criticality (vital, major, low, or unspecified). ASIO's Business Liaison Unit (BLU) interfaces with the business sector, briefing businesses and providing guidance on security planning. The BLU distributes sanitized intelligence from the National Threat Assessment Center (NTAC) to the private sector in the form of business security reports for specific industries and provides a range of other types of reports. Businesses receive BLU reports free of charge via a secure web site to which they must subscribe (Australian Security Intelligence Organisation, 2008, pp. 22–23). In 2003 the attorney-general's department (which also oversees ASIO and the Australian national police) created the Trusted Information Sharing Network for Critical Infrastructure Protection (TISN). The TISN, which is an advisory body rather than an operational one, is comprised of infrastructure advisory groups (IAAGs) each of which addresses a different sector, including banking and finance, communications, emergency services, energy,

FIGURE 7.4 Telstra communications tower, Canberra, Australia. This Wikipedia and Wikimedia Commons image is from the user Трансаэро and is freely available at http://commons.wikimedia.org/wiki/File:Telstra_Tower_Canberra_Australia.jpg under the Creative Commons Attribution-Share Alike 3.0 Unported license.

the food chain, health, icons and public venues, transport, and the water sector. The structure is open-ended, allowing additional groups to join as necessary. Each IAAG is comprised of owners and operators of critical infrastructure. Overseeing the IAAGs is a Critical Infrastructure Advisory Council (CIAC), which advises and reports to the attorney-general. The CIAC is comprised of selected members of each IAAG as well as representatives from each state and territory, relevant federal government agencies, and the National Counter-Terrorism Committee (Rothery, 2005, p. 48). Every company that joins the TISN and that will be privy to certain kinds of information is asked to sign a deed of confidentiality regarding handling information that is either commercially sensitive or security-related. Those participants who decline to sign the deed of confidentiality are allowed only limited access to information discussed within the TISN (Trusted Information Sharing Network for Critical Infrastructure Protection, 2007). In addition, most of the states and territories have their own critical infrastructure advisory and coordination organizations. The Business Government Advisory Group on National Security brings together senior ministers in the Australian government and CEOs of major Australian corporations to discuss security issues and to provide guidance to the business community with respect to the impact and potential impact of government policies on industry (Australian Government, 2006, p. 29). At the state level, state governments are expected to determine which critical infrastructure assets exist in their respective states or territories, maintain a state/territory database of critical infrastructure, develop emergency response procedures for each level of alert, provide security advice, and advise owners of critical infrastructure as to the relevant threat information (New South Wales Government, n.d., p. 2). State and territory police are responsible for the provision of security advice to critical infrastructure owners and operators, to maintain liaison with these private-sector entities, to disseminate appropriate intelligence information to them, to conduct and participate in exercises along with the critical infrastructure owners and operators, and to assist these in developing counterterrorism policies and strategies (Australian Attorney-General's Office, n.d., p. 3).

CRITICAL INFRASTRUCTURE PROTECTION INSTITUTIONS AND GOVERNMENTAL ENTITIES

United Kingdom

In the UK, the Center for the Protection of National Infrastructure (CPNI) advises critical infrastructure assets with respect to contingency planning, security measures, and business continuity (UK Center for the Protection of National Infrastructure, n.d.). NaCTSO, which supports the activities of CTSAs in the regional police forces, is a police unit colocated with the CPNI. NaCTSO provides the private sector with advice on securing explosives and precursor chemicals, pathogens, and toxic and radioactive materials and also provides guidance on business continuity, preventing vehicle-borne terrorism, and the protection of crowded places (UK National Counter Terrorism Office, n.d.).

Germany

In Germany, the Federal Office for Civil Protection and Disaster Response (Bundesamt für Bevölkerungsschutz und Katastrophenhilfe—BBK), operates a Center for the Protection of Critical Infrastructures. This center at the BBK is responsible for facilitating cooperation between industry and government, developing protection models, and proposing measures for protecting critical infrastructure (Germany, Federal Ministry of the Interior, n.d., pp. 50–51). In addition, the Federal Agency for Technical Relief (THW—Technisches Hilfs-werk), which is part of the Ministry of the Interior, provides technical relief in the areas of rescue, salvage, and the rehabilitation of infrastructure (Tech-nisches Hilfswerk, 2010).

Canada

Canada has a unique organizational approach to critical infrastructure protection in that in December 2003 it created a cabinet-level agency, the Department of Public Safety and Emergency Preparedness (informally known as PSEPC—Public Safety and Emergency Preparedness Canada) that includes intelligence (CSIS), federal law enforcement (RCMP), and the Office of Critical Infrastructure Protection and Emergency Preparedness. PSEPC has responsibility for coordinating the government's response to terrorist attacks and provides personnel to the National Emergency Response System (which is tasked with planning and organizing the federal response to major disasters) (Hay, 2006, p. 17). PSEPC operates a Government Operations Center (GOC), which acts as a hub for operations centers of various kinds dealing with terrorism, natural disasters, and other issues.

PRIVATE-SECTOR CRITICAL INFRASTRUCTURES RESPONSIBILITIES/LICENSING

Israel

In Israel, the Commercial Licensing Act of 1968 requires that approximately 60 types of business enterprises fulfill security requirements set by the police in order to receive licenses to operate their businesses. These requirements, which vary according to the nature of the business, its size, and other factors include following specific security procedures, installation of security equipment, and security vetting of employees. The Israel police carry out random checks on such businesses to ensure that they are complying with the licensing requirements (Geva, 1995).

Australia

In Australia, owners and operators of critical infrastructure are responsible for providing adequate security, applying risk management approaches to their planning, conducting regular reviews of risk management assessments and plans, reporting suspicious incidents to law enforcement, maintaining continuity of operations plans, and participating in government-run exercises (Rothery, 2005, p. 49). Moreover, operators are expected, under the Terrorism Community Protection Act of 2003, when it is possible to do so safely, to maintain continuity of operations in the wake of an attack (State of Victoria Auditor-General, 2009, p. 15).

Germany

In Germany, the German Stock Corporation Act requires that a number of sector-specific companies set up monitoring and risk management systems to proactively identify developments that threaten the functioning of their respective companies (Federal Ministry of the Interior, n.d., p. 9).

United Kingdom

In the UK, the National Counter Terrorism Security Office (NaCTSO) is part of the Center for the Protection of National Infrastructure (CPNI), which reports to the Association of Chief Police Officers (ACPO), a nongovernmental organization funded by the Home Office and local police authorities designed to provide policing and community interface guidance. The ACPO consists of chief constables and their deputy and assistant chief constables and is a prominent voice in effecting policy. NaCTSO publishes a range of guides for different sectors of the economy designed to assist those sectors in preparing contingency plans, business continuity plans, and security arrangements. NaCTSO also runs multimedia simulations for participants from different economic sectors under the framework of Project Argus. For example, NaCTSO produces a document for health institutions that takes managers and planners in those institutions through a four-step process involving identifying threats, determining vulnerabilities and protection needs, identifying

FIGURE 7.5 Cardiff Millennium Stadium, Wales, UK. This Wikipedia and Wikimedia Commons image is from the user Olivier Aumage and is freely available at http://commons.wikimedia. org/wiki/File:Cardiff_Millenium_Stadium.jpg under the Creative Commons Attribution-Share Alike 2.0 France license.

measures to reduce risk, and rehearsing and revising emergency plans, contingency plans, and security measures.

This guidance addresses issues ranging from threats to spotting hostile reconnaissance to business continuity, access control, evacuation measures, personnel security, and information security (National Counter Terrorism Security Office, 2009, pp. 10–70). NaCTSO produces similar publications for pubs and nightclubs, shopping centers, stadia, cinemas and theaters, hotels and restaurants, commercial centers, schools, and places of worship. CPNI also produces a planning guide designed to help industry sectors and individual businesses determine operational requirements for their security measures. It suggests that stakeholders first produce a "level 1 operational requirement (OR)," which defines the physical location, assets to be protected, perceived threats, the consequences ensuing from damage of the assets, criteria for security success, and possible security solutions and then produce a set of checklists based on the physical location, assets to be protected, threats, consequences, and so on. Subsequent to defining level 1 requirements, stakeholders are encouraged to produce a level 2 OR, which focuses in more detail on each area of concern and its possible solution. These will look at possible solutions for each of the areas addressed in the level 1 ORs. For example, if a level 1 OR assessment determined that a facility needed perimeter fencing with detection capabilities, a level 2 OR would address issues such as type of fending, type of perimeter intruder detection system, lighting, CCTV, and so on (Center for the Protection of National Infrastructure, 2007, p. 10). The British Security Service (MI5) also provides information to businesses on, among other things, how to carry out risk assessments, making security part of organizational culture, mail-handling procedures, proper computer security policies, and identity checking for employees (British Security Service, 2005).

The UK Cabinet Office issues a business continuity management toolkit designed to help business create a business continuity management (BCM) plan. Businesses are encouraged to follow a five-step process in order to analyze the impact of a disruption on their operations. Step 1 involves determining the nature of the key products and services produced by the enterprise and the probable effect on the business of the disruption in their production. Step 2 involves determining the maximum length of time that a disruption to key products and services can be managed before this threatens the enterprise's viability. Step 3 involves setting a recovery time objective (RTO) in which production and/or provision of services would need to be resumed after a disruption. Step 4 involves documenting the critical activities necessary to deliver key products or services, and step 5 involves quantifying the resources needed over time to maintain key activities at an acceptable level and meet the RTO (UK Cabinet Office, n.d.b., pp. 6–7).

AVIATION, MARITIME, AND SURFACE TRANSPORTATION SECURITY

In this section of the chapter we explore international policies with respect to security in the aviation, ground, and maritime transportation sectors. As with other parts of this book, we look at a mix of countries and policies based on the availability of information and highlight a few notable examples. We first address aviation security strategies and then move on to the rail and road transportation sectors, concluding with the maritime sector.

AVIATION SECURITY

The aviation subsector of the transport sector attracts the most attention. Airplane hijackings and attacks against airports have figured prominently in the history of modern terrorism, and, of course, the seminal act of modern terrorism, the attacks on September 11, 2001 against the World Trade Center and the Pentagon, were perpetrated to and with aircraft. The aviation sector is always likely to be an attractive target for terrorists because attacks in the aviation sector are certain to enjoy extensive media coverage. Moreover, the psychological impact of the fear of being victimized in a terrorist event when one is largely helpless and locked inside a metal tube at 36,000 feet is arguably greater than other forms of terrorism. In addition, aircraft are uniquely vulnerable to terrorist attacks, and explosions on aircraft or attacks with surface-to-air missiles guarantee large numbers of casualties—something that is not always assured with respect to attacks at public venues. Finally, as shown tragically in the 9/11 attacks, aircraft are modes of transportation that can be used as a weapon to the detriment of those onboard as well as those in targeted buildings. Consequently, aviation security is a critical area of ongoing homeland security concern.

Israel

In Israel, the aviation security strategy is based on a layered approach starting outside the airport and working in toward the aircraft. These security layers include the airport, passenger areas, catering and duty-free areas, aircraft access routes, and in-flight security, with each layer focused on intelligence, warning and observations, deterrence, and prevention. Travelers are queried by security guards as they drive into the airport and then pass another set of security agents trained to spot suspicious behavior as they access the terminal (Guttman, 2010). Much of the focus of aviation security measures is on observation of passenger behavior. This involves extensive interviews to verify that passengers' reports of their travel destinations, purpose of visit, lodging arrangements, and so on, dovetail with information obtained from the airlines and other sources as well as to afford security personnel the opportunity to observe the behavior of individual passengers as they respond to queries regarding their trip (Pickett, 2008, p. 12). Security personnel employed the airport authority and Israeli airlines such as El Al are well educated and highly trained and must speak at least two languages. Airport security also compares passenger manifests with intelligence lists to ensure that suspicious passengers are not scheduled to board flights (CNN, 2010). Israel airport authority personnel must operate according to security guidelines set by the Israel Security Agency (ISA; see Chapter 1). Ordinary law enforcement issues at airports (as well as arrests of suspected terrorist operatives) are handled by the Israel police. The country's primary international airport, Ben Gurion Airport, has its own police station.

Australia

In Australia, a number of agencies are involved in aviation security. In view of the fact that Australia has a federal system, many strategic aviation security policies must be reviewed and approved by the Council of Australian Governments (COAG), a decision-making and coordination body consisting of the commonwealth and state and territorial governments. In the commonwealth government, the Department of the Prime Minister and Cabinet (PM&C), the Attorney-General's Department, the Department of Transport and Regional Services, the Department of Justice and Customs, and the Department of Immigration and Multicultural and Indigenous Affairs are all involved in various aspects of aviation security policymaking (Wheeler, 2005, p. 13). The PM&C, through its National Security Division, provides the commonwealth government with strategic advice on aviation security, and the ministry's National Security Committee brings together commonwealth, state, and territory governmental and police agencies to oversee policy and the division of labor between the various entities involved in aviation security (Wheeler, 2005, p. 13). Two bodies attached to the Attorney-General's Department play a critical role in aviation security. The Australian Security Intelligence Organization (ASIO), as the country's domestic intelligence services, provides intelligence and threat assessments to the air sector and the Protective Security Coordination Center (PSCC). The PSCC maintains a round-the-clock operations center and is responsible for coordinating the response should a threat arise in the aviation sector. The Customs Service and the Australian Federal Police (AFP) have jurisdiction at airports, and the AFP also provides air marshals on selected flights. The Department of Transportation and Regional Services (DOTARS) deals with the private sector (airlines, airport operators, etc.) and, via its Office of Transport Security (OTS), provides policy advice and briefs industry on terrorism threats (Wheeler, 2005, pp. 15–16). The state governments also play a vital role in aviation security since they have the resources and the authority to respond to terrorism and other threats within their respective jurisdictions. States have policymaking and coordination bodies in their jurisdictions that roughly mirror those at the commonwealth level. In addition, states are responsible for responding to threats to aviation via their respective police services and to responding to attacks or other emergencies in that sector via their respective fire and EMS services.

Since Australia is very large and a considerable number of small towns and villages are located in very remote areas, air transportation is common and there are countless landing strips, some on private land, that are used for light planes. These strips are not regulated and do not have any security measures in place. There are 186 airports in the country that are regulated and required to have security measures, but these requirements differ depending on the type of airport and the perceived threat. There are essentially three categories of airports. The first tier of airports, of which there are 11, consists of the large airports and any other airports that serve international flights. These are required to have stringent security measures in place, including on-site AFP SWAT teams (known as the Federal Police Protective Service—FPPS). The second tier consists of 28 airports that service domestic flights. Most of these second-tier airports have some form of passenger screening and some

FIGURE 7.6 Alice Springs airport, Northern Territory, Australia. This Wikipedia and Wikimedia Commons image is from the user Stuart Edwards and is freely available at http://commons.wikimedia.org/wiki/File:ASP_12.jpg under the GNU Free Documentation License, Version 1.2.

security measures in place. Finally, the third tier of regulated airports, of which there are 147, serve regional and local air traffic, and screening of travelers and/or baggage is not compulsory. The Australian authorities do not consider it practical or desirable to expect high levels of security at regional airports (Wheeler, 2005, pp. 20, 50). Each of the tier 1 airports has an airport security committee, which includes the AFP, the state police force, representatives of the airlines fuel and telecommunications companies and cargo services, the Australian customs services, ASIO, and other interested governmental and industry partners. These committees serve as forums for the sharing of information and raising of security concerns. As far as international flights are concerned, Australia's use of advance passenger processing means that passengers cannot board flights to Australia without first being cleared for landing.

United Kingdom

In the UK, the Department for Transport (DT) is responsible for transportation security. Within the DT there is a Transportation Security Directorate (TRANSEC) responsible for developing appropriate security regimes in different airports or flights based on risk assessments, monitoring, and enforcing these regimes and providing security advice to industry. TRANSEC also funds research, development, and evaluation of new technologies that have the potential to improve transportation security. TRANSEC enjoys the authority to issue directives to airlines, airports, and other entities in the aviation sector (known as *directed parties*) that require them to carry out specific measures, such as passenger and baggage screening. Directives are couched in broad terms and identify minimum standards so that industry managers can determine the best way of implementing these general requirements and standards (House of Commons, Transport Committee, 2005, p. EV 7).

UK airports employ a multiagency threat and risk assessment (MATRA) process wherein stakeholders at the airport work together to create a risk register and identify actions required to reduce risks. MATRA helps to create greater mutual familiarity between stakeholders for each other's responsibilities and capabilities (House of Commons, Transport Committee, 2005, p. EV 7). Policing at the airports is the responsibility of the local police constabulary with Special Branch officers playing a key role.

France

In France, the country's armed forces continually monitor French airspace and the approximately 10,000 aircraft that fly within it on any given day with over 100 fixed radars. They maintain combat aircraft and attack helicopters to respond to air threats and maintain bilateral air security agreements with a range of neighboring countries that allow the French authorities to identify threats before they reach French airspace (France, Prime Minister's Office, 2007, p. 60). The armed forces also play an important role at the airports, as the Gendarmerie Nationale serves as the primary airport law enforcement agency.

Germany

Germany represents an interesting twist on one aspect of aviation security, the willingness to shoot down a hijacked passenger plane as an act of last resort to prevent it from being used by the hijackers to wreak additional destruction and loss of life, along the lines of the attacks on September 11, 2001. Most countries recognize that there is little choice and that shooting down a hijacked plane to prevent it from targeting people on the ground or in buildings essentially means saving additional lives because the lives of those on the planes can be considered already to have been lost. When the German parliament passed the Aviation Security Act in 2004, which authorized the German government to take this measure if necessary, this ignited a fierce debate over whether the government had the right to revoke the basic rights of the passengers on board the plane by instructing the military to shoot it down. The Federal Constitutional Court subsequently ruled that the parliament lacked the legislative authority to empower the government to make this decision and that the act was incompatible with the fundamental right to life and the human dignity of the passengers on board—given that they were not criminals but rather innocent bystanders in the event. The court rejected the argument that the lives of the passengers on board could already be considered forfeited and ruled that shooting down a plane would be an act of state murder (Mauer, 2007, p. 69).

Canada

In Canada, the Department of Transport (also known as Transport Canada) is the lead agency for aviation security. The RCMP and CSIS collect information, but Transport Canada must decide whether or not to refuse clearance to certain passengers. Transport Canada operates a no-fly list and has the authority to ask airlines for passenger information. The Aeronautics Act empowers Transport Canada to share passenger information with the minister of citizenship and immigration, the minister of public security, selected CSIS and RCMP personnel, and the CEO of the Canadian Air Transport Security Authority (CATSA) (Commission of Inquiry into the Actions of Canadian Officials in Relation to Maher Arar, 2006, p. 180). CATSA is a government-owned entity (Crown Corporation) that fulfills airport security functions similar to those of the Transportation Security Administration in the United States. CATSA screens air passengers, operates systems to detect explosives, and funds RCMP air marshals on certain flights but holds no responsibility for screening air cargo.

Policing services at Canadian airports are split between the local policing authorities (which may or may not be the RCMP on contract), which enforce the Canadian Criminal Code as well as provincial laws, and the RCMP, which enforces federal statutes. Consequently, while the RCMP may be the only law enforcement agency present at some Canadian airports (in those jurisdictions in which they provide provincial and/or local policing services), the RCMP does not enjoy overall responsibility for airport law enforcement (Canada, Standing Senate Committee on National Security and Defense, 2007, p. 9).

MARITIME AND PORT SECURITY

Israel

Israel has a long coastline, parts of which are close to areas of terrorist activity. The northern Israeli town of Nahariya is located only 9 kilometers south of the Lebanese border (and consequently, not far from Hizballah maritime bases) and the southern city of Ashkelon in just 15 kilometers north of the Gaza Strip (where Hamas maintains a naval presence). Moreover, Israel's largest port, located in the city of Ashdod, is merely 25 kilometers north of the Gaza Strip. Israel's southernmost city and third largest port, Eilat, is sandwiched between the Egyptian and Jordanian borders. With the exception of Jerusalem and a few smaller cities, most of Israel's urban areas are in proximity to the coastline. The short distances between potential terrorist threats and their coastal and port targets mean that the Israeli navy and maritime units of the Israel police have, in many cases, only minutes to evaluate and engage threats (Lorenz, 2007, p. 5). The first attempt to launch a maritime terrorism attack against Israel occurred in 1953, but the 1970s brought a wave of seaborne attacks to the country originating from both the coasts of neighboring countries and from mother ships out at sea and including maritime suicide bombing attacks (Lorenz, 2007, pp. 6–10).

The primary agency responsible for coastal defense in Israel is the Israeli navy. The Israeli navy also has the capability to conduct traditional naval battles (although these have never been a major factor in previous Arab–Israeli wars) and to project strategic power throughout the Mediterranean, Red Sea, and Persian Gulf areas. Nevertheless, the counterterrorism mission is one of the navy's primary missions. In addition to the navy, the Israeli air force, which is responsible for operating all military aircraft in the country (as well as antiaircraft and antiballistic missile systems), plays an important role

FIGURE 7.7 Israeli navy guided missile patrol boat. This Wikipedia and Wikimedia Commons image is from the U.S. military or Department of Defense and is freely available at http://commons.wikimedia.org/wiki/File:INS_Hetz.JPEG and has been released into the public domain as a work of the U.S. federal government.

since Israel's coastline is patrolled constantly by aerial reconnaissance aircraft (these are light planes with minimal defensive capabilities). Israel also employs a network of maritime radar stations along its Mediterranean and Red Sea coasts, and both aircraft and radar stations can provide intelligence to three local command centers (in Haifa, Ashdod, and Eilat) to dispatch fast patrol boats to respond to threats (Lorenz, 2007, p. 40). According to some reports, Israel is also building a maritime barrier off the Gaza coast to prevent terrorists from swimming up to the Israeli coast and small boats from accessing the Israeli coast. Israel also employs buoys with sensors at the maritime border with Lebanon (Lorenz, 2007, pp. 40–41). Finally, foreign ships accessing Israeli ports are escorted by naval ships when they are still 6 nautical miles away from the port, and some of the vessel hulls are examined by divers before they enter the port itself (Lorenz, 2007, p. 41). Other entities involved in maritime security include the Israel police's maritime units (which provide port security and also patrol in coastal waters), the Customs Authority, and the Ministry of Transportation (primarily in the realm of safety issues).

United Kingdom

In the UK, the Royal Navy, under Military Aid to the Civil Power (MACP) provisions (as noted in Chapter 5), can engage in a range of activities, from fishery protection patrols to counterterrorism operations in rivers, coastal areas, and littoral waters. The Royal Navy has been engaged increasingly in assisting the Serious and Organized Crimes Agency (SOCA), of which the former HM Customs Service is now a part, in countering drug trafficking. The Royal Navy also gathers intelligence and can carry out surveillance on port-bound ships in British waters (Dodd, 2009, p. 347). Maritime safety is the responsibility of the Maritime and Coast Guard Agency (MCA), which is an agency within the Department for Transport (DfT). Among other duties, the MCA is responsible for coordinating maritime search and rescue with Her Majesty's Coast Guard. Unlike the U.S. Coast Guard, HM Coast Guard does not have law enforcement or quasimilitary powers and focuses primarily on search and rescue and marine safety. The primary agency for regulating maritime security, as with aviation security, is TRANSEC. TRANSEC requires that ship operators and port facilities draw up security plans based on TRANSEC guidance, and these ships and ports are subject to compliance checks by the agency's inspectors. TRANSEC also delivers mandatory training for port and ship security officers. Finally, the agency also handles security alerts for shipping (in consultation with the JTAC), and these alerts range from level 1 (normal) to level 2 (heightened) to level 3 (exceptional), with each level being linked to specific preparedness measures (UK Department for Transport, n.d., p. 2).

FIGURE 7.8   Australian Customs Service ships dock at port. This Wikipedia and Wikimedia Commons image is from the user kenhodge13 and is freely available at http://commons.wikimedia. org/wiki/File:Darwin%27s_Stokes_Hill_Wharf_January_2010.jpg under the Creative Commons Attribution 2.0 Generic license.

Australia

In Australia, the Australian Customs Service monitors the entry and exit of ships, goods, and persons from or to ports and performs basic immigration services. The counterterrorism response within the boundaries of the port areas are the responsibility of the respective state or territorial governments. Beyond the territorial seawater line, the commonwealth government has jurisdiction over counterterrorism prevention and response activities, and the primary commonwealth agency tasked with this mission is the Joint Offshore Protection Command (JOPC). The JOPC is a military entity under joint civilian and military control. For missions involving civil surveillance, safety, or other regulatory functions, the JOPC operates under the authority of the Customs Department, whereas as part of its counterterrorism and security missions, it operates under the authority and command structure of the chief of the Australian Defense Force (Australian Government, 2005, p. 11).

Canada

Canada's maritime security approach is based on the concept of five concentric circles that expand outward from its national territory. The first circle is around a domestic port. The second is bounded by the 12-nautical mile territorial limit of Canadian waters. The third circle covers coastal and internal waters. The fourth circle covers waters between North America and Europe, and the last covers foreign ports such as Antwerp or London (Avis, 2003, p. 11). Each circle involves specific security activities related to protection, response, domain awareness, and collaboration. In coastal waters and seaways, the focus will be on domain awareness, collaboration, and response, whereas in the foreign ports circle, the focus in on collaboration with foreign authorities and domain awareness. Consequently, the outer circles involve primarily information-based activities, whereas physical response measures take an increasing role as one moves from the outer to the inner circles (Avis, 2003, p. 11).

Canada operates marine security operations centers (MSOCs), designed to detect, assess, and respond to maritime security threats. MSOCs collect raw intelligence and process it. MSOCs are led by the Canadian navy on Canada's Atlantic and Pacific coasts and by the Royal Canadian Mounted Police (RCMP) on the Great Lakes and St. Lawrence Seaway (Commission of Inquiry into the Actions of Canadian Officials in Relation to Maher Arar, 2006, p. 178).

SURFACE TRANSPORTATION SECURITY

United Kingdom

In the UK, the primary policing agency responsible for security and law enforcement in the railway and London Underground system is the British Transport Police (BTP). The BTP deploys close to 3000 police officers throughout the country. Naturally, the BTP, which is a national agency, cannot perform all security and policing duties over some 10,000 miles of track and, consequently, must rely on cooperation with local constabularies. The formal relationship between the BTP and the local police rests on agreements based on two types of frameworks: mutual aid and extended jurisdiction. Under mutual aid, BTP officers can assist the local constabulary when it so requests, in which cases BTP officers will come under the operational command of the local chief constable and will enjoy the same powers as officers of the local constabulary. Under the 2001 Anti-Terrorism, Crime and Security Act, BTP officers have the authority, in very limited circumstances, to operate outside the railway system in a non-mutual aid capacity in order to safeguard life (Scottish Executive, Justice Department, 2002, pp. 2–3). As with other British policing agencies, the BTP has the authority to erect cordons and implement stop and search procedures within those areas based on threat assessments.

The UK employs a surface transportation security program known as Crime Prevention Through Environmental Design (CPTED). Under CPTED,

FIGURE 7.9   British Transport Police motorcycle. This Wikipedia and Wikimedia Commons image is from the user Arpingstone and is freely available at http://commons.wikimedia.org/wiki/File:British.transport.police.arp.750pix.jpg and has been released into the public domain.

each police department has an architectural liaison officer (ALO) who develops expertise on blast effects on structures of various kinds. ALOs advise local companies on security issues relating to design and construction. CPTED guidelines are followed to ensure good visibility for passengers and CCTV systems when new stations are built or old ones are upgraded, and bomb shelter areas (BSAs) are identified so that people can take refuge there when physical evacuation from the structure is not an option (Jenkins and Gersten, 2001, p. 16). The Department for Transportation's Transportation Security Directorate (TRANSEC) is responsible for setting and enforcing rail security standards.

CONCLUSION

Addressing international CIP and transportation security policies clearly requires several volumes, and one chapter cannot do justice to the breadth of this topic. Nevertheless, this chapter provided a handful of examples across a few countries with respect to strategies and institutions involved in the provision of security to these clear potential targets. Israel has an extensive system of protocols and licensing for the public sector and can employ its emergency economy system to keep critical infrastructure providers operating in the event of an acute crisis. The British, Canadians, and Australians all have extensive contacts with the private sector, relying comparatively more on cooperation and persuasion than on legislation and regulation. With respect to aviation, maritime, and ground transport security, the Australians, with their wide variety of airports, follow a risk-management approach in terms of securing such facilities, and most of the countries surveyed rely on naval forces to ensure security on the access routes to their major ports.

ISSUES TO CONSIDER

- How does the Israeli system of emergency economy work, and should such a system be applied in other countries facing massive disruption and challenges to continuity of operations in the wake of a major natural disaster or similar threat?
- What are the security requirements across the countries surveyed in this chapter that are imposed on the private sector as a condition of licensing?
- What is the Australian approach to aviation security given the wide range of types of airports and flights in that vast and largely empty country?
- How do the British transport police provide mutual aid to local constabularies?

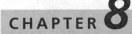

CHAPTER 8

EMERGENCY PREPAREDNESS, RESPONSE, AND MANAGEMENT

In this chapter we survey the emergency response and management policies and approaches taken by a number of countries with a particular though not exclusive focus on two countries with very well-developed response and management systems: Israel and the UK. Providing medical services in emergencies requires the capacity to provide services that include first aid and triage, transportation of the sick and injured, hospital surge capacity, disease surveillance and prevention, and mental health maintenance. These services overlap, particularly in large-scale disaster situations, with the provision of sanitation services, public hygiene, and the provision of shelter (Alexander, 2002, p. 189). In this chapter we look at emergency medical services and operations, incident response, command and management, response and planning strategies, and the provision of post-event social services.

EMERGENCY MEDICAL SERVICES AND THEIR ROLE

Israel

Israel has one primary emergency medical service (EMS), known as Magen David Adom (Red Shield of David—MDA). The MDA is a national organization designed both to serve as the primary EMS service in the country and to assist the military during wartime (the MDA's legal mandate requires that it train during peacetime in support operations for the military). The MDA is also responsible for running Israel's blood bank and for providing first-aid training. The organization operates 50 first-aid and emergency response

Comparative Homeland Security: Global Lessons, First Edition. Nadav Morag.
© 2011 John Wiley & Sons, Inc. Published 2011 by John Wiley & Sons, Inc.

FIGURE 8.1   MDA armored ambulance. This Wikipedia and Wikimedia Commons image is from the user Tewfik and is freely available at http://commons.wikimedia.org/wiki/File: MDA_Armoured_Ambulance.jpg under the Creative Commons Attribution-Share Alike 2.5 Generic license.

centers nationwide and has some 900 employees and approximately 5000 volunteers. The MDA's medical response emphasizes continuous triage and operates on the principle of "scoop and run," with only the very minimum treatment needed to preserve life provided on-site (Cole, 2007, p. 106). This approach is taken, among other reasons, because the EMS service has had to deal in the recent past with many injuries that were the result of terrorist attacks and in which there were well-placed fears of a secondary device or follow-up attack. When the first ambulance arrives on the scene, it acts as the EMS command post and does not provide treatment. The command ambulance reports to the Ministry of Health (and, in a mass-casualty incident, to the IDF Homefront Command) on the scope of the incident and estimates the number of casualties, thus allowing the Ministry of Health to direct resources to the site and notify hospitals (Marcus, 2002, p. 22). Severely injured persons will be evacuated to "close circle" (nearby) hospitals and given only the most necessary pre-hospital urgent care before arriving at the facility. Persons who are less critically injured will be evacuated to "second circle" hospitals— hospitals that are farther away (Leiba et al., 2005, p. 1). One of the interesting aspects of Israeli response strategies is the expectation that bystanders will assist in the emergency response. EMS personnel will typically solicit volunteers to drive less severely injured persons to the hospital and reserve the ambulances for those who are more critically injured. The MDA also relies on a large number of volunteers to staff its ambulances and provide other services.

The norm for response to and management of an event is one hour from the initial call until all the injured have been hospitalized and the dead have been removed from the scene. The emergency response process is assisted by the military's Homefront Command (HFC), which operates a round-the-clock medical operations center with communication links to MDA, all Israeli general hospitals, military command centers, fire services, police headquarters, HFC search and rescue units, military medical branch units, the Ministry of Health, the Ministry of the Environment, the air force (which conducts virtually all air medical transport), and the hazardous materials information unit (Raiter et al., 2008, p. 225). When a mass casualty incident occurs, hospitals in the area are informed by the MDA and the HFC medical department via beeper messages and phone calls from HFC officers to hospital emergency department heads. The hospitals are instructed to activate their mass casualty incident protocols and their national online casualty surveillance computer software (which creates a single database that allows tracking of all the injured arriving at hospitals or other medical centers). HFC medical officers are sent to the scene and to various hospitals in order to gather information and update the HFC's operational center as well as to get an overview of the situation and report to hospital managers and first-responders (Raiter et al., 2008, p. 226). An additional response organization that always arrives at the scene of terrorist attacks, accidents, and disasters is Zaka (Zihuy Korbanot Ason—Identification of Victims of Disaster). This volunteer agency of approximately 1500 personnel consists largely of ultra-orthodox Jewish men. Zaka assists the MDA in providing aid to victims and prepares the bodies of those who did not survive the event for transportation to the morgue. Jewish law requires that a person be buried whole, and consequently, Zaka volunteers can be seen at the site of terrorist attacks retrieving body parts and working to match these to the appropriate victim.

United Kingdom

In the UK, the role of the local ambulance service is to coordinate the on-site aspects of the National Health Service (NHS) response, liaise with other emergency services, identify resources needed, coordinate NHS communications at the scene, and, of course, treat and transport casualties (Macpherson, 1998, p. 13). In London, the first ambulance staff member who arrives at the scene acts as the Ambulance Incident Officer (AIO) until the arrival of personnel from the London Ambulance Service (LAS). The second ambulance crew arriving at the scene initiates command and control and prepares the triage plan (London Emergency Services Liaison Panel, 2007, p. 13). In London, a Medical Incident Officer (MIO) is also deployed by the LAS. He/she is a senior physician with the role of deploying physicians and nursing staff at the scene. The MIO provides professional guidance to the ambulance service, coordinates National Health Service clinical resources, determines the appropriate distribution of casualties to receiving hospitals, and provides a physician to confirm the deaths of those victims who did not survive the attack, to act as the clinical assessor for public health issues, and to deal with the media alongside the ambulance service incident commander (UK National Health Service, 2005, pp. 6–7). With the first report of an incident, the ambulance service will set up an Emergency Operations Center (EOC) and the senior officer in charge of the EOC will immediately dispatch a predetermined attendance (PDA) group of six ambulances and six personnel and prepare to deploy forward control vehicles. As the incident evolves, the EOC is responsible for mobilizing additional EMS resources, for liaison and briefing to the gold and/or silver commands depending on the degree of central government mobilization (a discussion of these commands follows later in the chapter), and for updating area hospitals with respect to victim numbers and condition (London Emergency Services Liaison Panel, 2007, p. 12).

FIGURE 8.2   London Ambulance Service bicycle. This Wikipedia and Wikimedia Commons image is from the user Oxyman and is freely available at http://commons.wikimedia.org/wiki/File: NHS_bicycle.jpg under the Creative Commons Attribution-Share Alike 3.0 Unported license.

At the scene, the LAS use two levels of triage: triage sieve and triage sort. In the *triage sieve* process, the injured are put into priority groups: immediate red, urgent yellow, delayed green, and expectant blue (the dead comprise the final priority category: dead or deceased white or black). They are then moved to a casualty clearing station for *triage sort*, in which patients are reclassified through more thorough clinical exams (London Emergency Services Liaison Panel, 2007, p. 33).

France

France has a broad range of fire services with roughly 250,000 personnel, of whom almost 80 percent are volunteers. One of the most curious and idiosyncratic elements of France's fire service is the 5 percent of fire personnel who belong to military units (army personnel in Paris and navy personnel in Marseilles). The Paris fire brigade is the largest fire service in Europe. The brigade is part of the French army's engineering divisions, hence these firefighters are referred to as sapeurs-pompiers (sapper-pumpers).

The primary EMS service in France is SAMU (Service d'Aide Medicale Urgente—Emergency Medical Assistance Service). Each French Département operates a SAMU service. SAMU operates on the basis of six principles:

1. Emergency medical assistance is a health care activity, and whoever provides critical health care (bystanders, the police, rescue workers, and of course, medical staff) acts as the first step in a process of medical care.

2. The emergency medical response must be proportionate to the need so that each victim receives the type of treatment that he/she needs without the unnecessary depletion of resources or tying up of personnel in noncritical situations.

3. The emergency response must take into account medical needs, operational needs (having to do with transportation, communications, etc.), and human needs (having to do with respect for the needs of the patient and reduction of the patient's anxiety).

4. The medical response system needs to be organized and coordinated so that personnel know their roles (as medics, transport teams, physicians, etc.) and coordinate their activities.

5. "Tradecraft" is critical, and the medical response personnel must be highly skilled.

6. EMS personnel should act as "observation platforms" for epidemiological data so that they can provide proactive information (SAMU de France, 2010).

HOSPITAL PREPAREDNESS AND RESPONSE

Israel

In Israel, the Ministry of Health (MOH) operates the Supreme Hospitalization Authority (SHA), which focuses on preparing the hospital system to cope with national emergencies (Israel Ministry of Health, 2007, p. 29). One of the subcommittees of the SHA drafts and distributes master mass-casualty plans to each hospital. These are then adapted by the preparedness committee of each hospital to its unique needs and capabilities and then translated into a set of standing orders relating to things such as personnel numbers and the prepositioning of equipment and supplies (Levi, Michaleson, Admi, Bergman, and Bar-Nahor, 2002, p. 13). Many hospitals maintain gas lines and other fixtures in corridors and basements so that beds can be brought in to provide surge capacity, which is generally required to be 20 percent of the total bed capacity. With respect to WMD threats, hospitals in Israel have chemical decontamination facilities at the entrance to the hospital because Israeli response doctrine emphasizes evacuation to the hospital for decontamination and treatment in chemical attack scenarios as opposed to decontamination at the scene of the attack (Association of State and Territorial Health Officials, 2006, p. 7). Moreover, as noted earlier, extensive citizen assistance in evacuating casualties means that fully 50 percent of the injured are evacuated to hospitals by private individuals; consequently, hospitals must have a decontamination capability (Briefing by MG Yair Golan, 2009).

All physicians working in hospitals, regardless of their specialty, are required to work shifts in the emergency department in order to create surge capacity. During a crisis situation, health care personnel work 12-hour shifts, thus increasing staff time by 50 percent. When patients arrive at the hospital, the surgeon in charge assesses their condition and triages them to one of three admission sites based on whether their condition is *critical*, *moderate*, or *mild* (Cole, 2007, p. 106). The hospitals closest to an incident often triage the more lightly wounded victims to other hospitals, while the Homefront Command coordinates air force helicopters to transport victims, and the Ministry of Health mobilizes hospitals to accept these patients (Marcus, 2002, p. 22). The overall goal of hospital care during mass-casualty incidents is to deliver an "acceptable" level of care in order to preserve lives and prevent complications. Finally, terrorist attack victims admitted to Israeli hospitals will typically be photographed before swelling disfigures a victim's face (more on this below). These photos are entered into a national database for ease of identification on the part of family members (Elliott, 2010, pp. 7–8).

United Kingdom

In the UK, planning for the health response to emergencies is managed by the National Health Service (NHS) via its Emergency Planning and Coordination Unit (EPCU). EPCU is responsible, among other things, for developing national health guidelines, facilitating training and coordinating the national health response to a major incident (Macpherson, 1998, p. 12). The role of the local ambulance service is to coordinate the on-site aspects of the NHS response, liaise with other emergency services, identify resources needed, coordinate NHS communications at the scene, and treat and transport casualties. The NHS is also responsible for ensuring that its hospitals, as well as those in private hands, engage in planning activities and are able to deploy agreed-upon resources in the event of an emergency. In England, the NHS operates locally via strategic health authorities (SHAs), which are responsible for mobilizing resources within an SHA's designated area and liaise with the Department of Health to provide for a regional response (UK Government b, 2005, p. 13). SHAs are brought in when a significant incident threatens to overwhelm the resources of more than one NHS organization, the incident crosses SHA jurisdictional areas, or the incident requires

FIGURE 8.3 The Royal London Hospital. This Wikipedia and Wikimedia Commons image is from the user LoopZilla and is freely available at http://commons.wikimedia.org/wiki/File: Royal_London_Hospital.jpg under the Creative Commons Attribution-Share Alike 3.0 Unported license.

communication with national health authorities. In the event of a highly complex and major incident in England, the Department of Health (DH) will take control of NHS resources via its emergency preparedness division coordinating center. The DH will also coordinate with the health departments in Scotland, Wales, and Northern Ireland.

In London, as is done at the scene of an emergency, each hospital also designates a Medical Incident Officer (MIO) to manage the arrival and treatment of casualties from a major incident. Hospitals are required to provide a trained and equipped Mobile Medical Team (MMT) to provide treatment at the site of a major incident. When the EMS service arrives at the scene, it selects and alerts the most appropriate hospitals. Initially, those hospitals are put on *major incident standby*. Once the EMS service determines that the hospitals have to make preparations to receive casualties, they alert the hospitals to activate their emergency plans, known as *major incident declared* status. Hospitals are then notified once all the casualties have been removed from the site of the incident (London Ambulance Service, 2007, p. 25). During a major incident, at least one area hospital is designated as a receiving hospital that will keep all its staff on site to receive the injured while other area hospitals send staff to the scene of the emergency.

INCIDENT RESPONSE, COMMAND, AND MANAGEMENT

Israel

In Israel, the police are in charge of incident command (with the exception of a WMD event or a war situation, in which case the Homefront Command runs the response; see Sidebar 8-1). The police incident commander directs all first-responders to the scene and the police handle access to the site. Incident scenes are divided into inner and outer perimeter areas. The inner perimeter area includes all the important response activities (firefighting, evacuation of victims, search for secondary devices, etc.) and the outer perimeter is used for crowd control, security, communications, and evacuation units and municipal departments with information on the wiring, water supply, and building plans at the site as well as municipal units that provide temporary housing and care of the other needs of evacuees (if there are such) (Merari, 2000, p. 16).

Once a report of a terrorist incident reaches the police from the public, they immediately distribute a report on the attack to response agencies as well as the IDF Operation's Branch and the HFC. The IDF Operations Branch, based on a preliminary assessment of the nature of the attack, can then activate the Homefront Command if the event requires an HFC response. Events will run a continuum from "ordinary" incidents (including terrorist attacks, traffic accidents, and the like) that do not involve massive destruction,

Sidebar 8-1: HFC Chemical Warfare Preparedness Levels

Prepared-ness Level	Description of the Preparedness Level	Significance of the Preparedness Level	Civilian Operations Activation Code Name	Impact on the Population
1	Calm	Normalcy	Moss	1. Normal life patterns 2. Gas mask/atropine kits closed and stored in civilian homes
2	Preliminary preparation	Warning of enemy operational activity	Bamboo	1. Mobilization of a limited number of HFC reservists 2. The beginnings of the provision of information to the public 3. Preparing bomb shelters and sealed rooms 4. Completion of distribution of gas mask/atropine kits 5. Coordination with the Israel police and other civilian response agencies
3	Intensive preparation	Potential for the use of chemical weapons	Bamboo	1. Opening of gas mask/atropine kits and attachment of the gas mask filters 2. Civilians instructed to take gas mask/atropine kits with them everywhere 3. Closing of schools (based on a government decision)
4	High alert	Warning of possible attack using chemical weapons or the use of chemical weapons in a neighboring area	Reed	1. Declaration of a state of emergency by the government 2. Reduction in civilian activity 3. Civilians told to remain at home, except for workers in critical industries

| 5 | Operations | Chemical weapons attack | Cereal | 1. Air raid sirens/ instructions via the media
2. Civilians directed to remain in bomb shelters and sealed rooms
3. Wearing of gas masks according to HFC directives
4. This stage will end by a stand-down siren call and/or messaging in the media |

Source: IDF Homefront Command (n.d.a., p. 56).

the use of WMD, or are otherwise extremely disruptive or ongoing to major disasters, WMD events, and finally, full-scale war situations in which the civilian sector is being bombarded. As noted above, the Israel police are given incident command in "ordinary" situations and are usually able to handle the response in cooperation with the EMS and fire services. On certain occasions in which the police must respond to a more significant disaster, the normal response agencies will lack the expertise and equipment to respond and thus will call on the HFC to assist in the response (although it will operate under police incident command in that context). Such scenarios have occurred, for example, during the collapse of buildings where the HFC was called upon to provide heavy equipment as well as extrication teams. In more extreme scenarios such as a WMD incident, it is likely that the government will use its statutory power to declare a state of emergency, thus putting the HFC, the only agency with a significant capacity to cope with such an attack, in charge of handling the incident. The same procedure will also occur in the event of a conventional conflict in which the Israeli civilian sector comes under rocket or missile attack. When the HFC is given authority to manage an event or series of events, the HFC will set up an outer perimeter operations center which includes representatives of all the response agencies. (Association of State and Territorial Health Officials, 2006, p. 3).

For historical reasons and contrary to the centralized nature of Israeli homeland security agencies, the fire service in Israel is divided into 24 regional fire services (four municipal and the rest shared by a number of local governments) that are overseen by the Ministry of the Interior. This regionalization has, however, proved inefficient, and at the time of this writing, the Israeli government is working on a bill to amalgamate the fire services into one national fire service. Firefighters in Israel are trained for every type of fire service role and there is no specialization of personnel.

United Kingdom

In London, a major incident at one of the above levels can be declared by any police officers arriving at the scene if the incident (1) involves large numbers of casualties; (2) requires the combined resources of the police, London fire brigade, and London ambulance services; (3) requires the mobilization of the emergency services and support services due to the large number of people involved or dislocated by the event and/or (4) because intense media involvement overwhelms police public communication resources (London Emergency Services Liaison Panel, 2007, p. 7). The officer who declares a major incident must then manage the scene and maintain contact with his/her dispatch center until relieved by a more senior officer. The police, in consultation with other first-responders, establish three cordons around the sight. An inner cordon, designed to provide immediate security at the incident scene and to maintain it as a potential crime scene, is established as well as an outer cordon, controlling access to the inner cordon and controlling the staging area, rendezvous points for emergency services (RVPs) and the staff responsible for authorizing access to the site (known as the Joint Emergency Services Control Center—JESCC), and a traffic cordon to limit access of unauthorized vehicles to the area (London Emergency Services Liaison Panel, 2007, p. 16).

The UK fire service consists of both full- and part-time firefighters (some of whom are professionals and others volunteers). Unlike police agencies, the fire services in the UK are funded and overseen by local government authorities. Fire services cooperate with other response agencies in planning via Local Resilience Forums (more will be said about these shortly).

France

In France, the Ministry of the Interior's Directorate of Civil Protection (DSC—La Direction de la Sécurité Civile) is the primary institution for the centralization of incident management. Among other things the DSC operates a fleet of helicopters, firefighting aircraft, and can call on engineering troops to reinforce firefighters in fighting forest fires, dealing with hazardous materials, and recovering people from collapsed structures. (France, Ministry of the Interior, 2009). The Ministry of the Interior also operates an Interministerial Crisis Management Center (COGIC—Centre Opérationnel de Gestion Interministérielle des Crisis) that coordinates strategic-level incident response

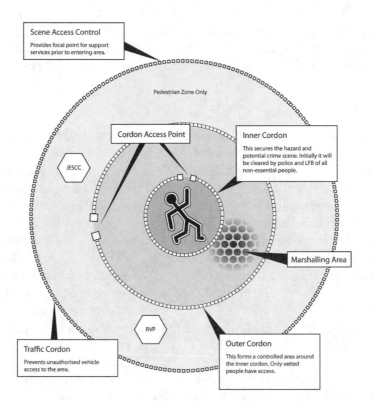

Scene Access Control

Provides focal point for support services prior to entering area.

Pedestrian Zone Only

Cordon Access Point

Inner Cordon

This secures the hazard and potential crime scene. Initially it will be cleared by police and LFB of all non-essential people.

JESCC

Marshalling Area

RVP

Traffic Cordon

Prevents unauthorised vehicle access to the area.

Outer Cordon

This forms a controlled area around the inner cordon. Only vetted people have access.

FIGURE 8.4   London Ambulance Service cordons and access control.

across the relevant ministries. At the regional level, prefects (and in Paris, the police prefect) are responsible for management of large-scale incidents, and these are handled at the local level by municipalities. Local government is mandated to prepare for emergencies through creating the appropriate response capabilities. Among other things, the municipality will create a Municipal Reserve for Civil Security (Les Réserves Communale), which consists of volunteers who report to the mayor and are responsible for assisting the municipality during crisis situations as well as undertaking activities in the area of preparedness, public information, and postaccident recovery. Municipal reserves are not designed to engage in actual rescue, firefighting, and other missions but, rather, to assist in auxiliary activities that free up the professionals to focus on their duties (France, Ministry of the Interior, 2006).

Canada

In Canada, the provincial ministers of public safety are expected to oversee provincial and local emergency response. Canadian legislation requires provinces and local governments to prepare emergency plans. In British Columbia, for example, the minister of public safety can require that these plans be activated and can require that emergency personnel respond to emergencies or disasters regardless of whether or not their agencies decide to do so. Moreover, the minister has the authority to order an evacuation; authorize entry into any building or land without a warrant in order to implement an emergency program; procure; fix prices for or ration food, clothing, fuel, equipment, medical supplies, or other essentially goods during a state of emergency; and control or prohibit travel to any part of the province [British Columbia Emergency Program Act, Chapter 111, Section 10(1), n.d.].

NATIONAL, REGIONAL, AND LOCAL RESPONSE STRATEGIES AND PLANNING

Israel

In Israel, the Homefront Command works with the health system (via the Ministry of Health, which brings the primary health actors together) through the creation of committees that draft recommendations on preparedness. The committees include a policy committee that makes final decisions and committees that deal with hospital preparedness, laboratories, medication dispensing, strategic communications, education, community health preparedness, and security (Marcus, 2002, p. 18). The HFC's preparedness efforts are focused on four goals:

1. The creation of common operational procedures and means of communication between first-responder agencies
2. The development of joint training protocols and the running of joint training exercises
3. Preparations for the setting up of a central command post under the command of the HFC to manage incidents
4. The development of clear protocols for who is responsible for what during and following an incident

In September 2007, the Israeli cabinet authorized the creation of the Reshut Herum Leumit (National Emergency Authority, also known as the National Emergency Management Authority—NEMA). This body is responsible for coordinating, at the strategic level, the military and civilian response to emergencies. NEMA is also responsible for developing strategies with respect to preparedness goals, logistical supply, civilian protection, and interagency cooperation as well as to prepare reports on the country's state of readiness, to support research and development in the area of civil defense, and to help draft legislation dealing with preparedness issues (Israel Ministry of Defense, Office of the Deputy Minister of Defense, 2008).

United Kingdom

The UK emergency response strategy is governed by the Civil Contingencies Act 2004 (CCA). The CCA divides local first-responders into category 1 and category 2 responders. As noted previously, category 1 responders are those organizations directly involved on the scene or in treating casualties and include the local police (or a national police force, such as the British transport police, where relevant), fire brigade, ambulance service, local hospitals, and National Health Service bodies responsible for hospital management and public health (UK Civil Contingencies Act, Schedule 1, Part 1, n.d.). Category 2 organizations include agencies not usually involved directly in coping with an emergency but, rather, in dealing in the aftermath of an incident. These include public utilities (electricity, natural gas, water, sewage, telecommunications, the transport system, airports and seaports, etc.) and safety bodies. Category 1 and 2 agencies coordinate their activities and information flow at the local level via Local Resilience Forums (LRFs) and Regional Resilience Forums (RRFs).

FIGURE 8.5 London Fire Brigade incident response unit vehicle. This Wikipedia and Wikimedia Commons image is from the user Jackus2008 and is freely available at http://commons.wikimedia.org/wiki/File:Incident_Response_unit.JPG and has been released into the public domain.

LRFs are geographically based on police districts and act as the platform for the joint planning of local response agencies, although they do not have command authority over any of the agencies represented. The role of the LRF includes compiling an agreed-upon risk profile for the area, planning joint approaches and operations on the part of the category 1 responders, plan a common strategic communications policy and common interface policies with the private sector, as well as supporting the preparation of exercises and coordinating their execution (UK Government a, 2005, p. 11).

The strategic approach taken to coping with emergency incidents is to handle these, in the first instance, at the local level. If local resources prove inadequate, local agencies will turn to neighboring agencies for mutual aid. The military can also sometimes be brought in, but this will be done, in the first instance, at the request of local authorities (UK Home Office, 2003, p. 3). Only if an incident is so serious that it proves beyond the capacity of local and regional resources will the central government be activated and, then, primarily to provide support to local first responders.

In the UK, serious emergency situations requiring central government resources are ranked on the basis of their level of seriousness (from level 3 to level 1). A *catastrophic emergency* (level 3) is one with a significant and widespread impact, or potential impact, requiring immediate central government direction and possibly the use of emergency powers. This would include attacks of the scope of 9/11 or major industrial accidents on the scale of the Chernobyl disaster. If a level 3 emergency is declared, the strategic response will be led by the Cabinet Office Civil Contingencies Committee (known as COBR—Cabinet Office Briefing Room—after the location in which the committee meets), often with the prime minister as the chair (UK Cabinet Office, 2005). In the case of a *serious emergency* (level 2) or *significant emergency* (level 1), the role of central government gradually diminishes with central government guidance handled at a lower decision-making echelon than level 3 incidents and with a concurrent increase in regional and local management. In level 2 crises, the COBR coordination is handled by the lead government department, and level 3 crises are generally not handled via COBR at all but, rather, in the premises of the lead government department (Cornish, 2007, p. 15). In all of the cases above, the government will appoint a Government Liaison Officer (GLO) to support the police incident commander and act as a single point of contact for central government services. As noted earlier, the central government emergency response process in the UK follows the *lead government department* (LGD) model, wherein one department or devolved administrative unit (such as Scotland) takes overall responsibility for coordinating the central government response. The LGD will be determined by COBR (except in Northern Ireland, which does not follow the LGD system) on the basis of the nature of the emergency, the nature and quality of the department's access to information (in terms of whether this department normally works closely with other departments), and the availability of facilities from which to run the operation (UK Cabinet Office, n.d.a., p. 66). For example, in cases of terrorism, the Home Office is the LGD. COBR will dispatch a government liaison team (GLT) led by a government liaison officer (GLO) to liaise between COBR and the responders at the scene of the event. Organizations that are likely to play an LGD role must prepare for this through consideration of their role during an emergency, developing contacts with other organizations likely to be involved in the response, identifying resources (staff, facilities, equipment, etc.) to be used during a crisis, and developing monitoring systems to provide information on a developing crisis (Northern Ireland Central Emergency Planning Unit, 2010, p. 23). During an emergency, the LGD will act as a communications and information focal point, facilitate discussion and decision-making among the response organizations, coordinate the provision of information to the media and the public, brief strategic-level decision-makers such as government ministers, oversee the process of movement from immediate response to recovery, reconstruction, and risk reduction, and finally, oversee the debriefing process in terms of lessons learned and the review of existing plans (Northern Ireland Central Emergency Planning Unit, 2010, pp. 23–25).

The UK also employs a three-tiered approach to dealing with emergencies known as *gold, silver*, and *bronze*. These designations have to do with roles in an emergency rather than rank and correspond to strategic, tactical, and operational approaches. The gold level is strategic and not on-scene, and it focuses on formulating the strategy for dealing with the major incident within each response organization at the senior management level. The gold command is only used in very large incidents (level 3 emergencies) to make strategic decisions regarding the deployment of resources, managing populations, providing information, and restoring the status quo. In cases where central government intervention becomes necessary, the gold command will act as the primary liaison body with the central government and will be based away from the scene, usually in the offices of the lead governmental department, which become the Strategic Coordination Center (SCC). The members of the gold command act as the Strategic Coordinating Group (SCG) and are usually made up of a senior member nominated from each of the response organizations. The military is also represented at the gold level via a military liaison officer posted to the SCC, who acts as a point of contact for requests for military aid (see Chapter 5). The gold-level person will have control over the resources of his/her own organization but will delegate the tactical response to his/her silver-level counterpart. The gold command is responsible for determining the strategic aim of the response, establishing a policy framework for the overall management of the event, prioritizing the requests from

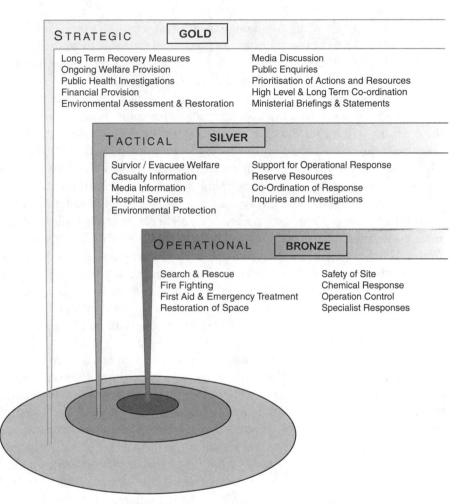

FIGURE 8.6   Gold, silver, and bronze tiers.

the silver command, formulating and implementing strategic communication policies vis-à-vis the media and directing operations focusing on the post-event recovery. The police will normally chair the SCG, but control can shift to other agencies, depending on the nature of the incident (UK Ministry of Defense, 2007, p. 10.5).

The silver level is tactical and involves control over the incident scene on behalf of the silver commander's organization based on the strategy set by the gold commander (in cases when the gold level has been activated). The silver commander will not be involved in the actual response activities but rather will set priorities, plan tasks, and obtain needed resources, and will typically be based at an incident control point near or adjacent to the scene of the event. The silver command provides the gold command with feedback and intelligence regarding the event based on information that it gathers as well as information of relevance to the gold level received from the bronze command (Pearce and Fortune, 1995, p. 183). The silver commander is responsible for allocating resources and obtaining additional resources, assessing risk, ensuring the safety of response personnel, and helping to determine if there is a need to activate the gold command. Finally, the bronze commander will be responsible for the operational response in terms of managing the scene based on the tactical guidance provided by his/her silver counterpart (the silver commander will be activated, in many cases, at the recommendation of the bronze commander, although all levels can be activated concurrently). As the police normally act as a coordinator of the overall response at the scene, the bronze police commander will normally focus on establishing cordons, maintaining security, controlling traffic, and generally managing the incident area. The bronze commander is responsible for assessing the extent of the incident, providing input on helping to determine whether additional levels of command (silver and/or gold) are needed, obtaining resources, and deciding on which tasks need to be executed. As additional resources arrive, managers will be assigned roles within the gold–silver–bronze structure.

One of the first roles of the police bronze commander at a scene is to establish an inner cordon. The boundaries of that cordon will be decided in consultation between the police, fire service, and ambulance service, and those entering or leaving the cordoned area must report to an egress point under the authority of the bronze commander. Personnel entering the corridor are checked to ensure that they have the proper personal protective equipment (PPE) and have been briefed on the situation and any potential hazards. The bronze commander will also set up forward control points (FCPs) for each agency and organizes regular interagency coordination committees. If possible, an outer cordon will be established as a staging area for the emergency services, where they can operate without having to deal with civilian vehicular and pedestrian traffic.

On the preparedness and planning side, the central government body responsible for resilience lies within the Cabinet Office Civil Contingency Secretariat (CCS), which operates to assess immediate needs, to establish possible scenarios for the event (including the possibility that things will get worse), to connect the lead government department with specialist advice and information, and to manage data flow between the response agencies (UK Cabinet Office, n.d.a., p. 68). The CCS is responsible for improving the UK's resilience through planning for emergency events, assessing risk, and enhancing preparedness. Its role is to develop an integrated response to threats, develop a greater understanding of critical networks and infrastructure, to ensure continuity of operations for government in the event of a crisis situation, and to develop policy and planning objectives of enhancing resiliency across the public sector (UK Government, 2003). The CCS is involved in three categories of activities: near-term planning and crisis management, medium-term planning, and training, which cover a range of activities, from developing policy with respect to continuity of government, strategic communications, provision of critical services, and public safety to suggesting changes in legislature, developing response plans, and training and exercising.

Regional Resilience Teams essentially act as the counterparts of the CCS. The role of regional resilience teams (RRTs) is to coordinate the governmental response to an emergency within each of the nine regional government offices in England (things work slightly differently in the devolved administrations of Scotland, Wales, and Northern Ireland) and to act as the main point of contact between government support agencies and local first-responders. In each of these regions, a Regional Resilience Forum (RRF) brings together the primary response actors to plan multiagency response, to facilitate information sharing, and to oversee the regional resilience teams (Cornish, 2007, p. 23). RRFs also produce regional risk maps, consider policy initiatives from local and central governmental authorities regarding civil protection, share lessons from other jurisdictions in the UK and overseas, and coordinate multiagency exercises (UK Ministry of Defense, 2007, p. 8.12). In other words, RRFs are planning and oversight entities, whereas RRTs are responsible for managing incidents. The same holds true for Local Resilience Teams and Local Resilience Forums within the limited geographic scope of the locality in which they operate.

Because the UK has quasi-federalist elements of government the non-English parts (i.e., Wales, Scotland, and Northern Ireland) enjoy differing degrees of devolved administration. In the case of emergencies, each devolved administration has its own civil protection plans and arrangements. If an incident occurs wholly outside England, coordination and management of the response would normally fall to the devolved administration (although in some areas, such as maritime security, the lead government agencies are all central government agencies, so in these areas the central government would manage the response) (UK Cabinet Office, n.d.a., p. 69).

France

In France, the government's plan for vigilance, prevention, and protection against terrorism is known as the Vigipirate Plan (which dates back to 1981 but has been updated a number of times). The present version of Vigipirate assumes that terrorism is a permanent threat and it defines a series of measures that must always be in force. Vigipirate also creates four levels of public alert—yellow, orange, red, and scarlet from lowest to highest—which are set by the prime minister in consultation with the president of the

FIGURE 8.7 French troops deployed to the Louvre Museum as part of a Plan Vigipirate exercise. This Wikipedia and Wikimedia Commons image is from the user Rama and is freely available at http://commons.wikimedia.org/wiki/File:Vigipirate_mp3h9208.jpg under the Creative Commons Attribution-Share Alike 2.0 France license.

republic and each of which carries a set of associated measures. These measures can include actions such as the deployment of troops to railway stations and airports, increased protection for schools, and searches at the entrance to department stores and shopping malls. Local versions of these plans determine the manner in which prefects, mayors, and critical infrastructure operators put measures into place to correspond with the appropriate national alert level (France, Prime Minister's Office, 2007, p. 68). There are also a series of specialized Pirate plans dealing with specific types of threats. PIRATOX deals with chemical attacks, BIOTOX deals with biological pathogen attacks, PIRATOME deals with nuclear or radiological attacks, PIRATAIR-INTRUSAIR deals with aircraft-borne attack scenarios, PIRATE-MER deals with maritime terrorism threats, and PIRATE-EXT deals with attacks against French nationals or interests overseas (France, Prime Minister's Office, 2007, p. 68).

To cope with a terrorist or disaster event that has already occurred, the French authorities devised the ORSEC Plan (Organization des Secours—Emergency Relief Organization). ORSEC is based on three geographic regions: departments, defense areas, and maritime zones, and focuses on risk analysis, designing common response protocols to major incidents, creating a single operational management organization to deal with major incidents, and managing training and exercises (France, Ministry of the Interior, 2007).

Australia

In Australia, the commonwealth government's disaster response plan, known as COMDISPLAN, empowers the Australian Emergency Management Agency (EMA) to coordinate the central government's response to large-scale emergencies. The EMA often deploys liaison officers to the states (states have overall jurisdiction over the emergency response) to help coordinate commonwealth assistance to state or territorial authorities. Each state level then has its own emergency structure and arrangements. The state of Victoria, for example, employs three tiers of agency command and control: incident command, area-of-operations command (with responsibility for a geographic area containing an emergency or group of emergencies), and a state command level (Government of Victoria, 2009, p. 3.4). Similar to the British lead government department approach, a single state agency must be clearly identifiable as having control in an emergency. In Victoria, the state controller serves as the strategic leader for the crisis, and this function is performed by a senior operational person in the controlling agency. That controller establishes a state emergency management team to handle the crisis at the state government level (Government of Victoria, 2009, p. 3.6). The area-of-operations command also has a controller, as does the incident command. Consequently, the Australian approach at the state level largely mirrors the gold, silver, and bronze structure in the UK.

POST-EVENT SOCIAL SERVICES

Provision of support in crisis situations extends not only to those who suffer physical injury but also to those who suffer psychological injury as well as the friends and family of those directly affected by terrorist attacks, disease outbreaks, or natural disasters. Moreover, in order for the general public to be resilient, psychological support must be provided as broadly as possible because social resilience often requires frequent and extensive addressing of psychological trauma. We look specifically at Israel in this section, as it has what is probably the most experience in dealing with these issues. In Israel, psychosocial support is significant and extensive (see Sidebar 8-2). The Tel Aviv Medical Center, for example, operates a program called Hosen (resilience), which focuses on treating patients at risk for posttraumatic stress disorder (PTSD). A "resilience scale" is used to assess patient risk for such complications, and a course of treatment is determined accordingly (including home visits and community referrals) (Inbar and Keren, 2002).

Sidebar 8-2: Hospital Response and Role of Social Workers in Israel—Case Study

In Tel Aviv, the first port of call to a terrorist incident is the hospital. Externally, a chain of relationships has been set up between Community Health Centers, Welfare Offices, Emergency Medical Teams, other Hospitals and Police and Military authorities. Certain types of wounded (e.g., head wounds) are taken to specified hospitals. Such hospitals have acute and specialized care facilities that can deal with those types of patients. Children may be taken to a specialized Children's' Hospital, for example. Hospitals also reorganize their internal systems to be ready in the event of terrorist attacks.

For instance, a few months ago, a bombing took place at the Dolphinarium, a seaside entertainment spot in Tel Aviv. Emergency police, military, and city ambulance services rushed people to Ichilov Hospital. Admissions at the hospital went on alert mode. Upon their arrival, patients were immediately numbered and sent to the Emergency Room. All through their stay they would stay numbered in order to protect confidentiality and to process them as terrorist victims. Depending on the numbers of wounded, this seems to be the most efficient process. Only later, admissions staff takes care of the paperwork and insurance.

Hospital social workers fulfill many functions during a terrorist alert. Medical social work staffs who are off duty know to come in to the hospital as soon as they hear of an incident over the radio or television. They are then organized to go to the Information Center that has been set up close to Admissions, to the Emergency Room and to the Operating Room. Some social workers go to their regularly assigned wards to wait for hospitalized victims to arrive. A phone and fax system are set up with outside lines going to the Emergency Medical services at the scene of the attack, to the police and military, and to the welfare offices. Inside the hospital, a centralized software program lists incoming patients by number, names if known, and any other critical details.

The Information Center is composed of a public reception area, a waiting area for patients' families, friends, and relatives, and a crisis line to answer public inquiries regarding loved ones. On the night of the Dolphinarium attack, hospital social workers did group work with the thirty or more youth that had been visiting the site of attack. A form of debriefing and mutual support took place that centered on encouraging participants to talk about what had happened to them, where had they been, how they were feeling, and what was going on. Group work is seen as one of the basic tools to be used with waiting area populations from an attack.

A multidisciplinary team sits inside the Information Center itself. The team is usually composed of a nurse, doctor, psychologist, hospital social workers, a police representative, translators, and Department of Social Services social workers from the community. The DSS people are skilled in going out into the community to find family members of patients who have been brought in from the attack for purposes of identification, and to aid family members of patients who have been killed at the Forensic facilities.

One of the main responsibilities of the Information Center is to keep people out of the Emergency Room. This allows medical staff to concentrate freely on treating the physical casualties. They deal with the public when they come in and man the phone lines. Their purpose is to provide reliable information. As Ms. Posen stated, depending on the size of the incident, people may come in from anywhere. Often former victims of prior terrorist attacks show up experiencing a "double-effect."

The hospital social worker in the Emergency Room's main function is to meet the concrete immediate needs of disabled victims, such as contacting a family member. Through talking, a main goal is to give support and a sense of control to the patient. This social worker also helps families who come inside the Emergency Room to aid in identification of unconscious patients, sometimes with great difficulty. Patients arrive covered with dust and wounds from the attack. As Ms. Posen stated, often hair color has become gray and one patient who was identified with a mustache turned out to have a burn instead.

In the Operating Room area, the hospital social worker works mainly to provide resources and support to waiting families. A telephone and computer line between this area and the Emergency Room, Admissions, and the Information Center are critical in providing a flow of information from and to the Operating Room area.

Response to a terrorist attack occurs in stages. After the initial reception of patients, the hospital after a few hours returns to its routines. However, the work of the hospital social workers and those social workers in the community continue in follow-up and support of victims that have returned home, providing resources to families. In the hospital, the social work supervisor must make decisions regarding how many social workers to have on at one time, when do they need to rotate out and go home and rest, and how to keep staff functioning. An automatic debriefing session takes place the day after an attack with social work staff and emergency room staff that basically cover the group work topics that were covered in the waiting area with victims and families along with an analysis of the effectiveness of the response network. Sometimes, two or three days later or three weeks later, a victim at the scene of the terrorist attack appears at the hospital requesting help, often with PTSD symptoms. (Levy, 2009).

In 2003, during the worst period in Israel's history in terms of the frequency and scope of terrorist attacks, the government launched a victim registry system (known as ADAM). Patients arriving at the hospital are photographed and the photos are entered into a computer along with other identifying information. All hospitals and emergency information centers have access to the ADAM system, and families may call any hospital to access information from this system. This helps prevent crowds of people from congregating at hospitals near the site of terrorist attacks (Association of State and Territorial Health Officials, 2006, p. 12).

One of the important features of Israel's emergency response approach is the effort to make a rapid return to normalcy after a terrorist event. This is deemed critical for the overall psychological resilience of society, so efforts are made not only to clean up the scene of the event quickly (in terms of the evacuation of the injured and dead and wrapping up the police investigation quickly) but also to provide municipal and national resources to repair damage at the scene of the attack and otherwise make every effort to remove any signs of an attack within a matter of days (and sometimes hours).

PUBLIC COMMUNICATION

In the final section of this chapter we focus on a handful of examples from a few countries in our survey with respect to public communications (also known as strategic communications) policies designed to communicate information to the public on preparing for and coping with crises and to enhance public resilience. If we accept the approach that the primary objective of homeland security is to maintain, to the greatest degree possible, the functioning of government and its services as well as the private sector and, more or less, normal economic and social activity in the face of large-scale threats (whether terrorism, natural disasters, pandemics or very large-scale crime that borders on insurgency), then homeland security policy should concern itself not only with the ability of government (and the private sector) to prevent such events and to bounce back effectively when such events occur, but also with making the public as resilient as possible. After all, the maintenance of economic and social activity and the functioning of government and industry depend essentially on the willingness of people to function under conditions of adversity. If people are paralyzed by fear and/or have no confidence in the ability of government and/or the economy to survive in the face of a massive event, government and the economy will, at least temporarily, be unable to function because they both depend on the active participation of the population. The public must therefore be "hardened" in the same sense that one might harden a potential target, by means creating a public that is psychologically resilient in the face of terrorism, disease out-breaks, natural disasters, and so on. This will allow the public to continue functioning despite attacks, disasters, or disease outbreaks in a manner similar to that of protecting a power plant and creating redundancies so that it can continue to provide electricity in the wake of attacks or disasters. Moreover, the public is often the source of critical information with respect to possible threats—particularly terrorism, significant crimes, and individual illness that could lead to pandemics. Consequently, public communications are a critical element of homeland security policy. If the public can be effectively hardened on the one hand (so that it can better survive attacks and disasters) and leveraged as a source of information with respect to threats on the other, this will contribute immeasurably to security.

CRISIS COMMUNICATION AND RESILIENCY PROMOTION

United Kingdom

The British authorities approach toward communicating risks and providing information to the public is founded on the principle that open communica-tion with the public enhances resiliency (see Sidebar 8-3). This is argued because the nature of risk has become more complex and uncertain and global interconnections increase exposure to previously remote risks, thus making it increasingly difficult for government to guarantee protection against

---

Sidebar 8-3: Principles That British Government Departments and Agencies Are Expected to Follow

Openness and transparency—both about their understanding of the nature of risks to the public and about the process they are following in handling them;

Engagement—Departments will be expected to involve a wide range of repre-sentative groups and the public from an early stage in the decision process;

Proportionality—action should be proportionate to the level of protection needed, consistent with other action, and targeted to the risk;

Evidence—Departments should ensure that all relevant factors, including public concerns and values, have been taken into account;

Responsibility and choice—where possible, people who willingly take risks should also accept the consequences and people who have risks imposed on them should have a say in how those risks are managed. (UK Cabinet Office, n.d.a., p. 9)

such risks. Moreover, increased public skepticism of government and its institutions coupled with greater public access to information from a wider range of sources means that government needs to work harder and provide more transparency to maintain public confidence (UK Cabinet Office, n.d.a., p. 8). As noted in a cabinet policy document, "communication which sets out to change or influence beliefs without recognizing the rational basis of those beliefs, or tries to divert attention away from people's real concerns, will almost certainly fail. A 'we know best' attitude is often a formula for disaster" (UK Cabinet Office, n.d.a., p. 14).

In light of these conclusions, the British government has changed its strategic communications approach to incorporate more openness about the nature of risks and the uncertainties of some situations, more transparency regarding decision-making processes in government, and more engagement with stakeholders and the broader public at an earlier stage in order to make the decision-making process more participatory.

Rather than relying largely on trying to restrict public access to information that could be disruptive, the British government recognizes that legislation such as the Freedom of Information Act as well as the proliferation of media outlets and information sources means that citizens will gain increasing access to risk and threat-related information. It therefore favors a proactive approach to providing people with information (as noted above). This does not mean, however, that quantity can replace quality. Providing more information does not necessarily lead to better communication with the public or with specific stakeholders and may actually cause confusion. Consequently, information has to be organized and edited so that the public can clearly see which information government agencies use and how they assess and analyze that information in order to make policy decisions (see Sidebar 8-4). Official bodies must also be able to explain clearly to the public why certain kinds of information are being kept private.

The British strategic communications approach also involves the use of market research techniques to ascertain potential public concerns about risk or risk-mitigation strategies, including the engagement of marginalized groups (UK Cabinet Office, n.d.a., p. 11). As the provision of reliable and accurate information is considered critical in maintaining public trust, the British cabinet has instructed government departments and agencies to (1) invite external subject-matter experts to review information provided to the public to ensure that it is accurate, (2) make it clear to the public which sources of information have been used in cases where there is a dispute over the nature of risk (including referencing conflicting sources of information) so that the public can judge by itself, (3) make greater use of trusted impartial sources of information such as independent agencies or academia, and (4) ensure that communication between experts and generalists in government

---

Sidebar 8-4: British Government's Six Guiding Principles of a Communication Strategy

Sound management systems—making sure communications experts, policy officials, operations staff come together quickly to deal with situations as soon as they arise. This means planned and rehearsed call-out arrangements, and being prepared to work on a 24 x 7 basis for a considerable period of time. It also means, if necessary, joining up with other departments.

Robustness—building flexibility into the planning process to allow for a variety of different and changing scenarios. Hours spent on producing ready-made solutions to predefined problems are unlikely to succeed in preventing all surprises. Agility of thinking, unfettered by pre-conceptions, is essential in fast-moving and unpredictable situations. Robust strategies should satisfy two criteria—First, initial statements and actions should appear sensible in a wide variety of possible scenarios. (This may rule out doing nothing or issuing completely anodyne statements). Second, they should as far as possible leave future options open, to be taken as more becomes known (communicating about risks to Public Health; Department of Health).

Speed—developing the ability to move quickly

- to agree and issue messages, latest information,
- to deal with rumors, speculation and misleading information.

Messages—getting out key information that is up-to-date, clear, coordinated, consistent, and actually satisfies public concerns, or if it is not possible to do this immediately, explaining why it is not possible.

Images—on the basis that pictures often speak louder than words, making sure that graphics, pictures and diagrams are used to provide impact, and to explain complex or unfamiliar concepts.

Intelligence—keeping fully in touch with latest developments, knowing what is going on. This means close monitoring of the media—particularly the broadcast media, and, wherever possible, contact with stakeholders. Your aim should be to be ahead of the game. (UK Cabinet Office, n.d.a., p. 46)

---

departments is enhanced to reduce the risk that technical information might be misrepresented to the public (UK Cabinet Office, n.d.a., p. 12).

The British authorities employ a four-stage risk communication and risk-management process. Stage 1 involves identifying areas of public concern (something that can be done via focus groups, public opinion surveys, or other means) with a focus both on potential risks of concern to the public and about specific communities within the larger public that may have special concerns. Stage 2 involves active discussion and exchange of information between stakeholder groups (i.e., those groups likely to be directly affected— this could include central and local governmental entities, private-sector entities, and community groups) to identify how various groups may react to risks. Stage 3 involves the authorities providing information about risks, assurance regarding the effectiveness of measures taken to address the risk, and a justification of the government's decisions. Finally, stage 4 involves obtaining information from the public regarding whether or not the risk management processes are working and risks are being adequately addressed (this may be via focus groups and surveys as well as follow-up discussions with the stakeholders involved in stage 2 (UK Cabinet Office, n.d.a., pp. 25–26).

Overall, it is recommended that the government and its constituent entities provide clear and simple messages to the public, avoid patronizing the public and using unnecessary jargon, and use different messages for different age groups and communities. The delivery of public messaging should also be consistent, allow people to know what they need to do to protect themselves, stick to facts and avoid speculation, and avoid over-optimistic assessments regarding the return to normalcy (UK Cabinet Office, n.d.a., pp. 49–50).

The Government News Network (GNN) serves as the UK's principal government information and communication service supporting the presentation of government policies, monitors the media, handles media arrangements for ministerial-level visits around the country, answers press queries, and issues communiqués. Use is also made of a national Media Emergency Forum (MEF), which is an ad hoc group of senior media editors, government representatives, local authority emergency planners, first responders, and private industry set up to consider media issues arising from terrorist attacks or other civil contingencies. The concept behind the MEF is that it will facilitate contacts and coordination during an incident to minimize rumor and inaccurate information being passed on by the media. By bringing senior media editors into the discussion, the authorities hope that they will be more cooperative during emergencies. Media debriefs are also routine and involve post-event discussion between media representatives and government and emergency services representatives to identify lessons learned regarding the management of information (UK Cabinet Office, n.d.a., p. 61).

The British Broadcasting Corporation (BBC) is the UK's public service broadcaster, funded by a tax on UK households. As a quasi-governmental broadcaster, it is required to support governmental preparedness initiatives (see Sidebar 8-5). The BBC thus operates a web site known as "Connecting in a Crisis," which provides information on crises, threats, and preparedness.

Sidebar 8-5: BBC Crisis Communications During the Floods of the Winter of 2000

In North Yorkshire, during the floods of the Winter of 2000, BBC North Yorkshire—Radio York became the main communication channel connecting the emergency services, local authorities and armed forces with the public.

In Cumbria during the flooding in January 2005 when large parts of the city of Carlisle were under 5–6 feet of water—BBC Radio Cumbria became the main outlet for emergency information connecting the emergency services, local authorities and responder organisations with the public. Due to the power cuts caused by the flooding most of the other media were unable to operate, however BBC Radio Cumbria's premises were powered by an emergency generator which enabled the station to broadcast for a 48 hour period through the height of the crisis.

Thousands of individual pieces of information were broadcast...

- Hourly updates from the Environment Agency on river levels.
- Updates from the electricity supplier on the power cuts.
- Updates on the bus services in the city following the flooding and loss of 65 buses.
- Police updates on roads and flooded areas and rescues.
- Business closures, school closures, hospital services status.
- Country Council and City Council emergency messages.
- Health warnings on polluted water. (BBC, 2010)

This web site is designed to complement radio and television transmissions that provide emergency information to people.

Another example of a web site that provides emergency information to the British public is that of the London Resilience Council. Its web site provides links to business continuity planning tools, the flu pandemic response plan, and information on preparing an emergency pack. The web site also provides information on swine flu and spotlights the role of various first-responders.

The British governmental web site *Directgov* provides general and specific advice on coping with emergencies, including bombings and chemical, biological, and radiological incidents (Directgov, 2010). The British Cabinet Office web site provides information to first-responders and recommendations for businesses with respect to establishing business continuity plans and conducting risk assessments. It also provides information on the government's concept of operations and the manner in which the central government responds to crises (UK Cabinet Office, 2010b).

Israel

Israeli strategic communications policies encompass both short- and long-term efforts. The longer-term efforts focus on educating the public through the school systems and educational television programs (for children) and via public service announcements on television and radio, and through web sites (an estimated 52 percent of the population regularly use the Internet). The authorities in Israel believe that the best way of strengthening psychological resilience is through reducing the public's fear of the unknown in the context of terrorism via the provision of information (Homeland Security Institute, 2009, p. 16). The Israel Defense Force's Homefront Command (HFC), the police, and the Ministry of Education (in the case of schoolchildren) engage in public relations campaigns providing a broad range of information on threats and how to cope with them. The Israel police will also typically provide information, via the media, at the site of an attack—along with information on casualties from the EMS commander on the scene and the director of the emergency medicine department at the hospital to which most of the casualties were taken, but if a major crisis, such as a war, occurs, the HFC will provide information to the public alongside the police and other emergency services.

The HFC runs both short- and long-term media campaigns to encourage preparedness and provide the public with information about terrorism and other threats. The HFC provides information via pamphlets as well as its web site that instruct people on planning for emergencies. It focuses, in particular, on family units and provides children with a role to play in family emergency preparedness activities. In fact, the HFC relies on children to enhance preparedness because it realizes that they are the population that is easiest to indoctrinate with preparedness messages broadcast by the media, on children's television programs, and communicated in the schools (Briefing by MG Yair Golan, 2009). Moreover, giving children, and adults, specific tasks in crisis situations helps enhance their psychological resilience because it reduces the sense of helplessness that can often be so psychological damaging in crisis situations. Children are also provided with information about the behavior of pets in emergencies and help empower them to take care of their pets as a way of encouraging them to take responsibility and to be active participants in coping with an emergency rather than being passive victims of it. The Ministry of Education, in conjunction with the HFC, also provides structured education in threats and emergency preparedness to schoolchildren in the fifth grade, and efforts are being made to create a preparedness curriculum that will be taught in every other grade during the course of a schoolchild's education. HFC conscript soldiers are also sent to schools periodically, at the request of the schools, to hold educational seminars and awareness sessions (Homeland Security Institute, 2009, p. 23).

The Israel police generally run both long- and short-term media campaigns in response to specific threats. In addition to providing the public with ongoing information about terrorist threats, the police will typically run public service announcements and provide information on their web site with respect to specific issues. For example, in the wake of Israel's 2006 war with Hizballah (the Second Lebanon War) and the firing of 3699 rockets into Israeli territory, the police ran a media campaign to warn citizens not to approach the remains of rockets, due to the threat of unexploded ordinance. During periods of heightened terrorism threats, the police have provided information to the media, which the latter broadcast immediately, with respect to intelligence indicating that a suicide terrorist was on the loose in a given Israeli city. In other words, the public were openly informed that there was a suicide bomber in their midst and that they should be in a state of heightened awareness in order to spot suspicious behavior and report to the police in the hopes that the attack could be thwarted (despite the fact that suicide bombers are virtually impossible to stop once they reach a population center). As with the case of HFC campaigns targeted at schoolchildren, the general consensus in Israel is that an informed and engaged public will not only be more effective in coping with adversity but will also ultimately be more psychologically resilient.

The HFC web site provides instruction on how to choose a secure room (for older apartment buildings that were constructed before building codes required homes to possess "safe rooms") for areas subject to rocket attacks from the Gaza Strip. It also provides information on types of protective kits (which include gas masks and atropine injections), preparing factories and other places of employment for civil defense measures, and suggestions for measures to be taken during terrorist attacks, earthquakes, fires, floods, and mortar, rocket, and missile attacks (IDF Homefront Command, n.d.a.).

Canada

The federal government in Canada has committed itself to a public communications approach based on a set of 10 principles, including the provision of timely, accurate, clear, objective, and complete information on government services and initiatives, high visibility and accessibility of government institutions, use of a variety of means to communicate information to the public, encouraging open communication with the public on the part of governmental institutions, and ensuring a common communications policy across federal agencies (Treasury Board of Canada, n.d., pp. 1–2) (see Sidebar 8-6). In addition to the above, the strategic framework that guides Canadian risk communications includes determining for each crisis which level of

Sidebar 8-6: Key Factors for Effective Risk Communication from the Canadian Public Health Association

Accessibility: Communicate clearly, concisely, with compassion and human appeal, in a respectful, adult tone. Use short sentences and everyday language. Define new or uncommon words. Be careful when quoting numbers—they are easily misinterpreted and misunderstood. Use graphics and visual aids, easy-to-read fonts, and leave lots of white space in written material.

Accuracy: Communicate information that is as accurate as possible in the circumstances. Refer to credible sources for situation updates.

Action: Help people to help themselves. Provide practical information with clear and consistent directions. Empower people to cope in an emergency.

Appropriateness: Respect the diversity and capabilities of your audience. Address them in an inclusive, representative and fair way.

Credibility: Refer to credible sources for situation updates. Cite these sources to strengthen your messages. In the case of a flu pandemic, refer to the World Health Organization, Public Health Agency of Canada, and your provincial, territorial, or local public health department for information.

Consistency: Follow the lead of federal, provincial/territorial and local public health officials for situation updates. Consistent and coordinated communications help build public trust in the information being disseminated.

Listen to concerns: Use communication channels that encourage listening, feedback and participation. Pay heed to the fears and concerns of your staff members and clients and respond to them in a respectful way.

Regularity: Repeat key messages to keep the issue visible, to make required actions more memorable, and to give them credibility through repetition.

Respond to rumors, misinformation and inaccuracies: Correct misinformation and quell rumor in a direct yet respectful way. (Canadian Public Health Association, c. 2007)

government or federal department has lead responsibility for public communications, which other organizations may be engaged in public communications, the degree of expected media interest, and who the media can be expected to approach for information, what the response time line is expected to be, and how this is likely to affect the public's perception of risk (Health Canada, Crisis Communications Unit, 2003, p. 9).

The Canadian government's GetPrepared.ca web site provides information for preparing a family emergency plan and provides instruction on when to call the emergency services, when and how to shelter in place, and how to respond to evacuation orders. The web site also provides links to detailed information on how to cope with a hurricane and a 72-hour emergency guide predicated on the assumption that families in disaster situations need to be self-sufficient for a minimum of 72 hours (Goverment of Canada, 2010).

Australia

In Australia, the country's Emergency Management Agency (EMA) produces preparedness publications such as *Preparing for the Unexpected* and *Tsunami Awareness* that provide practical safety information (see Sidebar 8-7).

Sidebar 8-7: Advice for Australian Public Health Communications Officers

- The aim of any public information campaign is to ensure public health and safety is the highest priority, to alleviate panic and anxiety, to provide a realistic expectation of the capability of authorities to manage a pandemic and its impacts, and to build community resilience.
- Public messages or statements should consider the following four principles as the basis for effective, transparent communication:
  - this is what we know
  - this is what we don't know
  - this is what we are doing
  - this is what we want you to do
- Public messages should be timely and appropriately targeted. Communications officers should consider the following questions when developing and distributing any public information:
  - WHY should this information be distributed?
  - To WHOM is this message intended (i.e. target audience)?
  - WHEN is the most appropriate time to distribute this message?
  - HOW should the information be presented or distributed?
  - WHO ELSE may need to be consulted on this information?
- Key messages should be developed and repeated regularly. Repetition of information is the best way to ensure messages are well received and understood.
- Messages should be accurate and supported by clear facts. Avoid speculation or generalisation.

- Avoid using technical jargon. Be as clear as possible, using simple language that is not overly scientific or specialised.
- Advice or instructions should be simple and practical, using checklists or clearly described steps where possible.
- Relevant spokespeople/information specialists should be identified early. Possible spokespeople include political leaders, subject experts and "on-the-ground" response representatives.
- Whilst it is important for the key spokesperson to be authoritative and credible in the public eye, the best representative may not always be the highest level government official. Spokespeople should possess an accurate and comprehensive understanding of the situation, be calm and appropriately trained/skilled.
- A record of all public information released should be kept by each agency.
- Jurisdictions should share information and keep relevant stakeholders informed of significant messages to ensure effective coordination of public messages. (Australian Department of Health, 2009, pp. 16–17).

MEDIA RELATIONS POLICIES

Israel

In Israel, the military and police have created a special system in which particular journalists (sometimes embedded within IDF units) are given access to information provided that they fulfill certain requirements. The IDF employs such a system with respect to what are known as "military journalists." These select journalists receive special identification and are given access to senior officers, military operations, and other activities provided that they follow directives with respect to sensitive information and can lose this status (thus creating a built-in tendency to fulfill the military's wishes with respect to the nature of information being reported on by these journalists). They were also required, in times of war, to clear their reports with the IDF spokesperson's office before publication or broadcast.

For immediate crisis situations, such as in the wake of a terrorist attack, the Israel police, which, as noted earlier in the book, is the lead response agency, will set up a communications center to provide information to the press, and senior commanders (including the regional commander and usually the police commissioner) will appear on scene to speak to the press.

United Kingdom

In crisis situations, the British government may authorize the Cabinet Office Communications Group (a permanent agency tasked with publicizing cabinet decisions and other matters) to create a news coordination center (NCC) for a specific crisis. The NCC may provide additional staff to work in the lead government department's operations center to help manage media relations, as that government ministry runs the response, and/or the NCC may act as a central press office to coordinate the government's public communications during the crisis (British Government, 2010, p. 181).

British policymakers recognize the reality of the "new media" in terms not only of 24-hour newscasts over cable, satellite, and the Internet, but also "citizen journalists" creating their own real-time content, including video, via the Web (see Sidebar 8-8). In the British view, this underscores the need to provide accurate and timely information in order to stave off rumors and misinformation (British Government, 2010, p. 187). Once an emergency occurs, the police, who as noted earlier have incident command, will usually dispatch an experienced media liaison officer (MLO) to brief the media at the scene and provide regular updates to the media in order to stave off conjecture, even if there is nothing new to report. Arrangements will also be made for the media to receive regular briefings and interviews with the key response agencies. When information needs to be withheld, such as information that may affect a potential criminal prosecution, the media are told why they cannot receive that information (British Government, 2010, pp. 190–191).

Sidebar 8-8: National Health Service of Scotland: Guidelines for Media Relations

Arrangements for Liaison with the Media

0.1 A major incident is news. Representatives of the media will arrive at the scene, at any casualty receiving hospital and at response control points very quickly, and in large numbers. News desks will also make repeated requests for information by telephone, especially in the early stages of an incident. Similar media pressure will accompany other types of major emergency. How the media is handled will affect how they report the emergency and the response to it. How the emergency response is reported can enhance the effectiveness of that response, both immediately and in the longer term. To this end, NHS managers responsible for emergency response must become familiar with media needs, methods and time schedules, and should prepare and train appropriate staff for media liaison duties.

0.2 Should the scale or circumstances of a major incident require it, initially the police will co-ordinate both the release of information to the media and the response to media enquiries. It is thus most important that police advice and assistance is sought by NHS managers when reviewing their arrangements for media liaison. Similarly, within the NHS, each Board must take the lead and establish a single focus for NHS liaison with the media, should more than one NHS organisation become involved in responding to an emergency in its area. The nature and scale of the emergency, and the nature of media interest, will determine the degree to which NHS Boards should act in concert with the police, local or other authorities. Plans for mutual support between Boards should be considered, and while assistance of the Scottish Government Communications Directorate may help a co-ordinating NHS Board in providing international, national and regional media with NHS information, Boards should not underestimate the demands on them which co-ordinating media liaison is likely to bring.

Preparation

O.3 While this Annex uses a scenario in which casualties are the primary focus of media interest, the principles and arrangements outlined should be adapted as necessary to other major emergencies.

These might include:

- An outbreak of disease in the community, among hospital patients or NHS staff
- Coping with the effects of bad weather or industrial action
- A quality control, equipment malfunction or other problem with a particular clinical procedure, screening process or pharmaceutical product

O.4 At every level, managers should ensure that arrangements for liaison with the media are integral to their major emergency plans. When these plans are exercised the media arrangements should also be exercised in as practical a way as possible e.g. participation by trained journalists.

O.5 As part of a NHS Board's emergency plan, a large room should be identified as a Media Centre. It should be sufficiently close to the Hospital Control Centre to facilitate authoritative briefings by members of the Hospital Control Team. If journalists there are provided with access to adequate communications, are supplied with refreshments, are regularly briefed and have questions answered, they will be less likely to wander elsewhere in the hospital. Ease of access to parked outside broadcast vehicles should be borne in mind.

O.6 A member of staff should be selected as Press Officer who can be dedicated to that role throughout an emergency. He or she should be of sufficient standing and personality to command respect and support within the hospital, particularly among nursing and medical staff. Additional staff should be earmarked to provide on a continuous basis the necessary administrative support which the preparing, typing and copying of news releases and state-ments will require. The advantages of designating a Press Officer include:

- Journalists and photographers will have a single point of contact for information and will tend to leave other staff free to work without interruption
- The Press Officer will quickly build a working relationship with the Press to the mutual benefit of hospital and media.
- Information communicated to the media can be more readily controlled.

O.7 Depending upon the circumstances surrounding the emergency NHS Boards may wish to consider joint arrangements with the police and/or local authorities.

Onset of Emergency

O.8 Emergency plan implementation should invariably include alerting the NHS Board's Press Officer. However, a telephoned media enquiry may well be the first intimation of emergency, and others may be received before a press officer is available. It is thus important that duty managers and/or on-call staff are permanently available and prepared to handle such calls.

O.9 The Press Officer's immediate action should be to ensure the Media Centre is functioning and that its communication facilities are operational. Hospital staff should be briefed that when the media arrive they should be escorted to the Media Centre. Film crews and photographers, after taking pictures of ambulances arriving with the injured, will want to take further pictures of the injured in hospital. All hospital staff should thus be on the look-out for strangers with cameras (apparent or concealed) and know of arrange-ments made to escort them back to the Media Centre.

O.10 Should the major emergency be caused, or suspected to be caused by an act of terrorism or crime, the police may impose a degree of security around casualties and hospitals treating them. Depending on the circumstances, the police may prevent anyone other than essential health personnel from entering the hospital grounds, including the media. Where this is necessary, it is likely that the police will co-ordinate media briefings and the preparation and release of information. However, an increase in telephone calls from the media to the hospital seeking information is to be expected. Thus it will be essential for the hospital press officer to make himself/herself known to the senior police officer present at the hospital, so that arrangements can be made to secure the co-operation of the media and which satisfy their needs, together with those of the police and the hospital.

O.11   If the emergency is such that more than one hospital receives casualties, has major public health implications or is otherwise likely to attract significant media scrutiny, its Central Co-ordinating Media Office should be set up by the NHS Board as soon as possible. All hospitals and other NHS organisations involved should keep that office fully up to date with information, including any proposed news releases etc., prior to issue. The staff of the Central Co-ordinating Media Office will require to liaise closely with police information staff, with those of other authorities or agencies involved in the emergency, and with the Scottish Government Communications Directorate. The earlier such co-ordination of media liaison is established, the less likely will it be that reporters and their enquiries will get in the way of those engaged in responding to the needs of the emergency itself.

First Media Briefing

O.12   Whatever the nature of a major emergency, a media briefing should be held as soon as possible. Press Officers must be aware of the need to maintain medical confidentiality and that any decision to release details of any individual patient must have prior medical and patient consent. At a hospital receiving casualties the Press Officer should, in preparation of briefing, seek to collate the following **factual information:**

- The time the hospital was told to expect casualties.
- The time the first casualties arrived.
- The number of injured received.
- General information about the casualties as to whether male or female, children under 16, the general nature of injuries, the general type of treatment being given, the numbers admitted or discharged after treatment
- Information about any patients transferred to other hospitals, either for specialist treatment e.g., burns/neurosurgery or to spread the load.
- A brief outline of the hospital's emergency plan, when it was activated and the effects on routine hospital work, normal visiting hours etc.
- Details of numbers of staff on duty, of specific specialist teams on standby or deployed, of routine operations cancelled and any other background information.

O.13   Based on such information, the hospital press officer should prepare a statement. Prior to issue, and after any scrutiny required by the Chief Executive, it should be agreed with the NHS Board's Central Co-ordinating Media Office and with the police, who may require to consult the Procurator Fiscal before agreement to release can be given. The text of the agreed statement must be furnished to the Scottish Government Communications Directorate in advance with details of time and place it is to be issued. The Chief Executive, ideally supported by a clinical director/senior consultant (in white coat) and senior nurse will be expected to:

- Read the statement to the media, answering questions arising from it.
- Be prepared to repeat the statement to radio interviewers/television reporters if required.
- Announce arrangements for further briefings on a regular and frequent basis.

O.14 The statement to the first media briefing should provide the basis for staff answering many subsequent telephoned media enquiries. Calls should be expected from local daily or evening newspapers or news agencies seeking information about the involvement in the emergency of people from within their circulation areas. The Press Officer should have sufficient support to allow such calls to be dealt with on a 24 hours a day basis, with all calls being logged. Where confirmation of information not previously released is sought, arrangements should be in place to check with senior staff that confirmation may be given or further, new, information released.

Subsequent Media Briefing

O.15 In preparing the second and subsequent media statements, the Press Officer will need to clear the release of information as before. The following might be given:

- Details of patients, giving names, addresses and ages bearing in mind the rules surrounding medical confidentiality, data protection laws, **and only where patient consent has been given. Patients should understand the possible impact release of their details might have on next of kin and that a media presence at their address might result**.
- Further details of the extent of injuries and of treatment.
- Details on the numbers of deaths, emergency operations, patients in intensive care; patients discharged home. Note that the police will not allow names of the dead to be confirmed until positive identification has been made and next of kin informed.

O.16 The number and frequency of media briefings will clearly depend on the development of the response to the emergency. Where there is nothing new to be said, then the Press Officer should make a statement to that effect, but promise further briefing when further information becomes available.

Media Interviews

O.17 As soon as patients arrive at a hospital the media will seek to interview, photograph or film both patients and staff treating them. The Press Officer should check with the police (who may in turn wish to consult the Procurator Fiscal) whether or not any individual patient might thereby be put at risk by such publicity in the context of criminal investigations. Media access to any patient should only be arranged with the consent of the consultant looking after the patient who can confirm that he or she is well enough. **No interview or photographs should be taken without the consent of the patient concerned**.

O.18   Press Officers should seek to identify a small number of doctors, nurses, ambulance and other staff directly involved in caring for patients who could be made available to give interviews to the media. The media will normally welcome such an opportunity. First-hand accounts of the health response to an emergency reduce the risk of wrong information being circulated and provide an opportunity to publicise what the hospital and the NHS have to face. However, any member of staff being interviewed should be carefully briefed beforehand.

Media Access to Patients

O.19   One method by which media access to patients and staff can be arranged is to organise a "short facility" with the media being admitted to a group of patients and/or staff under firm control.

O.20   Care must be taken to ensure that patients' wishes concerning interviews and photographs are made clear, that media activities do not take place without the patients' consent, nor in a manner which might cause them distress. The programme and time schedule for the facility should be agreed with the media beforehand. If necessary press, television and radio interviews should be done on a 'pool' basis to reduce disruption and stress for patients and staff.

O.21   Media deadlines (i.e., transmission/broadcast times or the times at which newspaper editions have to be finalised) may mean that the running order of groups above may need to be altered. Where there are space or time restrictions, pooling arrangements should be considered under which one reporter, one photographer, one radio reporter and one TV crew are admitted and subsequently share their reports/films with all.

VIP Visits

O.22   Members of the Royal Family and Government Ministers or other dignitaries will often visit the site of a major emergency and hospitals involved in response to it. The Scottish Government Communications Directorate in consultation with other press offices as appropriate will be responsible for providing advice on media coverage of such visits. (Scottish Government, 2010)

In the late 1990s, the UK authorities set up a Media Emergency Forum (MEF), which consists of an ad hoc and voluntary group of senior media editors, government representatives, local first responders, and private industry (see Sidebar 8-9). In 2003, regional MEFs were established based on the national MEF model. These forums are used essentially to float trial balloons in which the authorities try to gauge the possible media and community response, and they operate under Chatham House Rules (i.e., that discussions remain confidential) (British Government, 2010, p. 192).

Sidebar 8-9: Case Study of Media Relations During a Crisis in the UK

Background and Context

Avian Influenza was diagnosed in turkeys in rearing sheds at a multi-operation site at Holton, Suffolk. The site owner is Bernard Matthews, a high-profile multi-national operation. All turkeys at the site were culled. All persons coming into contact with the infected birds were offered health screening. Restriction and surveillance zones were enforced from 2 February to 12 March 2007. At the site, restrictions on the turkey rearing sheds were more prolonged than those on the meat processing operation. The source of the outbreak was not identified but the strain was identified as similar to that which caused an outbreak in Hungary, a country where Bernard Matthews also has commercial interests.

How the Topic Was Handled

The national media operation for the outbreak of avian flu and for any animal disease outbreak is managed centrally by press officers in Defra's Communications Directorate. Press officers at Defra work closely with GNN colleagues locally and with the Local Disease Control Centre (LDCC) to ensure that media are kept up to date consistently at the national and local level. In addition, at the local level, different agencies, including in this case Suffolk County Council and Suffolk police, play an important role in communicating messages to residents and keepers in the local area.

The return of the nation's media to Suffolk in February 2007 for the outbreak of H5N1 came just six weeks after the world's media focused on the county for the murder of five Ipswich prostitutes in December 2006.

As a result, the groundwork and firm foundations had already been laid between communication officers and national correspondents. Indeed for a number of national and local outlets, it was the same correspondents who covered both stories.

Having already established this rapport with media representatives, answering queries and setting up interviews was smoother. The close timing of two high profile cases also meant that a very strong relationship had been forged between communication specialists based in different agencies, especially Suffolk County Council and Suffolk Police.

Likewise, there was an awareness by elected members and officers of the importance of the media in terms of disseminating key messages, public reassurance and operational information in times of crisis and emergency.

In technical terms, providing for the demands of interview requests allowed the communications team to offer three different responses, depending on the nature of the query.

First, the assistant county trading standards officer was lined up to handle interview requests that would focus on operational procedure, such as the work of officers going door-to-door or from farm gate to farm gate. She also put across public reassurance messages. It was very useful to have one recognisable "talking head" that could be publicly identified with the operational messages. Her previous experience with the media stood her in good stead for the onslaught of questions and interview requests. However, we did find that placing one officer under such sustained level of media exposure over a number of days meant that this was a full time role for her during the period.

Second, the county's senior politician for public protection matters, also gave reassurance messages.

Third, a local county councillor, who also lived in the village of Holton, provided a public face and voice for the local community. She was a vital link with villages and the town council and gave information on the spread of Portuguese workers in the factory and surrounding area.

Throughout the bird flu outbreak there were three centres where communications specialists were based: near the site in Holton, at Gold command/Local Emergency Centre and at Suffolk County Council HQ at Endeavour House in Ipswich. Messages were co-ordinated between the three points.

This was particularly useful during the first 36 hours for the following reasons:

1 co-ordination of messages out of Gold to those doing live media interviews

2 co-ordination of media for pooled facilities, eg. photocall for putting up restriction zone signs, farm gate visits

3 co-ordination of live press conferences and avoiding the existence of an information vacuum for the media on site and at newsdesks.

As the story broke over a weekend, the most efficient way to inform the public was through the media. With this in mind, we made the assistant county trading standards officer, as the main spokesperson, as available as possible to the media. She disseminated key public and operational messages for the general public, poultry farmers, back-yard chicken keepers, partners and local business.

She undertook live interviews constantly at the scene over the first few days, explaining what had happened, what the restrictions meant and what the next steps would be.

Trading standards officers were filmed, but not interviewed, putting up signs and doing farm gate checks with local poultry owners. Highways staff were filmed putting up signs at key points entering the different disease zones that were implemented.

A live press conference was organised near the site within the first 48 hours on Sunday to cater for national news outlets.

The necessary information reached its target quickly and cost effectively this way. We then followed this up with a mail out of over 92,000 letters to local residents. The absence of panic among local people, while not completely removing all concerns, did show that fast, comprehensive information about the reality of the outbreak had reached those who needed it most and had reached them with enough detail to take effect.

The campaign also allowed the county's reputation to be restored quickly after the restriction zone was lifted.

Lessons Identified

During an outbreak, it is vital to manage the media. Whilst in this outbreak some media interviews were given at the site, media interviews are NOT encouraged at infected premises so as not to hamper operational matters by encouraging a large media presence. Media do head to sites where activity is occurring and Defra will send a Government News Network (GNN) press officer to supervise and disseminate information, but responding organisations should not encourage the media by providing a spokesperson at the scene. All media interviews should be conducted from the LDCC and no portable media facilities should be provided at the site.

Good established relationships with national press and with partner agencies ensured the media operation hit the ground running.

The story breaking over the weekend meant that media was the key outlet for informing the public.

It was invaluable having one recognisable operational expert, who was media trained, to front up most of the media interviews. However, this did take up all her time over the first few days, preventing her from undertaking more of an operational role.

Local knowledge on the ground helped us tailor the messages we were issuing, especially in terms of back-yard chicken keepers and the sizeable Portuguese community. (UK Cabinet Office, 2010a)

CONCLUSION

This chapter has provided some information and examples regarding international emergency response strategies and practices and differing international planning and response entities. Israel and the UK (the primary countries of focus in this chapter) both have highly developed emergency response systems, due primarily to having to cope with large-scale terrorism (and, in Israel's case, fairly frequent wars as well). Both countries put the police out front in terms of providing them with automatic incident command (with the caveat that Israel turns incident command over to the military in the event of an extremely severe and/or WMD attack). Both countries rely on their public hospitals to provide care, with the Israeli system focusing on bringing patients to the nearest hospital for triage, whereas in the UK, hospitals send out teams of doctors and nurses to treat the most severely injured at the scene. Both countries heavily emphasize planning and exercising response systems. The UK also enjoys a highly developed response strategy based on gold, silver, and bronze levels (corresponding to strategic, tactical, and operational response levels) that depending on the severity of the situation can involve everyone from the cabinet level down to the first responder at the scene. Israel's strategy also puts considerable emphasis on coping with the aftereffects of attacks in terms of psychological, economic, and social support for survivors of terrorist attacks and their families. All in all, a strong emergency response strategy should result in the mitigation of the immediate threat and saving of as many lives as possible, as well as allowing for a rapid recovery and return to normalcy. Crisis communications and public education also play a critical role in this problem, and all the countries surveyed engage in this to a certain degree, although Israel, arguably, does this with the most consistency and depth.

ISSUES TO CONSIDER

- What are the pros and cons of triaging and treating patients at the scene of an event (as the British do) as opposed to the scoop and run method (favored by the Israelis)?
- What role does the military play in WMD response in Israel?
- How does the British gold, silver, and bronze system work, and what are its benefits?
- Should social workers be considered first-responders, and how does this work in Israel?
- What are the approaches toward public education for preparedness and crisis communications across the countries surveyed, and can such approaches be implemented in the United States?

CHAPTER 9

PUBLIC HEALTH STRATEGIES AND INSTITUTIONS

In this chapter we focus on a handful of international public health strategies and their concomitant institutions. Public health, as a field, includes areas such as health protection (food and water safety and basic sanitation), disease prevention and the management of disease outbreaks, population health assessment, disease surveillance, and health promotion (Health Canada, 2003, p. 47). The field also covers the laws, institutions, and nature of the health system. We focus on those aspects that have a direct connection to homeland security issues: the nature and workings of a number of public health systems, how they handle disease surveillance, and how they plan for and handle outbreaks.

HEALTH SYSTEMS AND INSTITUTIONS OVERVIEW

Health systems around the world can be classified along a continuum from free market–based systems of private insurance to social insurance systems to government monopoly–based national health services. The private insurance model involves the smallest degree of state involvement in the supply of health services and is based primarily on private or employer-financed health care via private health care providers (although it may provide a government-funded health care option for the poor). The social insurance model is generally one of universal health coverage funded through social security payments through contributions by employers and employees via nonprofit health funds (large health care providers that offer all services). Finally, the national health service model is based on a state-administered (and frequently owned) health system funded through taxation that provides universal health coverage (Blank and Burau, 2007, pp. 11–12). As with other policy issues addressed in this book, the determination as to which health system will be adopted by a given country is the result of a range of factors. These include the nature of the political and legal system (particularly with respect to compensation claims and legal rights to health care), social values (emphasizing individual rights or collective goods and rights), and the demographic and cultural composition of the population (Blank and Burau, 2007, p. 31). These factors will thus affect the structure of the health system and the facilities and services made available.

Israel

Israel has a largely socialized health system. As of this writing, there are 24 general hospitals with some 23,000 beds in the country. Forty-five percent of hospitals are operated by the central government or local governments, and 30 percent are run by the largest HMO in the country (which is also government affiliated). Another 19 percent are run by nonprofit or religious organizations, and only 6 percent are in private hands. Israel has a ratio of 3.5 physicians per 1000 persons, one of the highest rates in the world (approximately 20 percent higher than the OECD average) but a comparatively

FIGURE 9.1 Schneider Children's Hospital in Petah Tikva, Israel. This Wikipedia and Wikimedia Commons image is from the user Dr. Avishai Teicher and is freely available at http://commons.wikimedia.org/wiki/File:PikiWiki_Israel_7694_schneider_children_hospital.jpg under the Creative Commons Attribution 2.5 Generic license.

low rate of nurses per person (approximately 5.8 per 1000 persons). Approximately 46 percent of acute hospitals beds are located in government-owned hospitals and another 30 percent in hospitals owned by the largest of the HMOs or "sick funds," the General Sick Fund (Kupat Holim Clalit) (Rosen, B., and Samuel, 2009, p. 90). Under the stipulations of the 1995 National Health Insurance Law, every Israeli resident must register with one of four HMOs (which are not allowed to deny membership on the grounds of age or health status), and these HMOs provide a basket of basic health services. Health insurance premiums are deducted from salaries based on a rate of 8 percent of one's salary, and the unemployed pay only a symbolic health premium fee (Israel Ministry of Foreign Affairs, 2009).

The Ministry of Health (MOH) has overall responsibility for the health care system. In addition to drafting health care bills to be brought before the Israeli parliament (Knesset), the ministry oversees the operation of government-run hospitals, the four HMOs, the certification of health care professionals (in cooperation with the Israel Medical Association), and the stockpiling of medications. The MOH is also responsible for preparing the health care system for emergencies (including terrorist attacks and war involving conventional weapons as well as WMDs) (Rosen, B., 2003, p. 13) (see Sidebar 9-1). The MOH ensures that hospitals develop a common response doctrine and that they are able to provide a surge capacity of at least 20 percent.

Sidebar 9-1: Israel's Health System and the Second Lebanon War

During Israel's Second Lebanon War (2006), Hezbollah carried out a massive and continued rocket attack on the entire northern part of the country (approximately 5000 km^2). Overall, 3970 rockets landed in Israel and halted the normal course of life for over 2 million people living in the areas affected. Workplaces shifted to limited emergency operations and the functioning of the social service system was greatly impaired. Israel's Home Front Command ordered the residents of the north to remain in or near underground shelters or other specially protected indoor spaces. Approximately 300 000 residents of the north chose to evacuate the area and stayed with family or friends in other parts of the country. This situation lasted until the ceasefire went into effect, about a month after the war began. As a result of the missile attacks, 42 civilians were killed, 1489 were physically injured and 2773 were identified as victims of stress.

During the course of the war, the health system had to operate emergency response models to meet the needs of three key groups of patients: soldiers injured in the fighting in Lebanon, civilians injured (physically or emotionally) by the missile attacks, and civilians in need of routine medical services unrelated to the war. Thus the war placed major demands not only on the hospitals (who take the lead in treating casualties injured in war or terror attacks) but also on the health plans (who play a major role in routine care and in trauma-related mental health care). All this took place under conditions in which it was dangerous for health care professionals and patients to travel to health care facilities, and the health care facilities themselves were open to attack.

Prior to the Second Lebanon War, the Israeli health system had already acquired substantial experience in addressing the needs of both injured soldiers and civilians injured in terror attacks. One of the unprecedented features of the Second Lebanon War was the need to provide routine care to a large population over an extended period of time under unsafe conditions. To some extent, there were precursors during the 1993 and 1996 missile attacks on Israel's most northerly cities and towns, but these were limited in duration and geographic scope.

Overall, the innovations in emergency response that were implemented during the missile attacks had two key sources: Israel's extensive prior experience in responding to major security-related public health emergencies, and the specific new challenges posed by the missile attacks.

All health systems need to prepare themselves for a wide range of public health emergencies. During times of peace, these include natural disasters, large-scale accidents and terrorist attacks; in wartime, there is a need to respond to both conventional and nonconventional attacks. Israel, because of its unique geopolitical situation, must invest substantial efforts and resources in preparing its health system for terrorist attacks and wars. These efforts include comprehensive contingency planning; development of national doctrines and protocols; the development of a wide variety of coordination, control and command mechanisms; extensive training at the individual, organizational and national levels; and construction of vital infrastructure and equipment.

A key feature of Israel's approach to emergency preparedness is national (governmental) control of preparedness for all mass casualty events and national coordination of the real-time response. This is not an insignificant challenge in a health system which, while largely publicly financed, relies primarily on nongovernmental hospitals and health plans for the provision of services. Under normal conditions, there is significant competition among health plans and among hospitals in an environment of regulated competition. During public health emergencies, all of the key providers are highly motivated by patriotism and humanitarianism to respond to national health emergencies, but are also interested in protecting institutional interests (including considerations of prestige and financing) while doing so. Cooperation is promoted via a mixture of legal stipulations, financial payments and guarantees, and ongoing consultative processes. Thus, in public health emergencies the system is transformed from managed competition to managed cooperation.

The health system's ongoing emergency preparedness efforts and its real time response to major emergencies are coordinated by the Ministry of Health's

Division for Emergency and Disaster Management (DEDM). Policy is set out by the Supreme Health Authority, which includes senior representatives of government, hospitals, health plans, MDA and other key players. The driving force behind this set-up is to combine government leadership with input from all the care providers, to develop effective and coordinated responses.

Since its founding in 1948, Israel has had to face a series of wars and waves of terror attacks. This led to substantial investment in preparedness of "the Home Front" in general, and within the health system in particular. Until the first *intifada* (1987–1991), the focus was on treating injured soldiers. In contrast, the terrorist attacks associated with the *intifadas*, and the missile attacks during the first Gulf War, focused attention on the need to respond to mass civilian casualties. Public health preparedness doctrines, training and exercises have been expanded accordingly. (Rosen and Samuel, 2009, pp. 182–184)

United Kingdom

In the UK, four separate health systems provide services in England, Wales, Scotland, and Northern Ireland, with each based on the National Health Service (NHS). The NHS was founded in 1948 on the principles that health care should be universal, standardized throughout the UK, and comprehensive, covering all health needs. Consequently, the NHS health care model consisted of health care that was free at the point of delivery, available to all on the basis of need, and funded through central taxation (income tax) (Talbot-Smith and Pollock, 2006, p. 2).

The NHS public health role consists of (1) improving health and reducing inequalities in health care, (2) providing preventive services and health care, and (3) safeguarding health, including the control of communicable diseases (Wanless, 2004, p. 41). The NHS is organized around Primary Care Trusts (PCTs), which provide most basic health services, and these trusts are, in turn, accountable to 28 Strategic Health Authorities (SHAs), the latter acting as the regional headquarters of the NHS. The main statutory functions of the PCTs are to improve communal health services and to commission primary, secondary, and community care. PCTs identify the health care needs of the populations in their jurisdiction and develop local delivery plans to provide this care (Talbot-Smith and Pollock, 2006, p. 37). SHAs are responsible for strategic planning and assessment of health system performance and act as the link between the strategic focus at the governmental level (via the Department of Health) and the point of service at the PCTs and hospital trusts (Talbot-Smith and Pollock, 2006, p. 19). In addition to the primary care trusts, which provide much of the primary and secondary care services for a particular region, other types of trusts perform specific types of medical services. Acute trusts manage hospitals and acute care centers, ambulance trusts provide ambulance services, and mental health trusts provide health and social services to mental health patients. The NHS also plays a health and disease monitoring role via public health observatories, which not only monitor health trends for the local authorities but also evaluate the implementation of health measures and work to identify future public health problems (Wanless, 2004, p. 43).

The central government's public health agency is the Department of Health (DH), which oversees the National Health Service in England and is responsible for coordinating a response to public health crises through its Emergency Planning and Coordination Unit (EPCU). The Health Protection Agency (HPA) is a nondepartmental public body that advises the DH and provides local health protection services in England while working closely with the health protection services in Wales, Scotland, and Northern Ireland. The HPA also provides training to medical personnel, carries out exercises, maintains disease surveillance, and organizes health arrangements in the country's ports (Health Protection Agency, 2010). The core functions of the DH with respect to disease outbreaks involve surveillance, monitoring, and analysis of such outbreaks, their investigation, and efforts to mitigate the effects of such outbreaks and the spread of disease.

Canada

Canada employs a socialized medical system in which nearly all Canadians are covered under the Medicare program (a health insurance system funded jointly by federal and provincial governments), with the remainder of the population provided with health care directly by the federal government

FIGURE 9.2 University Hospital in Edmonton, Alberta, Canada. This Wikipedia and Wikimedia Commons image is from the user WinterE229 and is freely available at http://commons.wikimedia.org/wiki/File:University_Hospital_Complex_University_of_Alberta_Edmonton_Alberta_Canada_02A.jpg and has been released into the public domain.

(Johnson and Stoskopf, 2010, p. 68). Canada has had a government-funded national health care system since the passage of the 1962 Canada Health Act (CHA). The CHA includes five program criteria that guide the approach behind the Canadian health care system:

1. *Public administration*—that the health care system be operated on a nonprofit basis by public agencies.
2. *Comprehensiveness*—that all insured health services provided by hospitals, physicians, or dentists are covered.
3. *Universality*—that health care services must be provided under uniform terms and conditions to all insured residents.
4. *Portability*—that residents moving from one province to another will be covered by their province of origin until the waiting period for eligibility for their new provincial or territorial health plan takes effect.
5. *Accessibility*—that insured persons must have reasonable access to medical services (Health Canada, 2008, pp. 4–5).

Consequently, the system in Canada is largely uniform, although there is some variance between provinces in terms of coverage for issues such as dental care, prescription drugs, optometry, and home care, Sections 92(13) and 92(16) of the Canadian constitution give the provinces responsibility for civil rights and matters of a private nature, and on the strength of this, provincial governments claim the right to pass public health legislation. At the same time, federal powers regarding "peace, order and good government," quarantine, national borders, and regulation of interprovincial trade and commerce also allow the federal government to legislate in public health matters (Health Canada, 2003, p. 48). Despite what might appear to be a clear dichotomy between the powers of the federal government and that of the provinces, the reality is considerably more opaque. For example, in the realm of epidemiology, the Statistics Act and the Department of Health Act give the federal government the power to collect information on public health threats that can affect the entire country. At the same time, provinces and territories are not legally required to share health surveillance data with each other and with the federal government (Health Canada, 2003, p. 457). The federal government is, however, also able to exercise influence via funding to provincial and territorial public health agencies (Health Canada, 2003, p. 2). As far as primary health care is concerned, this resides largely at the local level (with approximately 140 health departments and units). The next level is provincial/territorial and focuses on planning, the administration of budgets, and technical assistance. Finally, federal activity is coordinated by the Public Health Agency of Canada (PHAC), an agency of the federal public health agency, Health Canada. The PHAC includes centers for disease prevention and control, emergency preparedness, and response and surveillance coordination. Canada also has a Public Health Network Council with a mandate to develop and implement collaboration across Canadian health agencies and provides the auspices for a Council of Chief Medical Officers of Health (CCMOHs) from the provinces and territories, which act in an advisory role (Stachenko, Legowski, and Geneau, 2009, p. 129).

As noted above, Health Canada is the country's national public health agency. Its responsibilities include protecting the public from health and safety hazards associated with hazardous materials (including coordinating the federal response to major nuclear accidents), advising the government and public on food safety, carrying out national disease surveillance and risk assessments, and maintaining stockpiles of supplies and pharmaceuticals (Public Safety Canada, 2003, p. 3.10). The Canadian hospital system is made up primarily of facilities operated by local or provincial governments, and there are also hospitals operated by religiously affiliated institutions.

The Canadian constitutional framework is such that decision-making on public health matters of national significance requires that the federal government, the 13 provincial and territorial governments, and, often, local governments and regional health authorities work together on a consensual basis. Moreover, the role of the federal government is explicit in certain situations but is often negotiated with provinces/territories, regional authorities, and municipal governments and can vary from area to area (Stachenko, Legowski, and Geneau, 2009, p. 125).

France

The French health system provides 95 percent of the population with one of three primary health insurance systems (for employees in business and industry, for agricultural sector workers and for self-employed nonagricultural sector workers). These insurance systems are regulated and controlled by the French Ministry of Health, which handles their financial and operational management. The Ministry of Health also sets the prices of medical procedures and drugs, the number of hospital beds, and even determines the number of medical students to be admitted to medical school each year. According to the Universal Health Coverage Act of June 1999, all French citizens enjoy the right to health insurance coverage and those who live below the poverty line are entitled to free coverage (Sandier, Paris, and Polton, 2004, p. 20). The General Directorate of Health (DGS—Direction Générale de la Santé) is responsible for coordinating and evaluating public health policy, for ensuring preparedness, for limiting the scope of disease outbreaks, and for public communications (Bonin, 2007, p. 53). In France, about a quarter of hospitals are publicly owned, approximately one-third are owned by foundations, religious organizations, or other nonprofits and some 40 percent of hospitals are private, for-profit institutions (Sandier, Paris, and Polton, 2004, p. 70).

Australia

As with the other countries in this survey, Australia also has a largely socialized medical system. Australia introduced a comprehensive health care system, known as Medicare, in 1984 and this provides all Australian residents with free or low-cost medical care via funding from general tax revenue as well as a 1.5 percent health tax on income. Australians also have the option of seeking private health services via private clinics and hospitals, in which case Medicare pays for approximately 75 percent of the cost (unless they have private health insurance, in which case Medicare generally pays for services not covered by the health insurance policy). Australia's public hospitals (which provide approximately two-thirds of all hospital beds in the country) are funded jointly by the commonwealth government and the respective state governments and administered by state health departments (Australia, Department of Foreign Affairs and Trade, 2008). Private hospitals are licensed by state governments and medical practitioners are licensed and overseen by state health authorities.

Australia's health system is federated, with each state organizing public health services in a slightly different manner and with state health authorities given the primary role of safeguarding public health and providing health services. State governments determine health policy, provide budgeting for health services, engage in health care planning, determine standards of performance for the health care system, and provide capital investment for the building of health facilities (Wall, 1996, p. 32). The commonwealth government's role focuses more on oversight of pharmaceutical and other medical goods, supporting health research, financing public health campaigns, and regulation of the private health insurance industry. The commonwealth government is also responsible for food safety, immunizations, disease control and biosecurity, environmental monitoring, and various health educational programs (Australian Institute of Health and Welfare, 2002).

Germany

Germany, too, has a socialized health system. As of 2009, health care coverage is mandated by law. Around 75 percent of the population is covered by a publically financed health care scheme (for those earning up to approximately

€48,000 per year), an additional 10 percent of the population (which earns more than this) remains within the publically financed health care scheme, and the remainder of the population is covered by private health insurance. The public health insurance scheme is operated by approximately 190 health insurance funds that are autonomous, nonprofit bodies regulated by legislation, and these schemes are funded by a tax on earnings of close to 8 percent (with the employer or pension fund contributing another 7 percent). Almost all hospitals in Germany are nonprofit institutions, with about half publically owned and another third private (Busse, 2009, pp. 28–30).

EPIDEMIOLOGICAL SYSTEMS AND CONTINGENCY PLANNING IN EPIDEMICS AND PANDEMICS

One of the most important tools in coping with large-scale communicable disease outbreaks is an effective epidemiological system. The role of epidemiology (disease surveillance) is (1) to provide data on circulating strains of communicable diseases to help determine which vaccines may be effective; (2) to determine which populations are being affected and to identify high-risk groups; (3) to detect new and/or atypical strains, unusual syndromes or unexpected distribution of the disease within the population; and (4) to provide information of use in determining the appropriate response strategies (see Sidebar 9-2).

Sidebar 9-2: SARS in Canada

Severe Acute Respiratory Syndrome, now known worldwide by the acronym SARS, is considered to be the "first severe and readily transmissible new disease to emerge in the twenty-first century." In late February 2003, several guests at the Metropole Hotel in Hong Kong had come in contact with an ill doctor who had been involved in treating patients with an atypical form of pneumonia in Guangdong, China. Those guests continued their travels in Hong Kong and on to Canada, Singapore, and Vietnam. They fell ill, and began spreading the disease to others. Many of them died. This illness was soon identified as severe acute respiratory syndrome or SARS. As of July 11, 2003 in its daily summary, the World Health Organization [WHO] reported 8,437 probable cases of SARS and 813 deaths worldwide, and the toll has since risen to about 900 as some previously-ill individuals have succumbed. SARS was and remains a challenge to diagnose and manage because its symptoms resemble those of many other respiratory infections. Thus far, extensive research by a WHO-coordinated international network of research centres has identified a novel coronavirus as the presumed cause of SARS. The diagnostic tests available to test for the SARS coronavirus have limitations with respect to their reliability and sensitivity, and more research is needed to enable the rapid identification and characterization of this new coronavirus. SARS is spread through close contact with an individual who has SARS. The disease has an incubation period that typically ranges from 2 to 10 days. Affected individuals experience fever (>38°C) and later develop respiratory symptoms such as cough, shortness of breath, or difficulty breathing. Overall, case fatality from progressive respiratory failure ranges from less than 1% of cases for persons under 24 years of age to 15% of cases for persons aged 45 to 64 years of age; in persons over the age of 65, the fatality rate can exceed 50%. Diagnosis rests partly on the clinical syndrome, partly on a link to known cases of SARS, and partly on a process of exclusion. The virus can be isolated from respiratory secretions and stool; however, it is not always detected from these sources even in patients with probable SARS. Serological tests based on the body's immune response to SARS are also helpful, but these tests do not begin to yield useful information until a few weeks after the onset of symptoms. No vaccine or cure currently exists leaving clinicians to rely primarily on supportive measures and public health authorities to rely on isolation and quarantine as the predominant measures to control SARS. (Health Canada, 2003, p. 15)

Outside of Asia, Canada was the country most affected by SARS (Severe Acute Respiratory Syndrome). In Canada, by the time the outbreak had died down in Canada in the summer of 2003, there had been 438 suspected cases and 44 deaths. (Health Canada, 2003, p. 15)

Ontario Premier Ernie Eves declared SARS a provincial emergency on March 26, 2003. Under the *Emergency Management Act*, the Premier has the power to direct and control local governments and facilities to ensure that necessary services are provided. The same day, the province activated its multi-ministry Provincial Operations Centre for emergency response, situated on the 19th floor at 25 Grosvenor Street. All hospitals in the Greater Toronto Area [GTA] and Simcoe County were ordered to activate their "Code Orange" emergency plans by the OMHLTC. "Code Orange" meant that the involved hospitals suspended non-essential services. They were also required to limit visitors, create isolation units for potential SARS patients, and implement protective clothing for exposed staff (i.e., gowns, masks, and goggles). Four days later, provincial officials extended access restrictions to all Ontario hospitals. Later, the Committee heard mixed opinions about whether Code Orange was justified. Several interviewees noted the massive number of cancelled services, and suggested that the collateral casualties from the suspension of health care activities may never be fully measured. Other harms were more subtle, including hardship caused by restrictions on visits between families and patients hospitalized with conditions other than SARS. These informants claimed the activation of Code Orange demonstrated a "lack of understanding of the system." They suggested that The Scarborough Hospital could have been closed and converted into a dedicated SARS hospital, with staff support from other facilities, while selected other hospitals began urgent preparations to become SARS-care centres. The remainder of the system could then operate with increased infection control precautions. Other interviewees argued strenuously that the declaration of emergency and Code Orange were essential to galvanize infection control, and prevent unrecognized exposure by hospitals in the face of great uncertainty about the transmissibility of SARS. Dr. Jim Young, Ontario's Commissioner of Public Safety and Security, co-chaired the Provincial Operations Centre Executive Committee, and led the Executive and Scientific Advisory Committee in a lengthy and intense exercise to assess the pros and cons of designating one or more facilities as "SARS hospitals." Decision makers feared an outbreak would over-run any one or two designated SARS hospitals. The West Park experience suggested that the logistics of staffing a SARS specialty hospital would be extremely difficult. Concentrating SARS patients in a few institutions would put an enormous burden on these hospitals, and place their clinical personnel at great risk. Patients would still go to the emergency department nearest them, and language in current collective agreements constrained the ability of the system to move staff into new institutions. The team decided to build capacity for the management of SARS in multiple institutions. SARS patients were cared for at over 20 hospital sites scattered across the Greater Toronto Area. (Health Canada, 2003, p. 28)

Public health officials in York and Toronto continued to trace and quarantine contacts with good results. The outbreak management teams and leaders of the local public health units were identified by some interviewees as those who deserve greatest credit for containing the SARS outbreak. Nonetheless, concerns mounted that SARS was poised to spread into the community. Individuals who attended a funeral on April 3, 2003 were quarantined when some family members developed symptoms. An employee of a large information technology company defied quarantine, and returned to work while symptomatic; one co-worker contracted SARS, and nearly two hundred more were sent into isolation. A Scarborough school was closed by Toronto Public Health when one student, a nurse's child, exhibited SARS symptoms; four other schools would be closed by local school boards as a result of SARS concerns before the outbreak ended. Routine screening picked up a fever in a nurse caring for SARS patients—a hurried search to identify her fellow commuter train passengers ensued. Because Toronto was the only city outside Asia to be hit hard by SARS, the international media converged on the city like never before. The attention was not only unprecedented; it was unwanted. Despite the media attention, there was no evidence that the SARS epidemic was spreading through the community. The Amoy Gardens outbreak in Hong Kong, where the virus may have been transmitted through a defective sewer system, was an exception that proved the rule—the SARS virus was spread by either brief exposure to big doses of viral particles or close, prolonged contact. All but a few Canadian cases occurred in travelers, health care workers, and their immediate contacts. Using traditional surveillance, contact tracing, and quarantine, opportunities for community transmission were being identified and contained. The number of people quarantined grew daily. Very occasionally, someone would refuse to enter isolation, and public health officials had to resort to legal means to enforce compliance. But this was the exception; Torontonians were generally remarkably compliant with highly demanding strictures. Quarantined individuals lost income, suffered from boredom and loneliness, and most importantly, were fearful that they might develop SARS or that they may have spread SARS to family and friends. Committee informants commented that different public health units seemed to have different thresholds for the use of quarantine. A related issue is whether public health officials used quarantine too frequently. Some interviewees believed they did—one noted that while Beijing had 2,500 cases of SARS compared to Toronto with 250, both cities quarantined about 30,000 individuals. Beijing quarantined fewer people per SARS case because they focused on close contacts (e.g., household members, hospital visitors, and those who might have come in contact with bodily fluids). On the other hand, the higher caseload of probable and suspect SARS in Beijing might actually have been a result of too-limited use of quarantine. Perhaps the greatest scare of the Toronto outbreak occurred on April 12, 2003 when a cluster of SARS cases was identified in a close-knit religious community. Remarkably, it had begun with exposure back in mid-March of several members of a large extended family at the initial epicenter—The Scarborough Hospital, Grace Division. Over the ensuing weeks, the infection spread quietly through the extended family and some close friends, health care workers who cared for them, and then into a religious group. In all, 31 cases, including three health care workers, were associated with this cluster. Public health workers employed active surveillance and quarantine to control the spread of infection, and unchecked community transmission never materialized. As residents of their jurisdictions became exposed through the religious group cluster, public health units in the surrounding regions of Durham and Peel joined Toronto and York in trying to stop the outbreak. The various units collaborated, but there was no overarching coordination across jurisdictions. Hospitals later complained that they were sometimes contacted separately for information about the same patient by two public health units. Hospitals were also fielding requests for information from the OMHLTC Hospital Branch, the Public Health Branch, and the Provincial Operations Centre. Understandably, it appeared to those on the clinical front lines that public health officials were not communicating with each other. Meanwhile, in Toronto, local public health workers were nearing exhaustion—all non-SARS activity in infectious diseases and many other provincially-mandated programs had been suspended, and virtually all qualified employees were working on SARS full time. The monumental efforts of public health workers played a critical role in the containment of SARS. Toronto Public Health, for example, investigated 1,907 separate reports in addition to 220 cases of probable or suspect SARS, each of which involved several hours of investigative work, independent of contact tracing. A pair of papers later published in *Science* provided estimates of the "infectiousness" of the SARS virus. Both papers lead one toward the same conclusion: although SARS is only moderately transmissible, left unchecked it could have infected millions of people worldwide. Whether it would have done so before mutating into a more benign form is, fortunately, still unknown. (Health Canada, 2003, pp. 34–35).

Although most of the attention during the outbreak was directed toward hospitals, several instances of patients transmitting SARS to their family doctors produced apprehension. One academic family physician voiced concern as early as March 28, 2003: "Family physicians, just like hospitals, need precise and explicit directions for screening patients, and for contending with suspect or probable SARS patients who might make it past the screening system." They also "required full protective gear in the unlikely event that a SARS patient did make it into their offices." He suggested that family physicians could be used as sentinels—reporting cases of pneumonia to a central authority might pick up SARS clusters where there was no obvious epidemiologic link. Guidelines for family doctors were eventually issued on April 3, 2003 via the fax and e-mail network of the Ontario Medical Association. These instructions outlined three goals: first, to keep potential SARS patients out of doctors' offices using signs, pre-recorded telephone messages and screening questionnaires; second, to safely treat any SARS patients that did enter the office; and third, to protect physicians and staff from infection. Some informants later suggested that the guidelines were difficult to implement in community-based practices. More problematic was the lack of a system to distribute the necessary protective gear. The Ontario Medical Association proposed that the fastest strategy was for family doctors to buy their own supplies where they could, and apply for reimbursement later. A growing number of family physicians, however, were concerned by the lack of provincial support. On April 15, Drs. D'Cunha and Young convened a meeting of family doctors, hospital CEOs, and chiefs of emergency medicine at a downtown hotel. Family doctors left the meeting frustrated that the province had still not developed a plan to distribute protective equipment to physicians and their office staff. On April 21, 2003, almost four weeks after the Province of Ontario declared an emergency, the province finally used its vaccine distribution network to provide family doctors with protective equipment. (Health Canada, 2003, pp. 35–36)

Israel

In Israel, the Ministry of Health (MOH) is the primary public health agency and is responsible for disease surveillance. The MOH owns a network of family health centers (some of which are operated by the MOH and others by municipalities or HMOs) that are staffed primarily by public health nurses and provide services such as immunization and preventive health care for children. The MOH's regional offices maintain close contact with the family health centers as well as local physicians, clinics, and hospitals and receive regular reports on health issues (including communicable disease outbreaks) which physicians, clinics, and hospitals are required, by law, to report based on a specific list of reportable illnesses. The MOH operates an epidemiology unit, the Israel Center for Disease Control (ICDC—in Hebrew, Ha'mercaz Ha'leumi Le'bakarat Mahalot, also known by its acronym Malbam) that employs geographic information systems and other tools to identify and analyze outbreaks (Rosen and Samuel, 2009, p. 107).

Israel's pandemic strategy is based largely on the military. While the MOH engages in disease surveillance and its emergency division coordinates the medical aspects of preparedness and response (supporting the medical system, prioritizing medical and logistical goals, etc.), it is the Ministry of Defense's (MOD) responsibility to manage national preparedness and response during pandemics (Kohn et al., 2010, p. 259). As with other matters of homeland security in Israel, the military (and the MOD that oversees it) is often the only entity with the resources and expertise to cope with large-scale crises. The early stages of a disease outbreak would still be handled by the MOH, but should that outbreak overwhelm the resources of the civilian agencies or require a multiagency response of great rapidity, the government will authorize the MOD to manage the situation and would then delegate tasks to the Israel Defense Force (IDF) (Kohn et al., 2010, p. 260). Under Israel's pandemic plan, the military would be called in to

1. Provide equipment, food, fuel, and water.
2. Mobilize medical units and their equipment to manage field hospitals, conduct epidemiological investigations, and assist in the identification and burial of bodies.
3. Provide troops to replace workers in the public health sector who are ill or otherwise do not report for duty.
4. Communicate with the public, via the Homefront Command, in order to provide instructions for personal and family safety and to try and stave off panic.
5. Dispense vaccines and pharmaceuticals to the population (Kohn, et al., 2010, pp. 260–261).

United Kingdom

In the UK, the Health Protection Agency is responsible for the surveillance and analysis of communicable diseases via a center under its responsibility known as the Communicable Disease Surveillance Center (CDSC). The CDSC is responsible for monitoring and response to infections disease at the national level, provision of public health information to government and the public, and the provision of training in disease surveillance (Bonin, 2007, p. 193). The Department of Health's Emergency Planning Coordination Unit is tasked with the role of coordinating contingency planning for the National Health Service.

Australia

In Australia, the Commonwealth Department of Health and Aged Care coordinates the surveillance and management of communicable diseases through the National Center for Disease Control. In addition, Australia coordinates disease surveillance with its neighbor, New Zealand, via the Communicable Diseases Network of Australia and New Zealand (Hilless & Healy, 2001, p. 49). The Australian Department of Health also operates a national incident room to coordinate the department's emergency response capability and to coordinate with other government departments as well as state and territorial health authorities. Additionally, the Australian Health Protection Committee (AHPC) is tasked with coordinating the health system planning and response across jurisdictions. The AHPC is also tasked with determining the level of health preparedness and to plan and develop health emergency policy (Australian Government, 2006, p. 72).

France

In France, the French Institute for Public Health Surveillance (InVS—Institut de Veille Sanitaire), which reports to the Ministry of Health, is responsible for disease surveillance and alert. Its role is also to suggest measures to contain disease outbreaks and coordinate between the government departments responsible for dealing with public health threats (French Institute for Public Health Surveillance, 2010). France's syndromic surveillance system, which has been in operation since 1984 (albeit operating on different technological platforms), links to a representative sample of French general practitioners who remotely enter reports on 12 conditions on a weekly basis (Flahault, Blanchon, Dorleans, Vibert, and Valleron, 2006, p. 414).

Canada

In Canada, responsibility for coping with pandemics, as with other public health issues, is shared across the various levels of government. The federal government is responsible for nationwide coordination of the response, including surveillance, international liaison, and allocation and disbursement of vaccines. Provincial and territorial governments are responsible for mobilizing their respective contingency plans and resources and provide the first-response mechanism to a disease outbreak. Local public health authorities are responsible for planning the local response and liaising with local first-responders and other partners (hospitals, mortuary services, etc.), and they are often the first health authorities to detect a pandemic outbreak (Canadian Government, 2004, p. 9).

The declaration of a public health emergency in Canada occurs when the chief medical officer of the health department of a province or territory determines that there is a serious public health risk or when the chief public health officer (CPHO) of the federal government determines that such a risk exists. Once a health emergency is declared, under the terms of existing memoranda of understanding, federal, provincial, and territorial ministries of health will share information, including specific case information, laboratory results, information on the nature of the risk, number of cases and deaths, conditions affecting the disease, and health measures employed and also provide accurate information on the difficulties specific jurisdictions face in responding to the health emergency (Pan-Canadian Public Health Network, n.d.) (see Sidebar 9-3).

Sidebar 9-3: Pan-Canadian Public Health Network Proposed Criteria for Determining the Existence of a Public Health Emergency

Public Health Emergency:

An extraordinary, unexpected, or unusual health event...

Q.1) Is the health event extraordinary, or unusual?

The following are examples of extraordinary or unusual events:

- The event is caused by an unknown agent or the source, vehicle, route of transmission is unusual or unknown.
- Evolution of cases more severe than expected (including morbidity or case-fatality) or with unusual symptoms.
- Occurrence of the event itself is unusual for the area, season, or population.

Q.2) Is the health event unexpected from a public health perspective?

The following is an example of an unexpected event:

- Event caused by a Disease/agent that had already been eradicated from the jurisdiction or not previously reported.

If the answer to either Q1) or Q2) is yes, then the health event should be considered extraordinary, unexpected, or unusual.

II

... determined to i) constitute a Public Health Risk to other jurisdictions and/or internationally through the spread of Disease

Q.3) Is there a significant risk of spread to other Canadian provinces or territories and/or internationally?
Q.4) Is there evidence of an epidemiological link in other jurisdictions?
Q.5) Is there any factor that should alert health authorities to the potential for cross border movement of the agent, vehicle, or host?

The following are examples that may predispose Canadian provincial or territorial and/or international spread:

- Where there is evidence of local spread, an index case (or other linked cases) with a history within the previous month of:
- inter-jurisdictional and/or international travel (or time equivalent to the incubation period if the pathogen is known)
- participation in an inter-jurisdictional and/or international gathering (pilgrimage, sports event, conference, etc.)
- Close contact with an inter-jurisdictional and/or international traveller or a highly mobile population.
- Event caused by an environmental contamination that has the potential to spread across inter-jurisdictional and/or international borders.

If the answer to either Q3), Q4) or Q5) is yes, then the health event is determined to constitute a public health risk to other jurisdictions and/or internationally through the spread of Disease

III

... ii) have a serious adverse impact on the health of the population

Q.6) Is the number of cases and/or number of deaths for this type of health event large for the given place, time, or population?
Q.7) Has the event the potential to have an adverse impact on public health? The following are examples of circumstances that contribute to an adverse impact on public health:

- Event caused by a pathogen with high potential to cause epidemic.
- Indication of treatment failure.
- Event represents a significant public health risk even if no or very few human cases have been identified.
- Cases reported among health care staff
- The population at risk is extremely vulnerable (children, elderly, etc.)
- Event in an area with high population density.
- Concomitant factors that may hinder or delay the public health response (natural catastrophes, unfavourable weather conditions

If the answer to either Q6) or Q7) is yes, then the health event is determined to have a serious adverse impact on the health of the population

IV

... potentially require a coordinated response.

Q.8) Is assistance needed to detect, investigate, respond and control the health event, or prevent new cases?

The following are examples of when assistance may be required:

- Inadequate human, financial, material or technical resources, in particular:
- Insufficient laboratory or epidemiological capacity to investigate the event (equipment, personnel, financial resources)
- Insufficient antidotes, drugs and/or vaccine and/or protective equipment, decontamination equipment
- Existing surveillance is inadequate to detect new cases in a timely manner.

If the answer to Q8 is yes, then potential assistance and/or a coordinated response may be required.

In view of the foregoing:

A "yes" response by a jurisdiction to any one of Parts I–IV would indicate that a Public Health Risk exists.

A "yes" response to any two of Parts I–IV would indicate that: a Public Health Emergency exists. (Pan-Canadian Public Health Network, n.d.)

European Union

The European Union also plays an important role in epidemiological surveillance. In 2004, the EU established a European Center for Disease Prevention and Control (ECDC), based in Stockholm. The ECDC is designed to act in concert with national disease control centers in the member states. Its primary tasks are (1) to support epidemiological surveillance at the European level; (2) to provide scientific advice, to identify emerging health threats, and to disseminate real-time information to member states; (3) to engage in training of European epidemiologists; (4) to provide technical assistance to laboratories across Europe; and (5) to provide ongoing information through the publication of *Eurosurveillance*, a bulletin on disease prevention and surveillance circulated to European health agencies (McKee, Hervey, and Glimore, 2010, p. 253). The ECDC also coordinates other European information and monitoring networks, such as the European influenza surveillance scheme (EISS), the early warning and response system (EWRS), and the European Center for the Epidemiological Monitoring of AIDS (EuroHIV) (McKee, Hervey, and Glimore, 2010, p. 255).

QUARANTINE POWERS

Australia

In Australia, prevention of the introduction of diseases and restriction of their spread are governed by the Quarantine Act of 1908. The act lists a range of diseases that are subject to quarantine orders. Enforcement of the Quarantine Act is shared by the Minister for Health and Ageing and the Minister for Agriculture, Fisheries and Forestry, with a Department of Health and Aging unit known as the Australian Quarantine and Inspection Service (AQIS) given primary responsibility for quarantine activities at the country's air and maritime borders. The governor-general is empowered by the act to declare an epidemic caused by a quarantinable disease in any part of the commonwealth, and persons who fail to comply with the ensuring quarantine can be charged with an offence for which the maximum penalty is imprisonment for 10 years [Australia, Quarantine Act, Part 1, Section 3(3), 1908]. In addition to this commonwealth power, states and territories also enjoy public health powers to restrict movement of persons and goods. Quarantine officers are given broad search and seizure powers and can restrict the movements of persons or vehicles from any area declared to be under quarantine but must do so after obtaining a warrant from a magistrate (Australia, Quarantine Act, Part VIA, 1908). They can also require people to receive treatment. Even though quarantine powers are granted by law to the commonwealth

FIGURE 9.3   Quarantine port checkpoint, Tasmania, Australia. This Wikipedia and Wikimedia Commons image is from the user Bjørn Christian Tørrissen and is freely available at http://en.wikipedia.org/wiki/File:Tasmania_Fruit_Detector_Dog.jpg under the Creative Commons Attribution-Share Alike 3.0 Unported license.

government, Australian authorities take a "whole of government" approach involving the Council of Australian Governments (COAG), which includes representatives from commonwealth and state governments. As noted previously, state governments have the primary responsibility for emergency management within their districts and thus must be intimately involved in any pandemic response. In practice, this coordination is carried out at the level of commonwealth and state health ministers and chief health officers.

Generally speaking, the Australian authorities do not consider there to be a strong likelihood that Australia will serve as the origin of a global disease outbreak, due to its hygienic practices and geographic isolation. Consequently, the primary focus is on preventing or slowing the spread of a global pandemic to Australian territory through effective border control measures when the World Health Organization declares a phase 4 pandemic (sustained human-to-human transmission). At this point, pilots on inbound flights must notify authorities as to whether there are any unwell passengers on board, and passengers must complete a health declaration card answering questions regarding the risk of exposure and the presence of symptoms, and provide contact information in Australia. In addition, thermal scanners and border control nurses may be deployed at selected airports, and those with symptoms will be taken to hospital for further screening and evaluation (Government of Western Australia, 2009, pp. 14–15). In principle, the Australian Government has the authority to close its borders if the threat is deemed very extreme (Australia, Department of Health and Ageing, 2009, p. 19).

Canada

In Canada, the Ministry of Health (known as Health Canada) has overall responsibility for administering and enforcing the Quarantine Act. The Minister of Health is empowered by clauses 6 through 10 of the act to establish quarantine stations and to funnel all traffic into and from Canada through designated entry and departure points (Tiedemann and Norris, 2004, pp. 3–4). Travelers entering Canada can be required to undergo a health assessment, be disinfected, or have their clothing and personal belongings disinfected. Clause 28 allows for the detention of travelers if they refuse to be disinfected or undergo a health assessment, if they must be detained in order to undergo a medical examination, if they fail to comply with an order given them by the health authorities, and if they are believed to have a communicable disease or be infected or were in close proximity to someone who has a disease or is infected (Tiedemann and Norris, 2004, p. 6). Clauses 58 to 61 of the act empowers the governor-general, on recommendation of the cabinet, to prohibit the entry of persons from a specific foreign country (or area in that country) if it is believed that there is a communicable disease outbreak in that country/area and the spreading of the disease would post a severe and imminent risk to public health in Canada (Tiedemann and Norris, 2004, p. 12). The powers of the Quarantine Act do not, however, apply to persons already inside Canada. Quarantine powers for those already inside the country (citizens and noncitizens) are applied by the individual provinces according to their respective legislation. The public health authorities in the province of Ontario, for example, have the power to request a judicial order on the part of a judge of the Ontario Court of Justice to require the detention of persons who represent a public health threat (including instructing the police to apprehend such persons). That person must then be brought before the judge, who, if convinced by the health authorities that the person in question poses a public health risk, has the authority to extend the period of detention and treatment for up to six months in the first instance and, if necessary, can extend detention in additional increments of up to six months [Province of Ontario, Health Protection and Promotion Act, R.S.O. 1990, c. H.7, s. 35(11), n.d.].

Japan

In Japan, when suspected pandemic patients are identified coming into the country, they are transferred to designated medical institutions that treat infectious disease and can be detained there. In the case of suspected infection with pandemic influenza, for example, the period of detention at the medical facility is determined by subtracting the number of days from the patient's departure from the country of infection and his/her arrival in Japan from 10 days (Japan, Ministry of Health, Labor and Welfare, Pandemic Influenza Experts Advisory Committee, 2007, p. 8). In addition, people who came into intensive contact with patients can also be quarantined in the same manner.

European Union

Among EU member states, quarantine powers are seen as the prerogative of national governments and have not been transferred to EU institutions— despite the fact that disease outbreaks are likely to become European-wide problems in short order, due to the absence of internal borders and widespread cross-border activity. The current EU legislation, Directive 2004/38/ EC, allows for the restriction of movement of people between member states in cases of disease outbreaks that are considered by the World Health Organization to have epidemic potential or in cases in which the member state applies restrictions on movement to its own citizens as well (McKee, Hervey, and Glimore, 2010, p. 251). Enforcement of this restriction is, however, left to the respective national governments.

CONCLUSION

The countries surveyed in this chapter all benefit from public health systems that are comparatively centralized and thus more amenable to being mobilized by their respective governments in order to deal with public health emergencies. The Israeli health system is arguably the most centralized and most experienced in preparing for mass casualty events. The Canadians, in turn, have learned a great deal from their experience with the SARS pandemic of 2003. Quarantine powers, on the other hand, are arguably most robust in Australia, particularly with respect to those trying to enter the country.

ISSUES TO CONSIDER

- What are the different models for socialized medical and public health systems discussed in this chapter?
- How do quarantine powers and procedures differ across Australia, Canada, Japan, and the EU?
- What are some of the key lessons learned from Canada's brush with SARS?

CONCLUSION
International Homeland Security

In this book we have attempted to provide a partial survey of homeland security strategies and policies followed by a number of democratic countries. As this book has been written largely, although by no means exclusively, for an American audience, one of the primary objectives has been to introduce American students in the emerging discipline of homeland security (many of whom are seasoned practitioners in their own right) to overseas approaches and practices. As noted in the introduction to the book, studying and analyzing the policies of other countries that encompass the various subfields of homeland security (counterterrorism, emergency management, counter-radicalization, immigration, transportation security, etc.) is critical both in searching for effective models as well as in understanding the capabilities of a given country to cope with a threat that is soon likely to affect another country (such as international terrorism or global pandemics).

Although we have argued in the Introduction that homeland security is a uniquely American concept, there is a strong argument to be made for other countries to adopt the approach, if not terminology, of homeland security in the sense that it is really a "system of systems" that encompasses a range of seemingly disparate fields that all share a common objective: the maintenance of public safety and security, the stability of society and the economy, and most crucially, the continuity of government. Viewing, for example, the police, firefighters, paramedics, public health administrators, the military, and critical infrastructure operators as representatives of completely different disciplines and pursuits misses the point that they are all involved in trying to maintain the same objectives and that their respective missions dovetail in a considerable way. They are all in the business of coping with threats, protecting the citizenry, and maintaining the sovereignty and stability of

Comparative Homeland Security: Global Lessons, First Edition. Nadav Morag.
© 2011 John Wiley & Sons, Inc. Published 2011 by John Wiley & Sons, Inc.

the government over the national territory. Indeed, in many ways, homeland security is really about maintaining the functioning of the state and its ability to enforce its laws domestically—something that, of course, requires a minimal degree of social and economic stability. Threats to homeland security are accordingly threats that have the potential to undermine social and economic stability and hence the ability of government to function and to exercise its sovereignty. In many cases this is a question of scope rather than of the nature of a specific phenomenon. For example, a street gang operating in a particular neighborhood represents a law enforcement problem, not a homeland security issue. However, if that street gang were to morph into a large organization with branches all over the country with the capacity to terrorize large numbers of people, make particular areas considerably more lawless and challenge police authority in a significant way across a significant number of jurisdictions, it would then, in the estimation of this author, become a homeland security issue. Similarly, a localized tornado that creates a path of destruction in a small town in Kansas represents a tragedy and problem for the local area but is not a homeland security issue, whereas a major disaster like that produced by Hurricane Katrina, which crosses multiple state and jurisdictional boundaries, displaces large numbers of people, causes major economic damage and disruption and leaves large areas, at least temporarily, in a state of partial anarchy, is clearly a homeland security issue. Terrorism, although usually localized, is in its own category because terrorist threats, even if (unlike 9/11) they impact a small number of people, are a direct challenge to the government through their attempt to disrupt and produce a lack of confidence in the ability of government to provide stability and security.

In returning to the objectives of this book, and as noted above, it is important to understand the homeland security laws, policies, and strategies followed by other countries in order, first, to understand what they are able and are not able to do with respect to homeland security threats in their own countries. This is critical because many of those threats will cross borders, and other countries likely to be affected by those threats need to understand to what degree they can depend on a "forward defense" in terms of whether those threats will be effectively blocked, partially blocked, are likely to morph, or the country in question (because of either certain laws or the poor performance of certain institutions) will be unable to inhibit the movement of those threats in the direction of the homeland. If we do not understand how other countries operate, we cannot effectively develop our own laws, strategies, and policies to cope with these threats, either because, on one end of the spectrum, we leave ourselves wide open because we fail to recognize the inability of particular countries to deal with certain types of threats or because, on the other end of the spectrum, certain countries are extremely effective at what they do and there is no need to squander resources on duplication of efforts when our colleagues overseas know exactly what they are doing and have the necessary tools to do it.

Second, as noted previously, we must be able to learn from successful practices as well as mistakes made by other countries. This too requires delving into the laws, institutions, policies, and strategic approaches adopted by other countries to determine (1) what works and (2) what can be implemented in our country given different laws, institutions, cultures, strategic approaches, threat environments, and so on. The second parts sounds easier than it actually is. Determining the degree of applicability of another country's homeland security approaches is an extremely difficult undertaking, as it will be rare, at least at the strategic level, to be able to adopt a foreign practice wholesale without any sorts of modifications. Adopting a successful foreign model requires understanding and analyzing the differences in legal frameworks, institutional frameworks, culture/mentality (in terms of what is and what is not publically acceptable), and a range of other variables. While adopting a foreign strategy will require modifying it to fit into legal, institutional, cultural, and other realities of the country attempting to adopt the strategy, it may frequently also require some degree of change in "adopting" that country's laws, institutions, approaches, and ways of doing business. The difficulties involved in adopting strategic-level foreign practices are thus considerable but not unbridgeable. Laws, institutions, and even culture can evolve when necessary, and it does not appear that the types of threats that affect homeland security are likely to disappear, decrease, or even stay at the same level. The increased physical interdependence of the world through global travel and increasing human contact is likely to make pandemics an increasing threat—as pandemics in the past devastated populations due to increasing global contact between persons (as well as rats, mosquitoes, and other carriers). What appears to be a process of global warming may bring about greater natural calamities, as the planet's weather seems to be exhibiting increasing signs of instability and variability. Finally, there is absolutely no reason to believe that the phenomenon of terrorism will decrease. On the contrary, increasing technological capabilities to create chaos and destruction (including the greater ease in development of biological, chemical, and radiological weapons and, perhaps someday, nuclear weapons), coupled with increasing technological vulnerabilities associated with modern life (dependence on utilities, transportation networks, food supplies, financial markets, the cyber sphere, etc.), are guaranteed to ensure that groups of fanatics with no respect for human life (unfortunately, there will always be such people) will have the motivation, tools, and capacity to cause greater disruption.

This thus seems to suggest that we need to be open to foreign lessons driving the evolution of laws, institutions, and culture to better cope with a world in which such threats are only likely to increase. Failure to do so will increase our vulnerabilities and deny us the knowledge of what our foreign partners are able and unable to accomplish on their own and what they have discovered that works. The homeland security mission is thus a global one, and a homeland security approach that ends at a nation's borders is not a homeland security approach at all.

REFERENCES

Agreement on Australia's National Counter-Terrorism Arrangements. (2002, December).

Alexander, D. (2002). *Principles of Emergency Planning and Management*. Harpenden, UK: Terra Publishing.

Almandras, S. (2008). *Special Constables*. London: House of Commons.

Amaral, J. M. (2007). *Surveillance of Europe's Maritime Frontiers: Reply to the Annual Report of the Council*. Paris: Assembly of the Western European Union, The Interparliamentary European Security and Defence Assembly.

Amidror, Y. (2008). *Winning Counterinsurgency War: The Israeli Experience*. Jerusalem: Jerusalem Center for Public Affairs.

Amnesty, International., (2006). *United Kingdom, Human Rights: A Broken Promise*. London: Amnesty International.

Andeweg, R. B., and Irwin, G. A. (2009). *Governance and Politics of the Netherlands* (3rd ed.). New York: Palgrave Macmillan.

Anonymous v. Minister of Defense 54(1), Cr.A. 7048/97 (Supreme Court 1997).

Association of Chief Police Officers of England, Wales and Northern Ireland and The UK Border, Agency. (2008, April 3). Memorandum of Understanding: Police and United Kingdom Border Agency Engagement to Strengthen the UK Border. n.p.

Association of State and Territorial Health Officials. (2006). *Public Health Preparedness Infrastructures: Comparing Israel to the United States*. Arlington, VA: ASTHO.

Austen, I., and Johnston, D. (2006, June 4). *17 Held in Plot to Bomb Sites in Ontario*. Retrieved May 24, 2009, from New York Times: http://www.nytimes.com/2006/06/04/world/americas/04toronto.html?_r=2&hp&ex=1149393600&en=58798ee2c44c6f76&ei=5094&partner=homepage&oref=slogin.

Australia, Anti-Terrorism Act (No. 2), Section 105.42. (2005).

Australia, Criminal Code, Section 100.1. (2008).

Australia, Criminal Code, Section 104.5(3). (1995).

Australia, Criminal Code, Section 105.8. (1995).

Australia, Criminal Code, Sections 105.35–105.38. (1995).

Australia, Defence Act, Part IIIAAA, Division 1, Section 51A. (1903).

Comparative Homeland Security: Global Lessons, First Edition. Nadav Morag.
© 2011 John Wiley & Sons, Inc. Published 2011 by John Wiley & Sons, Inc.

Australia, Defence Act, Part IIIAAA, Division 1, Section 51E. (1903).

Australia, Department of Defense. (2009). *Defending Australia in the Asia Pacific Century: Force 2030*. Canberra: Department of Defence.

Australia, Department of Foreign Affairs and Trade. (2008, April). *Health Care in Australia*. Retrieved August 1, 2010, from About Australia: http://www.dfat.gov.au/facts/healthcare.html.

Australia, Department of Health and Ageing. (2009). *Fluborderplan, National Pandemic Influenza Airport Border Operations Plan*. Canberra: Australian Department of Health and Ageing.

Australia, Quarantine Act, Part 1, Section 3(3). (1908).

Australia, Quarantine Act, Part VIA. (1908).

Australian Attorney-General's Office. (n.d.). *National Counter-Terrorism Committee: Critical Infrastructure Protection in Australia*. Canberra: Australian Government.

Australian Department of Health. (2009). *National Influenza Pandemic Public Communications Guidelines*. Canberra: Australian Government.

Australian Department of the Prime Minister and Cabinet. (2010). *Counter-Terrorism White Paper: Securing Australia, Protecting Our Community*. Canberra: Australian Government.

Australian Federal Police. (2006). *Annual Report 2005–06*. Canberra: Commonwealth of Australia.

Australian Government. (2005). *National Counterterrorism Plan*. Canberra: Commonwealth of Australia.

Australian Government. (2006). *Protecting Australia Against Terrorism 2006*. Canberra: Commonwealth of Australia.

Australian Government. (n.d.). *Australian Laws to Combat Terrorism*. Retrieved May 25, 2009, from http://www.ag.gov.au/agd/WWW/nationalsecurity.nsf/AllDocs/A41A86E81E52A0B2CA25710A001A7EEA?OpenDocument

Australian Government, Attorney-General's Department. (n.d.). *National Guidelines for Protecting Critical Infrastructure from Terrorism*. Canberra: Commonwealth of Australia.

Australian Institute of Health and Welfare. (2002, June 27). *Australia's Health No. 8*. Retrieved August 1, 2010, from Australia's Health 2002: http://www.aihw.gov.au/publications/index.cfm/title/7637.

Australian Secret Intelligence Organization Act, Part III, Division 3, Section 34L. (1979).

Australian Security Intelligence Organisation. (2008). *ASIO Report to Parliament 2007–08*. Canberra: Commonwealth of Australia.

Avis, P. (2003, Spring). Surveillance and Canadian Maritime Domestic Security. *Canadian Military Journal*, 9–14.

Baldino, D. (2007). Good Instincts or Poor Judgement? Australia's Counter-Terrorism Response After 9-11. *American Political Science Association Conference*. Washington, DC.

Barnett, H. (2002). *Britain Unwrapped: Government and Constitution Explained*. London: Penguin Books.

Bassett, D., Haldenby, A., Thraves, L., and Truss, E. (2009). *A New Force*. London: Reform.

Bayley, D. H. (1991). *Forces of Order: Policing Modern Japan*. Berkeley, CA: University of California Press.

BBC. (2010, October 13). *Connecting in a Crisis*. Retrieved October 13, 2010, from Flooding: http://www.bbc.co.uk/connectinginacrisis/08.shtml.

BBC News. (2005, September 16). Retrieved November 5, 2009, from Police Forces No Longer Working: http://news.bbc.co.uk/2/hi/uk_news/4253138.stm.

Beckman, J. (2007). *Comparative Legal Approaches to Homeland Security and Anti-Terrorism*. Aldershot, UK: Ashgate.

Belanger, C. (2004). War Measures' Act. In *Readings in Quebec History*. Westmount, Quebec: Marianopolis College.

Benoit-Smullyan, E. (1938). *History of Political Theory, Part I: Plato to Locke*. Boston: Hymarx.

Blank, R. H., and Burau, V. (2007). *Comparative Health Policy* (2nd ed.). Houndmills, UK: Palgrave Macmillan.

Blum, S. C. (2008). *The Necessary Evil of Preventive Detention in the War on Terrorsim: A Plan for a More Moderate and Sustainable Solution*. Amherst, NY: Cambria Press.

Boender, W. (2008). Islam in the Netherlands: Expectations and Realities. *ISIM Review, 21*, 22–23.

Bonin, S. (2007). *International Biodefense Handbook 2007: An Inventory of National and International Biodefense Pratices and Policies*. Zurich: Center for Security Studies.

Brettfled, K., and Wetzels, P. (2006). *Muslims in Germany: Integration, Barriers to Integration, Religion and Attitudes Towards Democracy, the Rule of Law, and Politically/Religiously Motivated Violence*. Hamburg: University of Hamburg.

Briefing by MG Yair Golan. (2009, September 8). Herzeliya, Israel.

Briefing by senior German Federal Police official. (2010, January 27). Homeland Security in Germany. Monterey, CA.

British Columbia Emergency Program Act, Chapter 111, Section 10(1). (n.d.).

British Government. (2010). *Emergency Response and Recovery, Non Statutory Guidance Accompanying the Civil Contingencies Act 2004*. London: Stationery Office.

British Security Service. (2005). *Protecting Against Terrorism*. London: Stationery Office.

British Security Service. (2009). *What to Look For*. Retrieved August 17, 2010, from British Security Service MI5: https://www.mi5.gov.uk/output/what-to-look-for.html.

British Security Service. (2010). *Major Areas of Work*. Retrieved August 16, 2010, from Security Service MI5: https://www.mi5.gov.uk/output/major-areas-of-work.html.

British Security Service. (n.d.). Retrieved December 9, 2009, from https://www.mi5.gov.uk/output/national-intelligence-machinery.html.

Brogden, M., and Nijhar, P. (2005). *Community Policing: National and International Models and Approaches*. Portland, OR: Willan Publishing.

Brown, I. (2006, June 15). *UK Government Surveillance Powers*. Retrieved May 28, 2010, from Social Science Research Network: http://ssrn.com/abstract=1026974.

B'Tselem. (1997). *Prisoners of Peace: Administrative Detention During the Oslo Process*. Jerusalem: B'Tselem.

B'Tselem. (2009). *Statistics on Administrative Detention*. Jerusalem: B'Tselem.

Bull, M. J., and Newell, J. L. (2005). *Italian Politics: Adjustment Under Duress*. Cambridge, UK: Polity Press.

Burton, S. (n.d.). *Ramping Up: Reserve Response Force Established*. Retrieved July 12, 2009, from Army: http://www.defence.gov.au/news/armynews/editions/1075/topstories/story01.htm.

Bush, G. (2003). *The National Strategy for the Physical Protection of Critical Infrastructures and Key Assets*. Washington, DC: The White House.

Bush, G. W. (2007). *National Strategy for Homeland Security*. Washington, DC: The White House.

Busse, R. (2009). *The German Health Care System, 2009*. Retrieved August 1, 2010, from The Commonwealth Fund: http://www.commonwealthfund.org/Topics/International-Health-Policy-2009/Countries/~/media/Files/Publications/Other/2010/Jun/International%20Profiles/1417_Squires_Intl_Profiles_Germany.pdf.

Cabigiosu, C. (2006). The Role of Italy's Military in Supporting the Civil Authorities. In *Armies in Homeland Security: American and European Perspectives*. Washington, DC: National Defense University Press.

Caldwell, C. (2006, June 25). After Londonistan. *The New York Times*.

Canada, Criminal Code, Part 11.1, Section B(i) a–e. (1985).

Canada, Criminal Code, Part II, Section 83.3. (1985).

Canada, Emergencies Act, Section 3. (1988).

Canada, Ministry of Defense. (1994). *1994 White Paper on Defense*. Ottawa: Ministry of Defense.

Canada, Standing Senate Committee on National Security and Defense. (2007). *Canadian Security Guidebook: Airports*. Ottawa: Canadian Parliament.

Canadian Government. (2004). *Canadian Pandemic Influenza Plan*. Ottawa: Government of Canada.

Canadian Public Health Association. (c. 2007). *Communicating Risk: Pandemic Flu Preparedness*. Ottawa: Canadian Public Health Association.

Canadian Rangers. (n.d.). Retrieved July 24, 2009, from Canadian Ministry of Defense: http://www.armee.forces.gc.ca/land-terre/cr-rc/index-eng.asp

Canadian Security Intelligence Service. (2005, February). *Backgrounder No. 1 - The CSIS Mandate*. Retrieved October 21, 2009, from Canadian Security Intelligence Service: http://www.csis-scrs.gc.ca/nwsrm/bckgrndrs/bckgrndr01-eng.asp

Canadian Security Intelligence Service. (2008). *Public Report 2007-2008*. Ottawa: Public Works and Government Services Canada.

CBS News. (2003, September 7). *Marked for Death*. Retrieved October 15, 2008, from CBSNews.com: http://www.cbsnews.com/stories/2003/09/07/world/main571958.shtml.

Center for the Protection of National Infrastructure. (2007). *Guide to Producing Operational Requirements for Security Measures*. London: Center for the Protection of National Infrastructure.

Center for the Protection of National Infrastructure. (2010). *What We Do*. Retrieved April 1, 2010, from Center for the Protection of National Infrastructure: http://www.cpni.gov.uk/About/whatWeDo.aspx.

Chalk, P. (2009). Australia. In B. A. Jackson (Ed.), *Considering the Creation of a Domestic Intelligence Agency in the United States*. Santa Monica, CA: RAND.

Change Institute, Communities and Local Government: London. (2009). *Summary Report: Understanding Muslim Ethnic Communities*. London: Communities and Local Government.

Choudhury, T. (2007). *The Role of Muslim Identity Politics in Radicalisation*. London: Department for Communities and Local Government.

Clavier, S. M. (1997, July). *Perspectives on French Criminal Law*. Retrieved June 5, 2009, from http://userwww.sfsu.edu/~sclavier/research/frenchpenalsystem.pdf.

CNN. (2010, January 11). *How the Israelis Do Airport Security*. Retrieved July 25, 2010, from CNN: http://www.cnn.com/2010/OPINION/01/11/yeffet.air.security.israel/index.html.

Cole, L. (2007). *Terror: How Israel Has Coped and What American Can Learn*. Bloomington, IN: Indiana University Press.

Coleridge, S. T. (1833, April 8). *The Columbia Dictionary of Quotations by Robert Andrews*. Retrieved December 2009, from Google Books: http://books.google.com/books?id=4cl5c4T9LWkC&pg=PA668&lpg=PA668&dq=you + see + how + the + house + of + commons + has + begun + to + verify + all + the + ill + prophecies&source=bl&ots=87uDDbgAIU&sig=FrBCSDiiV9jbbRxrxOMYjFFgNFk&hl=en&ei=HE00S4-UMpL0sgPqgsDIBA&sa=X&oi=book_result&ct.

Colonel Zohar. (2006). Counterterrorism Intelligence: Hizballah and the Palestinians. *Israeli Intelligence and Heritage Commemoration Center Perspectives, 47*.

Commission of Inquiry into the Actions of Canadian Officials in Relation to Maher Arar. (2006). *A New Review Mechanism for the RCMP's National Security Activities*. Ottawa: Public Works and Government Services Canada.

Commonwealth of Australia, Incident Response Regiment. (May 2005). The Incident Response Regiment. *Australian Journal of Emergency Management*, 18–20.

Congressional Research Service. (2004). *CRS Report for Congress: Germany's Role in Fighting Terrorism, Implications for US Policy*. Washington, DC: Congressional Research Service.

Convention Implementing the Schengen Agreement, Chapter 1, Article 2. (1985).

Convention Implementing the Schengen Agreement, Chapter 2, Article 5. (1985).

Cornish, P. (2007). *Domestic Security, Civil Contingencies and Resilience in the United Kingdom: A Guide to Policy*. London: Chatham House.

Council of Europe. (1953, September 3). Convention for the Protection of Human Rights and Fundamental Freedoms, 213 UNTS 222.

Council of Europe. (2006). *Profiles on Counter-Terrorism Capacity: France*. Strasbourg: Council of Europe.

Council of Europe. (2008). *Codexter Profiles on Counter-Terrorism Capacity: Canada*. Strasbourg: Council of Europe.

Dammer, H. R., and Fairchild, E. (2006). *Comparative Criminal Justice Systems* (3rd ed.). Belmont, CA: Wadsworth.

Department of Justice, Canada. (2007). *Public Safety and Anti-Terrorism (PSAT) Initiative Summative Evaluation: Final Report*. Ottawa: Department of Justice Canada.

DHS. (2008a). *Brief Documentary History of the Department of Homeland Security 2001–2008*. Washington, DC: Department of Homeland Security.

DHS. (2008b). *One Team, One Mission, Security Our Homeland: US Department of Homeland Security Strategic Plan, Fiscal Years 2008–2013*. Washington, DC: Department of Homeland Security.

DHS. (n.d.). *What We Do and How We're Doing It*. Retrieved December 20, 2009, from Department of Homeland Security: http://www.dhs.gov/xabout/responsibilities.shtm#one.

Dichter, A., and Byman, D. L. (2006). *Israel's Lessons for Fighting Terrorists and Their Implications for the United States.* Washington, DC: The Saban Center at the Brookings Institution.

Directgov. (2010, October 13). *Preparing for Emergencies.* Retrieved October 13, 2010, from Directgov: http://www.direct.gov.uk/en/Governmentcitizensandrights/Dealing-withemergencies/Preparingforemergencies/index.htm.

Directorate-General for Maritime Affairs and Fisheries. (2008). *Non-Paper on Maritime Surveillance.* Brussels: European Commission.

Dodd, P. (2009). Do British Military Intelligence and Royal Navy Operations Have a Part to Play in the Fight Against Organized Crime in the Maritime Domain? In R. Herbert-Burns, S. Bateman, and P. Lehr (Eds.), *Lloyd's MIU Handbook of Maritime Security* (pp. 345–350). Boca Raton, FL: CRC Press.

Dupont, B. (2002). *Implementing Community Policing in a Centralised Criminal Justice System: Another French Paradox.* Working paper.

Dutheillet de Lamothe, O. (2006). *French Legislation Against Terrorism: Constitutional Issues.* Paris: The Constitutional Council.

Dyke, A. H. (2009). *Mosques Made in Britain.* London: Quilliam Foundation.

Elliott, R. (2010). Strategy for Improving EMS Response to Multi-Casualty Incidents. Unpublished paper.

European Commission. (2010). *Schengen Enlargement.* Retrieved July 24, 2010, from European Commission, Justice and Home Affairs: http://ec.europa.eu/justice_home/fsj/freetravel/schengen/fsj_freetravel_schengen_en.htm.

European Council Framework Decision 2002/475/JHA. (2002, June 13).

European Union. (2005, May). *European Arrest Warrrant Replaces Extradition Between EU Member States.* Retrieved April 25, 2010, from European Commission, Justice and Home Affairs: http://ec.europa.eu/justice_home/fsj/criminal/extradition/printer/fsj_criminal_extradition_en.htm.

European Union. (2009). *The Schengen Area and Cooperation.* Retrieved August 16, 2009, from Europa: Summaries of EU Legislaton: http://europa.eu/legislation_summaries/justice_freedom_security/free_movement_of_persons_asylum_immigration/l33020_en.htm.

European Union. (2010). *How Does the EU Work?* Retrieved July 24, 2010, from Europa: http://europa.eu/abc/12lessons/lesson_4/index_en.htm.

Explosive Substances Act, Sections 2 and 3. (1883). Retrieved May 21, 2009, from http://www.opsi.gov.uk/RevisedStatutes/Acts/ukpga/1883/plain/cukpga_18830003_en.

Facts About Germany. (2009). Retrieved December 27, 2009, from Political System: http://www.tatsachen-ueber-deutschland.de/en/political-system/main-content-04/the-basic-law.html.

Federal Ministry of the Interior. (n.d.). *Protection of Critical Infrastructures - Baseline Protection Concept, Recommendation for Companies.* Berlin: Federal Ministry of the Interior.

Federal Research Division of the Library of Congress. (1995). *Country Studies Series.* Retrieved December 26, 2009, from http://www.country-data.com/cgi-bin/query/r-4970.html.

Fetzer, J. S., and Soper, C. (2005). *Muslims and the State in Britain, France, and Germany.* Cambridge: Cambridge University Press.

Flahault, A., Blanchon, T., Dorleans, Y., Vibert, J., and Valleron, A. (2006). Virtual Surveillance of Communicable Diseases: A 20-Year Experience in France. *Statistical Methods in Medical Research, 15*, 413–421.

Flood, P. (2004). *Report of the Inquiry into Australian Intelligence Agencies.* Canberra: Commonwealth of Australia.

Foley, F. (2009). Reforming Counterterrorism: Institutions and Organizational Routines in Britain and France. *Security Studies, 18*(3), 435–478.

France, Code of Criminal Procedure, Article 145-2. (2004).

France, Code of Criminal Procedure, Article 421-3. (1997).

France, Code of Criminal Procedure, Articles 421-1 and 421-2. (1997).

France, Ministry of Defense. (n.d.). *The 30 Year Prospective Plan: A Summary.* Paris: Ministry of Defense.

France, Ministry of the Interior. (2006, March 28). *Civil Security.* Retrieved April 25, 2010, from Communal Reserves: http://translate.google.com/translate?hl=en&sl=fr& u=http://www.interieur.gouv.fr/sections/a_l_interieur/defense_et_securite_civiles/ dossiers/plan-orsec&ei=sdTUS52vAYXwsgPYyuz6CQ&sa=X&oi=translate&ct= result&resnum=3&ved=0CBcQ7gEwAg&prev=/search%3Fq%3Dorsec.

France, Ministry of the Interior. (2007, September 23). *The New Planning ORSEC.* Retrieved April 25, 2010, from Civil Security: http://translate.google.com/translate? hl=en&sl=fr&u=http://www.interieur.gouv.fr/sections/a_l_interieur/defense_et_ securite_civiles/dossiers/plan-orsec&ei=GNPUS8W7NobCsgPZxfiNCg&sa=X&oi= translate&ct=result&resnum=3&ved=0CBcQ7gEwAg&prev=/search%3Fq%3Dorsec.

France, Ministry of the Interior. (2009, August 12). *Civil Security.* Retrieved April 25, 2010, from Operational Services: http://translate.google.com/translate?hl=en&sl=fr&u= http://www.interieur.gouv.fr/sections/a_l_interieur/defense_et_securite_civiles/dos-siers/plan-orsec&ei=sdTUS52vAYXwsgPYyuz6CQ&sa=X&oi=translate&ct=result& resnum=3&ved=0CBcQ7gEwAg&prev=/search%3Fq%3Dorsec.

France, Prime Minister's Office. (2007). *Prevailing Against Terrorism.* Paris: French Government.

France, Senate. (2009). *The Legislative Process.* Retrieved December 28, 2009, from Welcome to the French Senate: http://www.senat.fr/lng/en/procedure_legislative.html.

Freilich, C. D. (2006). National Security Decision-Making in Israel: Processes, Pathologies, and Strengths. *Middle East Journal, 60*(4), 635–663.

French Institute for Public Health Surveillance. (2010). *Surveillance, Alert and Prevention.* Retrieved March 31, 2010, from French Institute for Public Health Surveillance: http:// www.invs.sante.fr/presentations/edito_en_.htm.

French Ministry of Foreign Affairs. (2010). *Other Means of Control and Protection.* Retrieved March 31, 2010, from France Diplomatie: http://www.diplomatie.gouv.fr/en/france-priorities_1/disarmament-arms-control-and-arms-trade-control_1109/france-and-non-proliferation-of-weapons-of-mass-destruction_7146/fight-against-biological-weapons-proliferation_7149/other-means-of-control-and-protection.

French President. (2008). *The French White Paper on Defence and National Security.* Paris: Office of the President of France.

FRONTEX. (2006). *FRONTEX Annual Report 2006.* Warsaw: FRONTEX.

Garapon, A. (2005). *Is There a French Advantage in the Fight Against Terrorism?* Madrid: Real Instituto Elcano.

German Criminal Procedure Code, Section 105. (1987).

German Federal Ministry of Defense. (2006). *White Paper 2006 on German Security and the Future of the Bundeswehr*. Berlin: Ministry of Defense.

German Federal Ministry of the Interior. (2005). *Annual Reports 2005 on the Protection of the Constitution*. Berlin: German Federal Ministry of the Interior.

German Federal Police. (2005). *Federal Police, Duties and Organization*. Berlin: Bundespolizeidirektion.

Germany, Basic Law, Article 35 (1–3). (n.d.).

Germany, Basic Law, Article 87a(3). (n.d.).

Germany, Basic Law, Article 87a(4). (n.d.).

Germany, Criminal Code, Section 129a. (2008).

Germany, Federal Ministry of the Interior. (2009). *National Strategy for Critical Infrastructure Protection*. Berlin: Federal Ministry of the Interior.

Germany, Federal Ministry of the Interior. (n.d.). *Protection of Critical Infrastructures— Baseline Protection Concept: Recommendation for Companies*. Berlin: Federal Ministry of the Interior.

Germany, Foreign Ministry. (2009). *The Immigration Act*. Retrieved September 17, 2009, from Germany Foreign Ministry http://www.auswaertiges-amt.de/diplo/en/ WillkommeninD/EinreiseUndAufenthalt/Zuwanderungsrecht.html.

Germany, Ministry of Defense. (2003). *The Bundeswehr Reservist Concept*. Berlin: Ministry of Defense.

Germany, Ministry of Defense. (2006). *White Paper 2006 on German Security Policy and the Future of the Bundeswehr*. Berlin: Ministry of Defense.

Geva, R., (1995, April 19). *Effective National and International Action Against Terrorism: The Israeli Experience*. Retrieved May 14, 2010, from Israel Ministry of Foreign Affairs: http:// www.mfa.gov.il/MFA/Archive/Communiques/1995/EFFECTIVE%20NATIONAL %20AND%20INTERNATIONAL%20ACTION%20AGAINS.

Gogou, D. (2006). Towards a European Approach on Border Management: Aspects Related to the Movement of Persons. In M. Caparini and O. Marenin (Eds.), *Borders and Security Governance: Managing Borders in a Globalised World*. Zurich: Lit Verlag.

Government of Canada. (2009). *Canada National Strategy for Critical Infrastructure*. Ottawa: Government of Canada.

Goverment of Canada. (2010, May–July). *Is Your Family Prepared?* Retrieved October 12, 2010, from GetPrepared.ca: http://www.getprepared.gc.ca/index-eng.aspx.

Government of Tasmania. (2006). *Tasmanian Emergency Management Plan, Issue 6*. Hobart: Government of Tasmania.

Government of Victoria. (2009). *State Emergency Response Plan, Part 3: Emergency Management Manual Victoria*. Melbourne: Government of Victoria.

Government of Western Australia. (2009). *Western Australian Health Management Plan for Pandemic Influenza*. Perth: Government of Western Australia, Department of Health.

Gregory, F. (2009). *UK Border Security: Issues, Systems and Recent Reforms*. London: Institute for Public Policy Research.

Gross, E. (2003). Democracy in the War Against Terrorism: The Israeli Experience. *Loyola of Los Angeles Law Review, 35*, 1161–1216.

Gross, O., and Aolain, F. N. (2006). *Law in Times of Crisis: Emergency Powers in Theory and Practice.* Cambridge, UK:.

Guild, E., Carrera, S., and Eggenschwiler, A. (2009). *Informing the Asylum Debate.* Brussels: Centre for European Policy Studies.

Guiora, A. N. (2006). *Transnational Comparative Analysis of Balancing Competing Interests in Counter-Terrorism.* Cleveland, OH: Case Western Reserve University, Case School of Law.

Gujer, E. (2010). Germany, the Long and Winding Road. In G. Schmitt (Ed.), *Safety, Liberty and Islamist Terrorism: American and European Approaches to Domestic Counterterrorism* (pp. 62–80). Washington, DC: AEI Press.

Guttman, N. (2010, January 6). *Israel's Airport Security, Object of Envy, Is Hard to Emulate Here.* Retrieved July 25, 2010, from Forward: http://www.forward.com/articles/122781/.

Haberfeld, M. R., and Herzog, S. (2000). The Criminal Justice System in Israel. In E. Obi, N.I. Ebbe, and O. N. Ebbe (Eds.), *Comparative and International Criminal Justice Systems* (2nd ed., pp. 55–78). Boston: Butterworth-Heinemann.

Hajjar, L. (2005). *Courting Conflict: The Israeli Military Court System in the West Bank and Gaza.* Berkeley, CA: University of California Press.

Hay, J. B. (2006). *Who Does What? Critical Infrastructure Protection in the Canadian Government.* Ottawa: Canadian Center of Intelligence and Security Studies, Carleton University.

Health Canada. (2003). *Learning from SARS: Renewal of Public Health in Canada.* Ottawa: Health Canada.

Health Canada. (2008). *Canada Health Act, Annual Report 2007–2008.* Ottawa: Health Canada.

Health Canada, Crisis Communications Unit. (2003). *Crisis/Emergency Communications Guidelines.* Ottawa: Health Canada.

Health Protection Agency. (2010). *What the Health Protection Agency Does.* Retrieved March 28, 2010, from Health Protection Agency: http://www.hpa.org.uk/web/HPA-web&Page&HPAwebAutoListName/Page/1153846674309.

Henry, S. (2009, April 16). Israel Supreme Court President Speaks at Princeton. Associated Press.

Hilless, M., and Healy, J. (2001). *Health Care Systems in Transition: Australia.* Copenhagen: European Observatory on Health Care Systems.

Hobbing, P. (2005). *Integrated Border Management at the EU Level.* Brussells: Centre for European Policy Studies.

Hodgson, J. (2006). *The Investigation and Prosecution of Terrorist Suspects in France.* London: UK Home Office.

Homeland Security Institute. (2009). *Public Role and Engagement in Counterterrorism Efforts: Implications of Israeli Practices for the US.* Arlington, VA: Homeland Security Institute.

House of Commons, Defense Committee. (2009). *The Defense Contribution to UK National Security and Resilience.* London: Stationery Office.

House of Commons, Home Affairs Committee. (2006). *Terrorism Detention Powers: Fourth Report of Session 2005–06,* Vol. 1 London: Stationery Office.

House of Commons, Transport Committee. (2005). *UK Transport Security: Preliminary Report.* London: Stationery Office.

Human Rights Watch. (2007). *In the Name of Prevention: Insufficient Safeguards in National Security Removals.* New York: Human Rights Watch.

Human Rights Watch. (2008). *Preempting Justice: Counterterrorism Laws and Procedures in France.* Retrieved June 5, 2009, from Human Rights Watch: http://www.hrw.org/en/node/62151/section/1.

IDF Homefront Command. (n.d.a). *Appendix 7: Doctrine for Dealing with Events in the Homefront.* Tel Aviv: IDF Homefront Command.

IDF Homefront Command. (n.d.b). *The Homefront Command.* Retrieved October 13, 2010, from IDF Homefront Command: http://www.oref.org.il/14-en/PAKAR.aspx.

Inbar, M., and Keren, U. (2002). *Identification of Casualties at a Mass Casualty Event: Our Responsibility to the Public.* Jerusalem: Israel Ministry of Health, Emergency Department.

International Crisis Group. (2007). *Islam and Identity in Germany.* Brussels: International Crisis Group.

Ireland, P. (2004). *Becoming Europe: Immigration, Integration, and the Welfare State.* Pittsburgh PA: University of Pittsburgh Press.

ISA Law, Section 7 (b1–b4). (2002).

ISA Law, Section 8(b). (2002).

Israel Defense Force Homefront Command. (n.d.a). *Annex C: Emergency Economy.*

Israel Defense Force Homefront Command. (n.d.b). *Chapter 3: Organizing the Civil Defense of the Factory.*

Israel Defense Force Homefront Command. (2007). *Summary of Civil Defense Doctrine in the Factory.* Tel Aviv: IDF Homefront Command.

Israel Defense Force, Order No. 1226, Section 9. (1988).

Israel Ministry of Defense, Office of the Deputy Minister of Defense. (2008, February 24). National Emergency Authority: Homefront Operational Approach.

Israel Ministry of Foreign Affairs. (2009). Ministry of Foreign Affairs Web site. Retrieved June 22, 2009, from The Health Care System in Israel—An Historical Perspective: http://www.mfa.gov.il/MFA/History/Modern%20History/Israel%20at%2050/The%20Health%20Care%20System%20in%20Israel-%20An%20Historical%20Pe.

Israel Ministry of Health. (2007). *Annual Report for 2007 Under the Freedom of Information Law.* Jerusalem: Ministry of Health.

Israel Police. (n.d.). *Annex B: Israel Police and the Emergency Services.* Jerusalem: Israel Police.

Israel Police, Police Order 14.01.08, Section 4(c). (1995, August 1). Soldiers and Citizens Eligible to Be Judged by Military Courts—Detention, Arrest, Searches, Investigations and Charging.

Israel Police, Police Order 14.01.34, Section 6(d) (1–2). (2008, January 18). Detention, Incarceration and Release.

Israel Supreme Court. (1999, September 6). Judgement on the Interrogation Methods Applied by the GSS.

Israel Supreme Court. (2005). *Judgements of the Israel Supreme Court: Fighting Terrorism Within the Law.* Jerusalem: Israel Supreme Court.

Israel, Criminal Procedure Law, Section 25(1). (1996).

Israel, Defense (Emergency) Regulations, Sections 58,59,62,64, and 66. (1945).

Israel, Defense (Emergency) Regulations, Section 84. (1945).

Israel, Emergency Powers Law, Section 2(a–b). (1979).

Israel, Emergency Powers Law, Section 4 (a). (1979).

Israel, Law and Administration Ordinance, Section 9(a–c). (1948).

Israel, National Emergency Economy Headquarters. (2001). *Planning and Organization of the Economy in a State of Emergency.* Jerusalem: National Emergency Economy Headquarters.

Israel, Penal Law, Section 34(11). (1977).

Israel, Penal Law, Section 128(1)(3). (1977).

Israel, Penal Law, Section 145. (1977).

Israel, Prevention of Terrorism Ordinance, No. 33, Section 7. (1948).

Israel, Prevention of Terrorism Ordinance, No. 33, Section 8. (1948).

Israel, Prevention of Terrorism Ordinance, No. 33, Section 11. (1948).

Israel, Prevention of Terrorism Ordinance, Section 12(a–c). (1948).

Israel, Prevention of Terrorism Ordinance, Section 15 (a–b). (1948).

Israel, State of Emergency Employment Law, Chapter 6, Section 34. (1967).

Italy, Law 690, Article 39. (1907).

Italy, Ministry of Defense. (2009). *Operation 'Forza Paris'.* Retrieved July 13, 2009, from Ministry of Defense http://www.esercito.difesa.it/English/History/Forza_Paris.asp.

Izbenberg, D. (2010, April 26). New Anti-Terrorist Bill Would Replace, Expand Current Laws. *Jerusalem Post.*

Japan, Ministry of Health, Labor and Welfare, Pandemic Influenza Experts Advisory Committee. (2007, March 26). *Guideline for Quarantine System of Pandemic Influenza.* Retrieved August 6, 2010, from Japan, Ministry of Health, Labor and Welfare: http://www.mhlw.go.jp/bunya/kenkou/kekkaku-kansenshou04/pdf/09-e02.pdf.

Jenkins, B. M. (2006). *Unconquerable Nation: Knowing Our Enemy, Strengthening Ourselves.* Santa Monica, CA: RAND Corporation.

Jenkins, B. M., and Gersten, L. N. (2001). *Protecting Public Surface Transporation Against Terrorism and Serious Crime: Continuing Research on Best Security Practices.* San Jose, CA: Mineta Transportation Institute, San Jose State University.

Johnson, J. A., and Stoskopf, C. H. (2010). *Comparative Health Systems: Global Perspectives.* Sudbury, MA: Jones & Bartlett Publishers.

Jones, A. A., and Wiseman, R. (2006). *Policing and Terrorism: The French (Dis)connection and the Lessons for America.* Retrieved from Jurawelt.com.

Jones, C. (2003). 'One Size Fits All': Israel, Intelligence, and the al-Aqsa Intifada. *Studies in Conflict and Terrorism, 26*

Jorry, H. (2007). *Construction of a European Institutional Model for Managing Operational Cooperation at the EU's External Borders: Is the FRONTEX Agency a Decisive Step Forward?* Brussels: Centre for European Policy Studies.

Klose, G. J. (2006). The Weight of History: Germany's Military and Domestic Security. In J. L. Clarke (Ed.), *Armies in Homeland Security: American and European Perspectives.* Washington, DC: National Defense University Press.

Kohn, S., Barnett, D. J., Leventhal, A., Reznikovich, S., Oren, M., Laor, D., et al. (2010). Pandemic Influenza Preparedness and Response in Israel: A Unique Model of Civilian-Defense Collaboration. *Journal of Public Health Policy, 256–269.*

Konze, A. (2009). Chapter 9: Germany. In M. Haberfeld, J. F. King, and C. A. Lieberman (Eds.), *Terrorism Within Comparative International Context: The Counter-Terrorism Response and Preparedness* (pp. 101–122). New York: Springer-Verlag.

Koopmans, R., Statham, P., Giugni, M., and Passy, F. (2005). Contested Citizenship: Immigration and Cultural Diversity in Europe. In *Social Movements, Protest, and Contention*. Minneapolis, MN: University of Minnesota Press.

Kuru, A. T. (n.d., April 23). *Muslims and the Secular State in France*. Retrieved 2009, from Kennedy Center for International Studies: http://europe.byu.edu/islam/pdfs/Kuru-paper.pdf.

Lackenbauer, P. W. (2005–2006, Winter). The Canadian Rangers: A 'Postmodern' Militia That Works. *Canadian Military Journal*, 49–60.

Laehnemann, J. (n.d.). The Training of Teachers for Islamic Religous Education at German Universities: The Erlangen-Nuremberg Project. Unpublished paper.

Larsson, P., Frisell, E. H., and Olsson, S. (2009). Understanding the Crisis Management System of the European Union. In S. Olsson (Ed.), *Crisis Management in the European Union: Cooperation in the Face of Emergencies* (pp. 1–16). Berlin: Springer-Verlag.

Laurence, J., and Vaisse, J. (2006). *Integrating Islam: Political and Religious Challenges in Contemporary France*. Washington, DC: Brookings Institution Press.

Lawday, D. (2001). *Policing in France and Brita in: Restoring Confidence Locally and Nationally*. London: Franco-British Council.

Leiba, A., Halpern, P., Kotler, D., Blumenfeld, A., Sofer, D., Weiss, G., et al. (2005). Case Study of the Terrorist Bombing in Tel Aviv Market: Putting All the Eggs in One Basket Might Save Lives. *International Journal of Disaster Medicine*, 3, 1–4.

Leith, K. H., Knott, A., Mayer, A. G., and Westermann, J. (2010). Germany. In J. A. Johnson and C. H. Stoskopf (Eds.), *Comparative Health Systems: Global Perspectives* (pp. 147–166). Sudbury, MA: Jones & Bartlett Publishers.

Lepsius, O. (2002). *The Relationship Between Security and Civil Liberties in the Federal Republic of Germany After September 11*. Baltimore: American Institute for Contemporary German Studies, Johns Hopkins University.

Lerhe, E. (2004, October 29–30). Civil Military Relations and Aid to the Civil Power in Canada: Implications for the War on Terror. *CDAICDFAI 7th Annual Graduate Student Symposium*.

Levi, L., Michaleson, M., Admi, H., Bergman, D., and Bar-Nahor, R. (2002). National Strategy for Mass Casualty Situations and Its Efects on the Hospital. *Prehospital and Disaster Medicine*, 17(1), 12–16.

Levy, P. (2009). *Terrorism and Social Work Practice: Memories of Terrorism in Israel*. Retrieved June 24, 2009, from BPD Update Online: http://bpdupdateonline.bizland.com/bpdupdateonlinespring2002/id1.html.

Lewis, P. (2003). Beyond Victimhood—From the Global to the Local: A British Case Study. In J. Cesari (Ed.), *European Muslims and the Secular State in a Comparative Perspective*.

Lijphart, A. (1971). Comparative Politics and the Comparative Method. *American Political Science Review*, 65(3), 682–693.

Lindsay, J. (2003). The Crime Prevention Continuum: A Community Policing Perspective on Crime Prevention in Canada. In S. P. Lab and D. K. Das (Eds.), *International Perspectives on Community Policing and Crime Prevention* (pp. 42–57). Upper Saddle River, NJ: Pearson Education.

London Ambulance Service. (2007). *Major Incident Plan*. London: London Ambulance Service, Emergency Preparedness Unit.

London Emergency Services Liaison Panel. (2007). *Major Incident Procedure Manual* (7th ed.). London: Stationery Office.

London Metropolitan Police. (2010). *Specialist Operations*. Retrieved August 16, 2010, from Metropolitan Police: http://www.met.police.uk/so/counter_terrorism.htm.

Lord, Carlile. (2007). *The Definition of Terrorism: A Report by Lord Carlile of Berriew Q.C., Independent Reviewer of Terrorism Legislation*. London: Stationery Office.

Lorenz, A. J. (2007). *TheThreat of Maritime Terrorism to Israel*. Retrieved April 11, 2010, from MaritimeTerrorism.com: http://www.maritimeterrorism.com/2007/09/19/the-threat-of-maritime-terrorism-to-israel/.

Macpherson, D. (1998). *Good Practice*. London: Emergency Planning Coordination Unit, NHS.

Malcolmson, P., and Myers, R. (2005). *The Canadian Regime: An Introduction to Parliamentary Government in Canada* (3rd ed.). Peterborough, Ontario, Canada: Broadview Press.

Marcus, L. J. (2002). *Israel's Preparedness for Responding to the Health Requirements of Its Civilian Population in the Event of Deployment of a Nuclear, Biological or Chemical Weapon of Mass Destruction*. Boston: Harvard School of Public Health.

Mauer, V. (2007). Germany's Counterterrorism Policy. In D. Zimmerman and A. Wenger (Eds.), *How States Fight Terrorism: Policy Dynamics in the West* (pp. 59–78). Boulder, CO: Lynne Rienner.

McBride, M., and Collins, G. (2002). *The UK Police: A Pocket Guide 2002–2003*. Barnsley, UK: Pen and Sword Books.

McKee, M., Hervey, T., and Glimore, A. (2010). Public Health Policies. In E. Mossialos, G. Permanand, R. Baeten, and T. K. Hervey (Eds.), *Health Systems Governance in Europe: The Role of European Union Law and Policy* (pp. 231–281). Cambridge: Cambridge University Press.

Mellis, C. (n.d.). *Amsterdam and Radicalization: The Municipal Approach*. Amsterdam: City of Amsterdam.

Merari, A. (2000). *Israel's Preparedness for High Consequence Terrorism*. Boston: Belfer Center for Science and International Affairs, Harvard University. "The Met," (January 2008) *Specials Magazine: The Voice of the Special Constabulary*, pp. 9–10.

Meyr, E. (1999). A New Strategy for the Israeli Police. *Law and Order, 47*(9), 103–106.

Miko, F. T., and Froehlich, C. (2004). *Germany's Role in Fighting Terrorism*. Washington, DC: Congressional Research Service.

Mill, J. S. (1998). *On Liberty and Other Writings*. Cambridge: Cambridge University Press.

Mirza, M., Senthilkumaran, A., and Ja'far, Z. (2007). *Living Apart Together: British Muslims and the Paradox of Multiculturalism*. London: Policy Exchange.

Monar, J. (2006). In M. Caparini and O. Marenin (Eds.), *Borders and Security Governance: Managing Borders in a Globalised World*. Geneva: Centre for the Democratic Control of Armed Forces.

Morag, N. (2010). Foreign Intelligence and Counterterrorism: An Israeli Perspective. In K. G. Logan, (Ed.), *Homeland Security and Intelligence* (pp. 146–167). Santa Barbara, CA: Praeger Security International.

Mosque and Imams National Advisory Board. (n.d.). *Draft Constitution of the Minab*. Retrieved April 19, 2009, from Mosque and Imams National Advisory Board: http://www.minab.org.uk/essential-documents/draft-constitution-of-the-minab.

Naqshbandi, M. (2006). *Problems and Practical Solutions to Tackle Extremism: and Muslim Youth and Community Issues.* Shrivenham, UK: The Defence Academy of the United Kingdom.

National Coordinator for Counter-Terrorism. (2008). *Salafism in the Netherlands: A Passing Phenomenon or a Persistent Factor of Significance?* The Hague: Netherlands Ministry of the Interior and Kingdom Relations.

National Counter Terrorism Security Office. (2009). *Counter Terrorism Protective Security Advice for Health.* London: Association of Chief Police Officers.

National Policing Improvement Agency. (2008). *PCSO Review.* London: NPIA.

Netherlands General Intelligence and Security Service. (2004). *From Dawa to Jihad: The Various Threats from Radical Islam to the Democratic Legal Order.* The Hague: Netherlands Ministry of Interior and Kingdom Relations.

Netherlands House of Representatives. (2009). *Operational Action Plan on Polarisation and Radicalisation, 2009.* Amsterdam: Netherlands Parliament.

New South Wales Government. (n.d.). *Countering Terrorism.* Sydney: New South Wales Government.

New South Wales Government Counter Terrorism, Plan. (2008). Sydney: New South Wales Government.

Northern Ireland Central Emergency Planning Unit. (2010). *A Guide to Emergency Planning Arrangements in Northern Ireland.* Retrieved May 30, 2010, from Office of the First Minister and Deputy First Minister (Northern Ireland): http://cepu.nics.gov.uk/pubs/emerplanarrange.pdf.

Northern Ireland Special Powers Act, 2(4). (1922).

Nuriel, N. (n.d.). IDF Home Front Command: Structure and Aim of the HFC.

Open Society Institute. (2007a). *Muslims in the EU—Cities Report: Germany.* New York: Open Society Institute, EU Monitoring and Advocacy Program.

Open Society Institute. (2007b). *Muslims in the EU—Cities Report: The Netherlands.* New York: Open Society Institute, EU Monitoring and Advocacy Program.

Open Society Institute. (2007c). *Muslims in the EU—Cities Report: France.* New York: Open Society Institute, EU Monitoring and Advocacy Program.

Pan-Canadian Public Health Network. (n.d.). *Federal/Provincial/Territorial Memorandum of Understanding (MOU) on the Sharing of Information During a Public Health Emergency.* Retrieved August 24, 2010, from Pan-Canadian Public Health Newtork: http://phn-rsp.ca/pubs/mou-is-pe-pr/index.html#annd.

Paris, J. (2007). Discussion Paper on Approaches to Anti-Radicalization and Community Policing in the Transatlantic Space. *Weidenfeld Institute/Migration Policy Institute Conference* (pp. 1–18). Washington, DC: Weidenfeld Institute for Strategic Dialogue.

Pearce, T., and Fortune, J. (1995). Command and Control in Policing: A Systems Assessment of the Gold, Silver and Bronze Structure. *Agency Report, 3*(3), 181–187.

Pearlman, J. (2009, August 20). *Australian Spooks Face FBI-Style Makeover.* Retrieved May 16, 2010, from Sydney Morning Herald: http://www.smh.com.au/national/australian-spooks-face-fbistyle-makeover-20090819-eqky.html.

Pedahzur, A. (2009). *The Israeli Secret Services and the Struggle Against Terrorism.* New York: Columbia University Press.

Pickett, S. (2008). *Aviation Security: A Comparision of the Aviation Security Approaches by the United States and Israel.* Unpublished paper.

Police National Legal Database and Andrew Stainforth. (2009). *Blackstone's Counter-Terrorism Handbook*. Oxford: Oxford University Press.

Ponenti, A. M. (2007). *An Examination of the United Kingdom's Counter Terrorism Security Advisor (CTSA) Program*. Unpublished.

Post, D., and Niemann, A. (2007). *The Europeanisation of Germany Asylum Policy and the "Germanisation" of European Asylum Policy: The Case of the "Safe Third Country" Concept*. Montreal: European Union Studies Association.

Prevention of Terrorism Act, Section 12(3)(a). (1984).

Prime Minister and Home Secretary. (2009). *The United Kingdom's Strategy for Countering International Terrorism*. London: Stationery Office.

Prince, R., (2008, October 13). *Terror Bill: 42-Day Detention Rejected by House of Lords*. Retrieved May 31, 2009, from Telegraph.co.uk: http://www.telegraph.co.uk/news/newstopics/politics/3191241/Terror-bill-42-day-detention-rejected-by-House-of-Lords.html.

Province of Ontario, Health Protection and Promotion Act, R.S.O. 1990, c. H.7, s. 35 (11). (n.d.).

Public Safety and Emergency Preparedness Canada. (2004). *Government of Canada Position Paper on a National Strategy for Critical Infrastructure Protection*. Ottawa: Government of Canada.

Public Safety Canada. (2003, February 19). *An Overview of Canada's Counter-Terrorism Arrangements*. Retrieved May 23, 2010, from Public Safety Canada: http://ww2.ps-sp.gc.ca/publications/national_security/terrorism_arrangements_e.asp.

Angel Rabasa, Stacie L. Pettyjohn, Jeremy J. Ghez, Christopher Boucek (2010). *Deradicalizing Islamist Extremists*. Santa Monica: RAND.

Raiter, Y., Farfel, A., Lehavi, O., Goren, O. B., Shamiss, A., Priel, Z., et al. (2008). Mass Casualty Incident Management, Triage, Injury Distribution of Casualties and Rate of Arrival of Casualties at the Hospitals: Lessons from a Suicide Bomber Attack in Downtown Tel Aviv. *Emergency Medicine Journal*, 25, 225–229.

Reed, J. (2005). *Young Muslims in the UK: Education or Integration*. London: IPPR.

Reichel, P. L. (1999). *Comparative Criminal Justice Systems: A Topical Approach* (2nd ed.). Upper Saddle River, NJ: Prentice Hall.

Roberts, G. K. (2000). *German Politics Today*. Manchester, UK: Manchester University Press.

Rollins, J. (2005). The United Kingdom: Military Assistance to Civil Authorities. *Foreign Studies, Doctrine #06*.

Rosen, B. (2003). Israel. In S. T. Mossialos (Ed.), *Health Care Systems in Transition*. Copenhagen: European Observatory on Health Care Systems.

Rosen, B., and Samuel, H. (2009). *Israel: Health System Review*. Copenhagen: World Health Organization, Regional Office for Europe.

Rosen, P. (2000). *The Canadian Security Intelligence Service, Current Issue Review 84-27E*. Ottawa: Parliamentary Research Branch.

Rothery, M. (2005). Critical Infrastructure Protection and the Role of Emergency Services. *Australian Journal of Emergency Management*, 20(2), 45–50.

Roy, O. (2004). *Globalized Islam: The Search for a New Ummah*. New York: Columbia University Press.

Russell, J. (2007). *Terrorism Pre-Charge Detention Comparative Law Study*. London: National Council for Civil Liberties.

Safe Cities Project. (2004). *Hard Won Lessons: How Police Fight Terrorism in the United Kingdom.* New York: Manhattan Institute for Policy Research.

SAMU de France. (2010). *Historical Background.* Retrieved August 15, 2010, from SAMU de France: http://www.samu-de-france.fr/en/System_of_Emergency_in_France_MG_0607#2.

Sandier, S., Paris, V., and Polton, D. (2004). *Healthcare Systems in Transition: France.* Copenhagen: World Health Organization Regional Office for Europe.

Savage, T. M. (2004). Europe and Islam: Crescent Waxing, *Cultures Clashing. The Washington Quarterly,* 27(3), 25–50.

Schmitt, G. J., and Gerecht, R. M. (2007). *France: Europe's Counterterrorist Powerhouse.* Washington, DC: American Enterprise Institute.

Schoenholtz, A. I., and Hojaiban, J. (2008). *International Migration and Anti-Terrorism Laws and Policies: Balancing Security and Refugee Protection.* Washington, DC: Institue for the Study of International Migration, Georgetown University.

Sciolino, E. (2004, October 18). Europe Struggling to Train a New Breed of Muslim Clerics. *The New York Times.*

Scottish Executive, Justice Department. (2002, December). *Protocol Between the British Transport Police and the Scottish Police Service.* Edinburgh: Scottish Executive, Justice Department.

Scottish, Government. (2010). *National Health Service in Scotland Manual of Guidance: Responding to Emergencies.* Retrieved May 22, 2010, from NHS Scotland: http://www.sehd.scot.nhs.uk/emergencyplanning/documents/annex_o.htm.

Secretariat of the Judicial Reform Council. (1999, July). *The Japanese Judicial System.* Retrieved March 18, 2010, from http://www.kantei.go.jp/foreign/judiciary/0620system.html

Serino, P. (2003). *The Italian Army's Role in Homeland Security.* Carlisle, PA: US Army War College.

Serious and Organized Crime Agency. (2008). *Annual Plan 2008–09.* London: SOCA.

Shapiro, J., and Suzan, B. (2003). The French Experience of Counter-terrorism. *Survival,* 45 (1), 67–98.

Shore, Z. (2006). *Breeding Bin Ladens: America, Islam and the Future of Europe.* Baltimore: Johns Hopkins University Press.

Siegel, P. C. (2007, August). An Insider's Look at France's Mosque Surveillance Program. *Terrorism Monitor,* 5(6).

Silber, M. D., and Bhatt, A. (2007). *Radicalization in the West: The Homegrown Threat.* New York: New York City Police Department.

Skurka, S., and Pringle, L. (1999, April). Northern Lights. *The Champion.*

Slapper, G., and Kelly, D. (2009). *The English Legal System 2009/2010* (10th ed.). London: Routledge-Cavendish.

Somerset Local Authorities' Civil Contingencies Partnership. (2006). *Major Incidents: A Guide for Local Businesses.* Somerset, UK: Somerset Local Authorites' Civil Contingencies Unit.

Spiegel Online. (2010, June 22). *Had It with Jihad: Germany to Launch Exit Program for Militant Islamists.* Retrieved July 24, 2010, from Spiegel Online: http://www.spiegel.de/international/germany/0,1518,702103,00.html.

Stachenko, S., Legowski, B., and Geneau, R. (2009). Improving Canada's Response to Public Health Challenges: The Creation of a New Public Health Agency. In R. Beaglehole and R. Bonita (Eds.), *Global Public Health: A New Era* (2nd ed., pp. 123–137). Oxford: Oxford University Press.

State of Israel. (2001). *Israeli Report to the UN Committee on Counterterrorism*. Retrieved May 20, 2009, from http://www.mfa.gov.il/MFA/MFAArchive/2000_2009/2001/12/Israeli%20Report%20to%20the%20UN%20Committee%20on%20Counterterrorism.

State of Victoria Auditor-General. (2009). *Preparedness to Respond to Terrorism Incidents: Essential Services and Critical Infrastructure*. Melbourne: Victoria Government Printer.

Stemmann, J. J. (2006). Middle East Salafism's Influence and the Radicalization of Muslim Communities in Europe. *The Middle East Review of International Affairs*.

Stevens, A. (2003). *Government and Politics in France*. Houndmills, UK: Palgrave Macmillan.

Stevens, A. (2008, December 29). *A Variety of Taxes Funds Japan's Prefectures and Municipalities*. Retrieved March 18, 2010, from City Mayors Government: http://www.citymayors.com/government/jap_locgov.html.

Stevenson, J. (2006). The Role of the Armed Forces of the United Kingdom in Security the State Against Terrorism. In J. L. Clarke (Ed.), *Armies in Homeland Security: American and European Perspectives*. Washington, DC: National Defense University Press.

Stoddart, J. (2009). *Surveillance, Search or Seizure Powers Extended by Recent Legislation in Canada, Britain, France and the United States*. Ottawa: Privacy Commissioner of Canada.

St-Pierre, Y. (2008). Caught in the Storm: Canada and the Netehrlands as Barometers for the West's Changing Attitude Towards Security and Human Rights After 9/11. *The International Human Rights Regime Since 9/11: Trans-Atlantic Perspectives, April 17–19, 2008*. Pittsburgh, PA: University of Pittsburgh.

Talbot-Smith, A., and Pollock, A. M. (2006). *The New NHS: A Guide*. New York: Routledge.

Tatchell, M., Tatchell, R., and Tatchell, T. (2010). Australia. In J. A. Johnson and C. H. Stoskopf (Eds.), *Comparative Health Systems, Global Perspectives* (pp. 187–201). Sudbury, MA: Jones & Bartlett Publishers.

Technisches Hilfswerk. (2010). *Technisches Hilfswerk*. Retrieved April 3, 2010, from The German Federal Agency for Technical Relief (THW): http://www.thw-ffm.de/english/.

Terrorism Act Section 1. (2000). Retrieved May 21, 2009, from http://www.opsi.gov.uk/acts/acts2000/plain/ukpga_20000011_en.

Terrorism Act Section 5. (2000). Retrieved May 21, 2009, from http://www.opsi.gov.uk/acts/acts2000/plain/ukpga_20000011_en.

TFEU, Title 1, Article 2. (2008).

TFEU, Title 1, Articles 4 & 5. (2008).

TFEU, Title 1, Article 6. (2008).

TFEU, Title IV, Articles 78 and 79. (n.d.).

The Economist. (2002, July 23). Anger and Assassination. *The Economist*.

The Economist. (2009, September 12). French Criminal Justice: A Delicate Judgement. *The Economist*.

The Economist. (2009, March 28). Getting Metaphysical: The Government Is Targeting Islamist Ideas as Well as Violence. *The Economist*.

Tiedemann, L., and Norris, S. (2004). *Legislative Summary: Bill C-12, The Quarantine Act*. Ottawa: Parliamentary Information and Research Service, Library of Parliament.

Tiefenbrun, S. (2003). A Semiotic Approach to a Legal Definition of Terrorism. *ILSA Journal of International and Comparative Law, 9*(357), 357–402.

Tirosh, Y. (2003). The Legal Framework for Military Activities in Israel from the Comparative Legal Perspective. *Law and the Military, 17,* 289–349.

Tol, G. (2008). Institutionalization of Islam in Germany and the Netherlands: Beyond EU Jurisdiction. *European Diversity and Integration Conference,* Miami, FL.

Treasury Board of Canada. (n.d.). *Communications Policy of the Government of Canada: Executive Summary.* Retrieved August 6, 2010, from www.tbs-sct.gc.ca/pubs_pol/sipubs/comm/cpgces-pcgcr-eng.rtf.

Trusted Information Sharing Network for Critical Infrastructure Protection. (2007). *Fact Sheet: TISN Deed of Confidentiality.*

Turnbull, A. (2004, April). *Draft Report on Young Muslims and Extremism.* Retrieved April 26, 2010, from GlobalSecurity.org: http://www.globalsecurity.org/security/library/report/2004/muslimext-uk.htm.

UK Borders Act, Chapter 30, Section 23. (2007).

UK Cabinet Office. (2005). *Central Government Arrangements for Responding to an Emergency.* London: UK Cabinet Office.

UK Cabinet Office. (2010a, May). *Case Study.* Retrieved August 18, 2010, from UK Resilience: http://www.cabinetoffice.gov.uk/ukresilience/response/recovery_guidance/case_studies/y5_bernard_matthews.aspx.

UK Cabinet Office. (2010b, May 12). *UK Resilience.* Retrieved October 13, 2010, from UK Cabinet Office: http://www.cabinetoffice.gov.uk/ukresilience.aspx

UsK Cabinet Office. (n.d.a). *Communicating Risk.* Retrieved May 22, 2010, from UK Reslience: http://www.cabinetoffice.gov.uk/media/132679/communicatingrisk.pdf

UK Cabinet Office. (n.d.b). *Dealing with Disaster* (rev. 3rd ed.) London: UK Cabinet Office.

UK Cabinet Office. (n.d.c). *How Prepared Are You? Business Continuity Management Toolkit.* London: UK Cabinet Office.

UK Center for the Protection of National, Infrastructure. (n.d.). Retrieved December 11, 2009, from UK Center for the Protection of National Infrastructure: http://www.cpni.gov.uk/About/whatWeDo.aspx.

UK Civil Contingencies Act, Part 2, Sections 20–23. (n.d.).

UK Civil Contingencies Act, Schedule 1, Part 1. (n.d.).

UK Department for Transport. (n.d.). *A Brief Overview of the United Kingdom National Maritime Security Programme.* London: Department for Transport.

UK Department of Business Innovation and Skills. (n.d.). *National Emergency Plan for the Telecommunications Sector.* London: Stationery Office.

UK Directorate of Reserve Forces and Cadets. (2005). *Future Use of the UK's Reserve Forces.* London: Directorate of Reserve Forces and Cadets.

UK Foreign Office. (2005). *Counter-Terrorism Legislation and Practice: A Survey of Selected Countries.* London: UK Foreign Office.

UK Government. (2003). *Emergency Preparedness and Response, Report Prepared for the Meeting of the States Parties to the Convention on the Prohibition of the Development, Production and Stockpiling of Bacteriological (Biological) and Toxin Weapons and Their Destruction.* Geneva.

UK Government. (2005a). *Emergency Preparedness: Guidance on Part 1 of the Civil Contingencies Act 2004, Its Associated Regulations and Non-Statutory Arrangements.* London: Stationery Office.

UK Government. (2005b). *Emergency Response and Recovery: Non-Statutory Guidance to Complement Emergency Preparedness.* London: UK Cabinet Office.

UK Government. (2006). *Countering International Terrorism: The United Kingdom's Strategy, July 2006*. London: Stationery Office.

UK Government. (2008). *The Prevent Strategy: A Guide for Local Partners in England*. London: Stationery Office.

UK Government (2011). *Prevent Strategy*. London: Stationery Office.

UK Government. (2009). *The United Kingdom's Strategy for Countering International Terrorism*. London: Stationery Office.

UK Government. (2010). *Pursue Prevent Protect Prepare: The United Kingdom's Strategy for Countering International Terrorism, Annual Report*. London: UK Home Office.

UK Home Office. (2002). *Secure Borders, Safe Haven: Integration with Diversity in Modern Britain*. London: UK Home Office.

UK Home Office. (2003). *The Release of Chemical, Biological, Radiological or Nuclear (CBRN) Substances or Material: Guidance for Local Authorities*. London: UK Home Office.

UK Home Office. (2008). *From the Neighbourhood to the National: Policing Our Communities Together*. London: UK Home Office.

UK Home Office. (n.d.a). Retrieved December 9, 2009, from Office for Security and Counter-terrorism: http://www.security.homeoffice.gov.uk/about-us/about-the-directorate/?version=1.

UK Home Office. (n.d.b). *Young Muslims and Extremism*. London: UK Home Office.

UK Home Office and Northern Ireland Office. (1998, December). *Legislation Against Terrorism: A Consultation Paper*. Retrieved January 13, 2010, from The Stationery Office: http://www.archive.official-documents.co.uk/document/cm41/4178/chap-08a.htm.

UK Home Secretary. (2004). *Counter-Terrorism Powers: Reconciling Security and Liberty in an Open Society: A Discussion Paper*. London: Stationery Office.

UK House of Commons. (n.d.). *Civil Contingency Planning to Deal with Terrorist Attack*. Retrieved July 6, 2009, from UK House of Commons Proceedings: http://press.home-office.gov.uk/documents/civilcontingencies.pdf?view=Binary.

UK House of Commons Intelligence and Security Committee. (2009). *Could 7/7 Have Been Prevented? Review of the Intelligence on the London Terrorist Attacks on 7 July 2005*. London: Stationery Office.

UK Inspectorate of Constabulary. (2008). *Preventing Violent Extremism: Learning and Development Exercise, Report to the Home Office and Communities and Local Government*. London: HM Inspectorate of Constabulary.

UK Investigatory Powers Act, Section 6(2) (n.d.).

UK Ministry of Defense. (2007). *Operations in the UK: the Defence Contribution to Resilience*. London: Ministry of Defense.

UK National Counter Terrorism Office. (n.d.). *NaCTSO: Who We Are, What We Do and How We Do It*. Retrieved December 11, 2009, from UK National Counter Terrorism Office: http://www.nactso.gov.uk.

UK National Health Service. (2005). *The NHS Emergency Planning Guidance 2005: Underpinning Materials*. London: National Health Service.

UK Nationality, Immigration and Asylum Act, Section 4/40. (2002).

UK Office of National Statistics. (2009). *Labour Market: Muslim Unemployment Rate Highest*. London: Office of National Statistics.

UK Regulation of Investigatory Powers Act, Section 22(2). (n.d.).

UK Terrorism Act 2006, Part 1, Section 8. (n.d.).

UK Civil Contingencies Act, Schedule 1, Part 1. (n.d.).

UK Terrorism Act, Sections 33-36. (2000).

United Nations Office on Drugs and Crime. (2010). *Digest of Terrorist Cases*. Vienna: United Nations.

United Nations, Office of the High Commissioner for Human Rights. (2001). *Concluding Observations of the Committee Against Torture: Israel*. Geneva: Office of the Higher Commissioner for Human Rights.

van Selm, J. (2005). *The Hague Program Reflects New European Realities*. Washington, DC: Migration Policy Institute.

Vaultier, D. (2006). The Military's Role in Homeland Security in France. In J. L. Clarke (Ed.), *Armies in Homeland Security: American and European Perspectives*. Washington, DC: National Defense University Press.

Walker, C. (2002). *The Blackstone's Guide to the Anti-Terrorism Legislation*. Oxford: Oxford University Press.

Wall, A. (1996). Australia. In A. Wall (Ed.), *Health Care Systems in Liberal Democracies* (pp. 12–46). London: Routledge.

Wanless, D. (2004). *Securing Good Health for the Whole Population: Final Report*. London: Stationery Office.

Warnes, R. (2009). Germany. In E. B. A. Jackson (Ed.), *Considering the Creation of a Domestic Intelligence Agency in the United States*. Santa Monica, CA: RAND.

Watts, D. (2006). *British Government and Politics: A Comparative Guide*. Edinburgh: Edinburgh University Press.

Watts, R. L. (1999). *Comparing Federal Systems* (2nd ed.). Montreal: McGill-Queen's University Press.

Weisburd, D., Amir, M., and Shalev, O. (2001). *Community Policing in Israel: A National Evaluation*. Jerusalem: Ministry of Public Security.

Weitz, R. (2007). *The Reserve Policies of Nations: A Comparative Analysis*. Carlisle, PA: US Army War College, Strategic Studies Institute.

Western Australia Police Service. (n.d.). *Watching Brief on the War on Terrorism*. Unpublished report.

Wheeler, J. (2005). *An Independent Review of Airport Security and Policing for the Government of Australia*. Canberra: Commonwealth of Australia.

Wilkinson, P. (1979). *Terrorism and the Liberal State*. New York: New York University Press.

Williams, G. (2006). Australia's Legal Response to Terrorism: Where Will It End? *Arena Journal*.

Yates, A. (2003). *Engineering a Safer Australia: Security Critical Infrastructure and the Build Environment*. Barton: Institution of Engineers.

Yates, A., and Bergin, A. (2010). *Here to Help: Strengthening the Defense Role in Australian Disaster Management*. Barton: Australian Strategic Policy Institute.

Yehezkeli, P. (2001). Between 'National Police' and 'Municipal Police'. *Journal of Police and Society*, 5, 71–106.

Yoaz, Y. (2005, June 19). A State in Emergency. *Ha'aretz*.

Zoller, V. (2004). Liberty Dies by Inches: German Counter-Terrorism Measures and Human Rights. *German Law Journal*, 5(5), 469–494.

INDEX

Comparative Homeland Security: Global Lessons, First Edition. Nadav Morag.
© 2011 John Wiley & Sons, Inc. Published 2011 by John Wiley & Sons, Inc.